D1265897

YOUR PRIVATE SKY

R. BUCKMINSTER FULLER

YOUR PRIVATE SKY

R. BUCKMINSTER FULLER
THE ART OF DESIGN SCIENCE

Edited by
Joachim Krausse
Claude Lichtenstein

Lars Müller Publishers

Those taking part in this project express their thanks to those responsible at the Buckminster Fuller Archive and the Buckminster Fuller Institute (Santa Barbara, Ca.) for their ever greater interest and participation in our project and for their readiness to grant us access to the holdings of the Fuller Estate.

We thank the firm CargoLifter AG for their openness towards a theme, of which we are convinced, that, in its present and future capacity holds still exciting possibilities.

"Scientific design is linked to the stars far more directly than to the earth. Star-gazing? Admittedly. But it is essential to accentuate the real source of energy and change in contrast to the emphasis that has always been placed on keeping man 'down to earth'."

Nine Chains to the Moon (1938), 67

Most of the documents depicted in this book are small-format (up to letter-format). The sketches from Fuller's early work *4D House* are inventoried as *Sketch Series.* The majority of the manuscripts and sketches from the time since 1945 come from the *Manuscript Files* and *Synergetic Papers* archive collections. Numerous photographs are prints from 18×24 cm; sometimes they are smaller images; larger-formats are the exception. For the photographs from Fuller's childhood years, they are sometimes from contact sheets. Some of the images in question come from the *Dymaxion Chronofile.* Another portion are photographs after 1945, particularly the color prints small-format slide images, whose color quality has remained astonishingly high in the time since the photographs were taken (up to fifty years in some cases). Some of these come from Fuller himself, who enjoyed taking pictures and did so quite often, though the photographs are not signed. A large section consists of black and white photographs, that Fuller received from numerous participants in his workshops and projects. Only sporadically are the photographs stamped on the back-side. See the illustration credits at the end of the book.

YOUR PRIVATE SKY

R. BUCKMINSTER FULLER

Introduction 11

Chronology 26

NAVIGATOR 42

NEW ENGLAND 66

LIGHTFUL HOUSES 80

DYMAXION HOUSE 122

GREENWICH VILLAGE 146

T-SQUARE/SHELTER 156

DYMAXION CAR 178

DYMAXION BATHROOM 202

DYMAXION DEPLOYMENT UNIT 212

WICHITA HOUSE 228

DYMAXION WORLD MAP 250

FULLER RESEARCH 276

ARCHITECTURE OUT OF THE LABORATORY 314

GEOSCOPE 344

GEODESIC DOMES 354

TENSEGRITY 392

GARDEN OF EDEN 412

HOW NATURE BUILDS 442

WORLD GAME 464

TETRASCROLL 500

Appendix 522

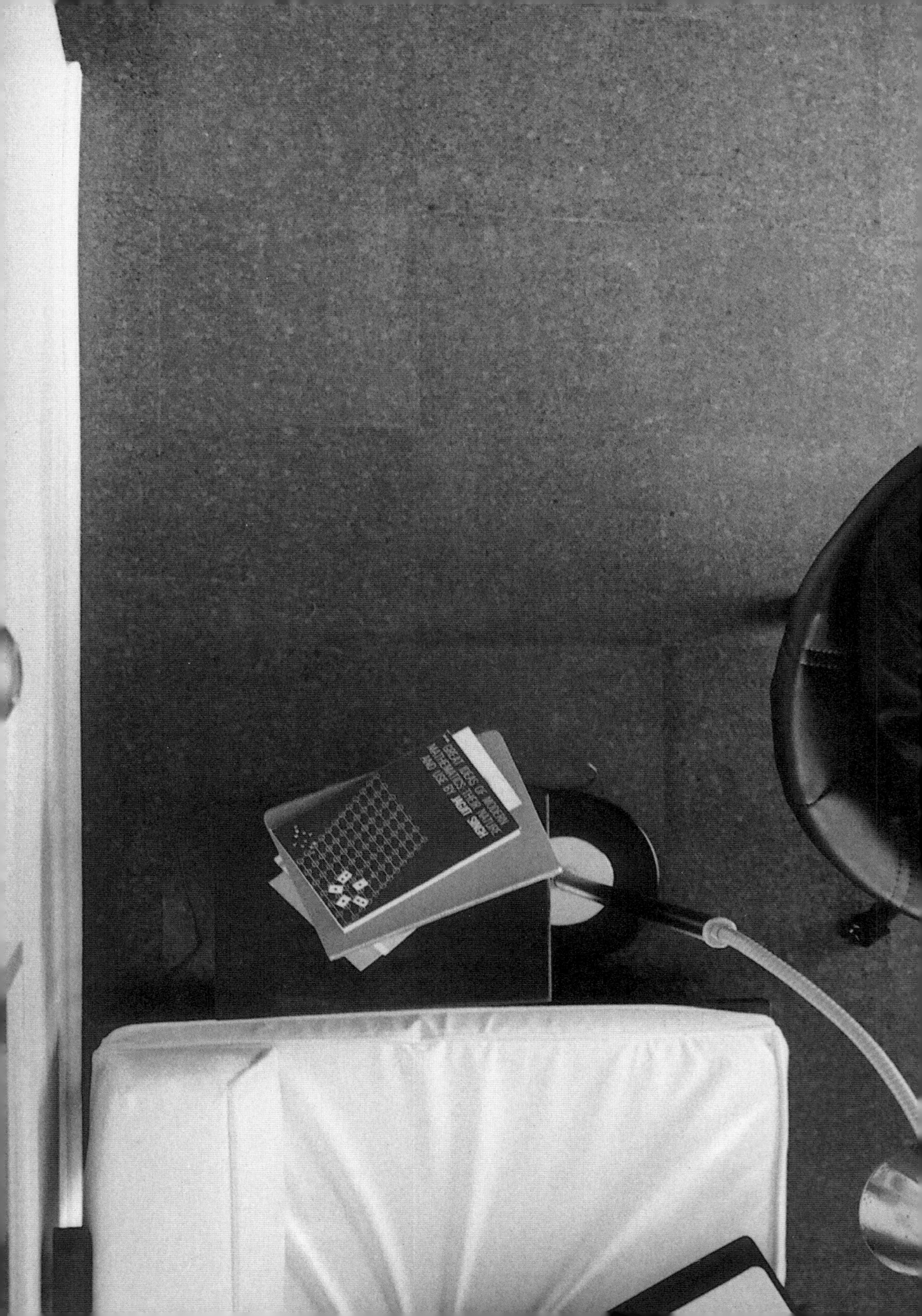

Claude Lichtenstein, Joachim Krausse

"How to Make the World Work"

Over the course of his life, Buckminster Fuller increasingly framed the justifications and explanations of his work in terms of the entire universe. Was he, then, a universalist? Fuller's many careers, callings, and activities would seem to suggest he was. In 1980, when the book *R. Buckminster Fuller: An Autobiographical Monologue/Scenario* was published–edited by his son-in-law, the well-known documentary filmmaker Robert Snyder–Fuller's many careers, fields of activities, awards, and titles were listed on the cover: "A sailor, a machinist, a comprehensive generalist, a doer, a new former, a student of trends, a technical editor, a businessman, an angel, a quarterback, a lecturer, a critic, an experimental seminarist, a random element, a verb, a comprehensive designer, an inventor, an engineer, an architect, a cartographer, a philosopher, a poet, a cosmogonist, a choreographer, a visionary, a scientist, a valuable unit, a mathematician, an air pilot, a Navy lieutenant, an affable genie, a geometer, a maverick thinker, a gentle revolutionist, a lovable genius, an anti-academician, doctor of science, doctor of arts, doctor of design, doctor of humanities, an amiable lunatic, a prophet, the custodian of a vital resource."

An imposing figure! This enumeration is indispensable, if one wants to do Fuller justice, because he became all of these things on his own strength. (The rest of us can still console ourselves that at least some of these terms apply to us as well.) Measured by the usual standards of a successful life, Fuller set off on a false start in his youth. He did receive a high school education, but he was not successful as a student and had no certain prospects for a respectable profession. His life demonstrated that these fears were unjustified: what range of thinking and areas of activity, indeed, what a life in general! That suffices to answer the question asked above. Yes, he was a universalist. If a different answer is given here, then an explanation is necessary.

He was not a universalist if a universalist is understood to be someone who in the course of his or her activities always reaches out to other increasingly distant disciplines, trying to find his or her self and achievements there. Universalists are expansive (admittedly, Fuller was as well, and how!), but universalists also accumulate. For all the admiration we might have for universalists, there is an element of sport in it, the drive to demonstrate multiple talents: a kind of intellectual decathlon. That was not Fuller's intention at all. It is true that as he grew older he referred, with fewer and fewer qualms, to the whole of the universe, but he discussed it as if describing something around the corner, discussing the "wisdom in the design of the universe." Of course, such far-reaching references placed excessive demands on his audience. But he had to look out into the world from the outside, as it were, because he could not be satisfied with that which lay before him. So, even if it is impossible not to see him in the context of the universal, it is precisely because Fuller did not strive centrifugally from the earth toward the universe, but rather sought the center of knowledge from outside. Naturally, he had to take a standpoint "from outside" the earth in order to present proposals that conformed to reality.

This characteristic of a viewpoint from "outside" explains why Fuller is so fascinating, but also why he is so unsettling. When, in a lecture in 1951, he spontaneously hit upon the phrase "Spaceship Earth," it seemed so direct and surprising. It seemed as if he were looking from "outside" at our blue planet–almost two decades before the public saw the inspiring photographs from the NASA missions to the moon–and characterized it as a closed ecological system. Already as a thirty year old, he had seen the populated land masses of our planet and realized that the continents, including Antarctica, formed an inhabitable archipelago: the "one-world island" in the "one-world ocean." No wonder, then, that he also saw the political activity on the archipelago "from outside" and viewed it as an irrational turmoil of court intrigues, of representatives of particular interests, of self-important people who purported to offer insight but who could not, in fact, see at all. Against all this he would protest, usually polemically but sometimes sadly. He also regarded academic disciplines "from outside," less with admiration than with curiosity. And in his own activities regarding the constitution of matter and the question whether the fields of mathematics, geometry, physics, and chemistry, which are not separate in nature, should not, perhaps, simply be *one* field: "how Nature builds."

Fuller first became known as an architect; his contributions were recognized and honored within the world of architecture. But these appraisals were often condescending. In the USA, where architects must be licensed, he was not permitted to construct buildings under his name alone until well into old age. When Fuller presented his *Dymaxion House,* most of the important architects from American cities who were present were not unimpressed; they envied Fuller for his intellectual freedom, but at the same time they smirked at his evident lack of "realism." At best, they regarded Fuller's concept as a refreshing questioning of standard ideas, brashly presented but not intended as a concrete proposal. Given all of this, it is no surprise that Fuller also saw the world of architecture and building "from outside." Clearances, height limitations, length of escape routes, access rights–parameters that determine the design agenda of an architectural office–did not exist for him. Architecture without the neat parcels that underlie it is a very dubious prospect in the eyes of professionals, as if someone were mischievously cutting off the circulation of the unquestioned necessities.

Fuller's early sketches–inexpert drawings when compared to the designs of others but highly forthright for precisely that reason–clearly reveal the background of his proposals in architecture. The individual building serves as a base in a concept that spans the globe, and "spanning the globe" does not simply mean that Fuller intended for many such homes to be built throughout the world. He show us how he intends to do it: the globe is a construction site for lightweight towers that reach up to the skies. The construction site is indeed a sphere. To say "reach up to the skies" is the wrong phrase, as the sketches show, because there is no absolute up and down, but only the inside of the Earth and the outside of space. The towers reach out to space. Who else saw architecture as Fuller does here? Not Tatlin, not El Lissitzky, not Le Corbusier. The 4D towers create a network of bases, signal flares of the colonizing activity of mankind. Building becomes a strategic act of colonizing the Earth, taking possession of land as a "claim," very much in the American tradition of a strategy for settling a place. But there was one important difference with respect to earlier history. For Fuller, the possession of land is free of its mythical claim on proprietary ownership of land and soil, a claim that has always resulted in discord, war, and destruction. For Fuller, ownership means to live in one place for a period, to use the land, no matter as a source of food or as the site for a house. This also means that the land remains available to others, just as the sea is available to others when we cross it in a ship. On the constantly moving sea, no one crossing it aspires to "possess" the water beneath the ship's hull. Only on static land does anyone get this sort of idea, as Fuller remarks sarcastically, polemically, but ultimately in the firm belief that sooner or later it will become clear how absurd this practice really is.

Everything that Fuller thought of, that interested him, and that he researched was in some way dynamic. The child of the industrial age is mobile, living now here, now there. He or she occupies a house for a time, then moves out, and someone else moves in; or he or she moves and takes the house along. His aphorism "In architecture, form is a noun; in industry, form is a verb" was incomparably concise. He needed to imagine things in interdependent relationships, in dynamic patterns; any idea of a static order a priori was deeply implausible to him. Visible reality is always just a snapshot, never an inventory, for the reality behind the visible cannot be recognized from an inventory–because there is no static reserve–but only from a dynamic and critical comparison of various snapshots.

It becomes clear that Fuller chose the far more difficult path over that of those who operate within static certainty. He himself always emphasized the life-long influence his experience in the Navy had had on his way of thinking. Working at sea, in a group of people working together in an intense way, in a form of transport subject to swaying, where every wave makes the assault of outside forces palpable and transforms them into motion; exposed to the elements of nature but capable of accepting it, navigation as a necessary activity and the expression of a consciousness of space and time that had been gained over the centuries by those who needed it to survive: these are for Fuller the existential bases for a perception of reality that is both realistic and creative. This book begins with the theme of navigation and "sea travel" less for their biographical significance than for their deeply metaphorical significance for Fuller's entire oeuvre.

The *Dymaxion Chronofile*

One astonishing sign of Fuller's "consciousness of himself" is the decision he made early in life to conceive of his own life as an experiment. This is a very radical position. An experiment contains not only a chance of success but also of failure. The experiment comprises both the way in which one's own life can fail and the way in which this can be avoided. These are the rules of a game that may have something of the typically robust Yankee style. As for the nature of this experiment, it goes beyond the usual "You can make it" of a competitive attitude to life and leads to the pioneering exploration of a true life project. The nature of this project is, to put it mildly, not without pathos: to do everything that can benefit humanity. Fuller: "I said, 'What can a little man effect toward such realizations in the face of the formidable power of great corporations, great states, and all their know-how, guns, monies, armies, tools and information?' Then, self-answering: 'The individual can take initiatives without anybody's permission.' Only individuals can think, and can look for the principles manifest in their experiences that others may be overlooking because they are too preoccupied with how to please some boss or with how to earn money, how to take care of today's bills. Only the individual disregards his fears and commits himself exclusively to reforming the human environment by developing tools that deal more effectively and economically with evolutionary challenges. Humans can participate–consciously and competently–in fundamental ways, to changes that are more favorable to human life. It became evident that the individual was the only one that could deliberately find the time to think in a cosmically adequate manner." *(Influences on My Work, 1963)*. This was the task he set himself. Fuller saw himself as "Guinea Pig B." He subjected himself to an experiment and made himself an object of observation. The instrument for checking up on this task was the *Chronofile,* which Fuller kept from about 1915 on and had bound in 1930 under the title *Dymaxion Chronofile.* It consisted of a kind of "central diary," which did not simply record and document events that had already happened, but also formulated or "crystallized" projects, as Fuller called it. Even so, the Dymaxion Chronofile was not the usual documentation of works; it combined life and work, daily life and events, the important and unimportant, which made it possible later to reevaluate the decision whether something was important or unimportant. The abscissa of the Chronofile is the time line, and only the time line, a strictly linear concept, in which there are no separated rubrics of "life" and "work," "private" and "public." The time line is one of the geometric coordinates that enable us to find what we seek. The other geometric coordinate, the ordinate, does not exist; more precisely, it would be the knowledge of the event, and that means nothing other than the knowledge of Guinea Pig B himself. One has to be familiar with both the event and its significance in order to find it within the time line of the Chronofile. Either that or one needs to *know* it. Or one has to comb through the records and find the significance on one's own. Precisely because it is meant to give information about the current position at every moment, Fuller intended for the Chronofile to be like the logbook of a sea traveler. Fuller committed himself to making his life project testable at every stage, both instantaneously and retrospectively by himself as well as historically by those who would come later. This is, it must be said, a very presumptuous plan. Nevertheless, Fuller held to the archival model of the Chronofile all his life, and he invested large amounts of money, time, and strength of will in the production and continuation of this monumental journal. Up to the late thirties, he would collect the folders into leather-bound volumes; later, confronted with a rising flood of data, activities, transcripts, letters, and notes, the documents were stored in boxes. One can only imagine the quantity of letters we are discussing. There is no inventory of the correspondence; there is no index of senders and recipients. The idea of the Chronofile did not permit it. (In reality, however, the pressure of circumstances meant that an index of the late correspondence, dating from about 1970, was prepared.) The Chronofile swelled to encompass some seven hundred fifty units by the time of Fuller's death, a "diary" or a life's transcript that stretches more than three hundred linear feet. As the model of a radically linear method of notation, the Chronofile soon reached its limits; the analogy to a nautical crossing and the incomparably immodest attempt to document fully a life journey proved to be illusory when confronted with reality. It is not without irony that, of all places, in Fuller's lifework, or rather work life, the connecting thread of his own personality remained at the center–the "soul" as it were–but from the time of Bucky's forties it was increasingly populated by highly talented and appropriately motivated assistants who boldly crisscrossed the principle of linearity. To continue the metaphor, Fuller was now commanding an entire squadron, and the idea of an exactly determinable intersection of space and time increasingly became a grossly simplifying illusion, even for himself, that over the course of time threatened to lose perspective on the before and after, the cause and the effect. Still, Fuller clung to the idea of the *Dymaxion Chronofile,* even though it probably only became practicable with the age of the computer and the existence of computer disk drives.

Preventative Designer or Charlatan?

An article in *The Saturday Evening Post* of 18 October 1944, in which Fuller and his Dymaxion World Map were presented, began: "Some people say the inventor of the 3-wheel automobile and the 'dwelling machine' is an authentic 18-carat genius. Others just laugh and laugh." Buckminster Fuller was, no doubt, not the silently serving spirit in the background; he polarized opinions, and a portion of the public saw him as a charlatan. And even some early colleagues who had fallen out with him.

He was no charlatan, this has been clear for some time, but it is understandable how someone could come to this view. Fuller characterized his work as "anticipatory design science." If one reads how Buckminster Fuller in important presentations would speak matter-of-factly about things as if they existed, when they most certainly did not, it is easy to imagine how that might have come across to somewhat less imaginative audiences: as a bluff or, worse, as a scam. To give just one example, when Fuller presented his Dymaxion House in a lecture for the Architectural League in New York (presented in detail in *Your Private Sky: Discourse*), he described, among other things, the contents of the appliance room. He described a washing and ironing machine as if it already existed: you put the dirty clothes in on one side, and they come out the other side, freshly washed and steaming from the iron, and land in the appropriate basket. Inventions are made of this stuff; wish dream and imagination are the parents of innovation. (What Fuller was imagining here was quite simply the transformation of the mechanical means of processing of, say, a wheat-harvesting machine into an advanced household apparatus; it is the inventor's task to transpose the process of turning grain into flour into the transformation of dirty laundry to clean.) Fuller probably just neglected to declare the *Vor-Schein* (pre-appearance) as such, to use Ernst Bloch's word. And why should he when everything is so easily anticipated? The anticipation must have been the greatest intellectual pleasure of all for him: to imagine what was not yet visible. To imagine means to make an image of something. He seemed to be on the intimate terms with the "not yet"; he spoke of it as if the doors were already opening and it was entering the present. The confusion that this caused is understandable. Before the background of the cautious and circumspect design theories of the American consumer society, which then as now was essentially concerned with the degree of innovation that could be tolerated by society, a preventative designer like Buckminster Fuller was like a stone fallen from the sky.

It speaks for itself that Fuller did not encounter such difficulties at advanced teaching institutions, where the public hung spellbound on his every word. At the Design Institute in Chicago or at Black Mountain College, it was not the demonstrable reality that counted but rather the shared path to the conceivable. The *Zukunftsmusik* (literally, "future music" but meaning "dreams of the future") was not what is usually called that in German, namely, something that lies outside the realm of one's own range of hearing. The students wanted to make music for themselves, and Fuller simply gave the cues. It can be assumed that the years before and after 1950 were the most imaginative ones, both for him and for his public, and he himself had an unheard-of presence. On photographs from this period, Fuller looks younger and noticeably less staid than during the previous period in Wichita.

Inventor Bucky

Buckminster Fuller is known as the inventor of the geodesic dome. But that is not quite right. He invented it for himself, it is true, but roughly thirty years after the engineer Walter Bauersfeld had sought a construction for the Zeiss Planetarium in Jena that would have an almost egg-shell-thin concrete dome, and found the solution in a great-circle supporting truss. Fuller knew nothing about this.

Nor did Fuller "invent" the octahedral-tetrahedral truss. It was Alexander Graham Bell who, also decades before Fuller, had experimented with this structure for kites and an observation tower. Bell–the inventor of the telephone, among other things–was certainly a first-class eighteen-carat genius, as was Bauersfeld.

The problem of "closest packing of spheres" was already explored by Johannes Kepler, three hundred years before Fuller. The Platonic bodies are called that because the greatest philosopher of antiquity was interested in their properties, and not because of Fuller's geometric research. And the C-60 carbon molecule was named "buckminsterfullerene" after him, but he did not discover it.

Finally, the principle of the multipolar tensegrity structure was discovered by Kenneth Snelson, one of Fuller's young students at Black Mountain College. It was a very significant and stimulating discovery that Fuller had been anticipating in his thought, but which he did not achieve himself.

Why make this list? (It could be continued: Fuller is, of course, not the inventor of the three-wheel automobile either, as stated above.) Because it is necessary to confront one of the greatest misunderstandings, namely, the notion that the value of Fuller's work lies above all in the particular inventions and discoveries. If this were the case, then one would have to ask: did Fuller discover anything in the field of chemistry, yes or no? In the field of geometry or mathematics? Or in civil engineering? Questions like these, however, miss the essence of Fuller's significance. What interested him was not the question of priority but that of plausibility. The former is an absolute value; the latter has to do with context, with a concrete problem. It presumes a newly invented context; it is a matter of innovative application. Fuller's greatest significance lay in the combination of various disciplines with an eye to solving a concrete application. Thanks to this correlative interest, he brought things to light that would otherwise have remained hidden.

His principle of the geodesic dome is not just a matter of geometry. It is a constructional problem: How, on the basis of a geodesic network, can a supporting structure be created that is suited to industrial production, permits rapid and simple assembly, uses a minimum of materials of the lightest possible weight, requires no supports, is cost effective, is capable of being disassembled and reused, can be transported by area, is suited for extreme climatic conditions, and so on? These sorts of questions about optimal use make the differences between Fuller's systematic thinking and the Zeiss Planetarium, a special case for the here and now, especially evident.

It is particularly ironic that what is perhaps Fuller's most important discovery, the Jitterbug Transformation, is also the one that for decades had no such parameters for application. Fuller took the name from a popular dance of the forties. (Is there any other discovery of such significance for science and the humanities that explicitly took its name from the so-called trivial world of popular culture?) In the Jitterbug Transformation, a cuboctahedron is placed over an icosahedron to form an octahedron (which can be folded into a triangle). For two thousand years, the Platonic bodies stood statically and proudly next to one another, and then Fuller discovered them and used the empirical model experiment of his energetic-synergetic geometry to show that they were closely related in structure, and that they could be understood as the phase transitions of a transformative "loop," specifically, a periodic swinging back and forth between two "end" phases. The demonstration of this was striking and fascinatingly manifest, but the significance of the discovery went unrecognized for some time. Only after Fuller's death did the discovery fall under the spotlight in the context of research into "dirty crystals" and superconductors. This comprised another contribution to the increasing "ephemeralization" of technical resources that Fuller had been discussing since the thirties. Ephemeralization is the principle of "Doing the Most with the Least," that is, greater

effects with decreasing expenditure of materials, constant improvements in the ratio of input to output. The example of structural research makes Fuller's astonishing obstinacy especially evident. It is difficult to imagine someone today who would be prepared on his or her own initiative to explore the unfathomable depths of hidden knowledge in the way that Fuller did. Today that would be considered infinitely naive–it was probably already somewhat naive at the time Fuller began (as a twenty year old during his Navy service), but less so than today. Today it is claimed that all essential questions are, if not already answered, at least exhaustively framed and divided into highly specialized sciences that are suspicious of the other specialized sciences and shielded from any intruder from outside the profession, so that a "random element," as Fuller jokingly referred to himself, would not have a chance of obtaining the goal at all. *Your Private Sky:* the self-confidence with which Fuller said the sky under which he stood was his sky, and encouraged us to do the same, is also the attitude of an artist who produces something, not that of a scientist who recognizes only what already exists. How else could one begin a project with the title *Everything I Know,* a video recording of a forty-two-hour-long extemporaneous speech from 1972? (It is also another example of the speed with which Fuller and his circle responded to technological development: the transportable and inexpensive medium of the video cassette was a prerequisite for a project like *Everything I Know.*) In its faith in the greatest possible presence of mind and the authenticity of the moment, this work is an artistic act, so much so that it could never have been at home in the world of science.
Bucky as an epic poet who elaborates on his life and experiences. If this is taken literally, it is easy to criticize the monomania of such an enterprise. Still, it is an open question whether this perception is the right one. Everything I know! Isn't this an imaginative question that anyone could ask? (Can it really be that the question was first posed by Guinea Pig B? Socrates must have known that he did not know nothing when he said that he knew nothing; his followers were quick to interpret his coquetry as a modesty from which they themselves could profit.) Everything I know! The question is relevant to the *Chronofile,* but in an inverted relationship. The latter is diachronic and accumulative, a chain of immeasurably many constituent parts that only gets its meaning when seen as a whole. *Everything I Know,* by contrast, recapitulates and radically synthesizes. Fuller calls this imparting of information "thinking out loud." There are two ways of approaching language: as information about thoughts and as provocation of thoughts. The latter is poetic, and this is what Buckminster Fuller does.
As a successive production of thoughts, speaking is related to industrial production in surprising ways, even though it is not the production of a repetitive series of the same thing. Speaking means the constant "sub-assembly" of words into sentences and their "assembly" into meaningful statements that are both more than and something other than the sum of their constituent parts. This is the essence of synergetics, the true driving force behind technical and civilizing progress, and thus the progress in the circumstances of human life.
All this begins with language. The decisive term *comprehensiveness* is founded in this. What Fuller meant by *comprehensive,* a word he used often and repeatedly and tirelessly insisted upon, is essentially about understanding that does not remain stuck in the rational but becomes a comprehensive and inclusive grasping of the whole person, becoming a physical experience. These were also questions that received isolated and local attention in the pedagogical reform movements in Europe and the United States in the early twentieth century. Fuller's childhood had been affected by them.

At this point, it is necessary to discuss the "architect" Fuller again. His projects and realizations are among the most independent contributions to architecture. If he had not himself presented his concept of the energetic-synergetic geometry of geodesic structures in an unusually active lifetime of lectures around the world, then one would also have to speak of one of the most unmediated contributions to architecture. The theme of the geodesic dome touches on spatial thinking and feeling more deeply than one would think if the dome is seen as simply a screen stretched over a plot of land. Here, too, seen again "from outside" (and that means peering into Fuller's thought) the conclusion is obvious. It is a unique achievement that (and how) he was able to generate an "architectonic" concept that developed into a totally different conception of space than our usual, familiar, and self-evident one. In a long process, Fuller moved away from the Euclidean premises of space structured as rectangular. The first step was the choice of the hexagonal floor plan of the 4D House, and twenty years later the second, far more radical step of the use of the energetic-synergetic geometry with great-circle arches as supports. Architecture's entire connection to the "supremacy of verticals" (Sigfried Giedion) had to be exploded, and thus also the connection between architecture and the upright stance of human beings. Fuller did not shy away from challenging the validity of the image of the surface of the earth as a planar ground plan: why speak of a "plan" when it is really a spherical (if only slightly so) section! Plans are evidence of a "radial" way of thinking, said Fuller, so and so many meters out from the corner and then at right over to the side and so an so many meters high. He did not make fun of it, but the fundamental difference to his theme was completely clear to him. What interested him were the energetic events in the limitations of space, that is, the "circumferential" moment. Lest there be any misunderstandings: The approach resulted in many beautiful spaces, breathing, rising, round, translucent, and firmly fastened, and spacious, colorful, airy, and protective. Fuller almost always spoke of technology, mentioned distinguishing quantities like weight tolerance and degree of effectiveness. But he certainly possessed a highly developed sense for such aesthetic and sensory characteristics, as the statements found in this book demonstrate.

The World as a Game

The image of Spaceship Earth rushing through space, with no operating instructions included, is one of Fuller's countless graphic and palpable metaphors *(Operating Manual for Spaceship Earth, 1969)*. Is it an image of catastrophe? No, but it plays skillfully with the possibility that it could become one. It is our task to avoid that possibility. The title of Fuller's first proper book, *Nine Chains to the Moon* (1938), set the tone for the significance of this planetary worldview against a universal measuring stick. Even before that, his privately printed *4D Time Lock* (1928) did the same: confidence, faith in progress, specifying the tasks, but no panic under any circumstances, exactly as he had learned as a sailor. "Emergency" means a serious or threatening circumstance, it does not mean accident. Is it permissible to include in the same trajectory two books separated by the greatest catastrophe yet to befall mankind? The confidence of *Nine Chains to the Moon* is expressed in the image that the whole of mankind could make a chain that would reach from the earth to the moon nine times, if we all just joined hands. In 1969, moon travel had become a reality. But it was just a technological fact and not yet an anthropological one, which is evident today in that this immense technological achievement has remained equally immense in its abstractness. Military technology was indispensable in achieving this goal, even if it was not exactly the same technology that had produced the atom bomb, armed it, and dropped it on cities. Nine chains that reach to the moon. Some believe the image became old-fashioned in 1969. We don't believe it was so for Fuller.

Fuller's thinking is bipolar. On the one hand, there is his faith in technology, quantification, the proof of inches, feet, and pounds, of meters and tons, "man hours" and transportation times. He was a passionate statistician who knew how to read the statements in a curve relating mass and number, about quantities, about increases and decreases, acceleration or deceleration of increase, acceleration or deceleration of decrease as a trend or a pattern of development. He had a lifelong, vital interest in quantifiable facts. But that was not all. He always subjected statistics to the question of their significance. An early example: During his work on the 4D House, in June 1928, he wrote his father-in-law, J. M. Hewlett: "We observe the even greater capital purchasing ability of the increasingly available (both as to time and space) and more attractive fresh fruits and vegetables, as against the diminishing capital equivalent of meat purchases, and realize that with the atmospheric control of the 4D House the necessity and desire for meat will vanish." (*Chronofile,* vol. 34 [1928]). This reveals the second pole of Buckminster Fuller's relationship to statistics, which was nothing less than sensuous: a strongly metaphysical faith in the meaning of life and in its unquestionable right to succeed. The sickness and death of his first daughter, Alexandra, was for him and his wife, Anne, clearly a consequence of living circumstances that were hostile to human life. His ambitions for an architecture that was up to date, for a chance to grow up healthy and life a beneficial life in its buildings, were as valid as ever. There could be no separation between architecture for people and for the universe. The latter would be the highest instance, and he saw his task as recognizing its will: in this way, he came to want to research "how nature builds." Although, as we noted, when naming properties or objects, Fuller himself preferred to derive in to derive them from verbs rather than nouns, he chooses not to use the verbal noun *creation,* but speaks rather of universe and of the wisdom of its design. His way of expressing himself was like that of Albert Einstein, whom he greatly admired and whose religiosity sublimated by science was quite close to his own stance.

Fuller believed he recognized the driving force of the development of civilized societies, and thought that the impetus was right, but not the course. There is a tendency from the elementary to the complex; Buckminster Fuller analyzed the development of civilized mankind and in a path through different ages became convinced that there was a regularity function at the deepest levels. Mankind's energy for design had long focused on the question of how to make the solution lighter, faster, and better: lighter, faster, more comfortable–no matter whether it is getting from here to there or making a utensil, a tool, or a weapon. Already in 1938, Fuller referred to this process in his book *Nine Chains to the Moon* as *ephemeralization;* he analyzed its essence and significance far more deeply than the almost inflationary use of this concept, now worn to a mere cliché, would have one believe (see *Your Private Sky: Discourse*). The triumph over feudal structures was in his eyes the political prerequisite for the welfare of society. But it was just the first step that would bring the emancipation of modern societies and industrial nations. The second step was still to be taken, namely, the emancipation of world society from the feudal relationships among the countries and the elimination of the unbelievable inequities in the standards of development. When Fuller said that politics was incapable of filling this task, this was just one side. What was incapable was the politics of self-interest, that is, the so-called realpolitik employed by all governments in this world. Fuller's critique of the political self-conception was itself highly political, even if he did not say so or perhaps even realize it. He called it "making the world work" or, more precisely *How to make the world work.* This is not the cry of warning of the cult leader who wants to urge his followers by predicting the end of the world. Fuller's working life began with an apprenticeship as a machine fitter in Québec, assembling textile machines that were delivered from Europe in parts; sometimes he had to repair them and make them run again. "How to make the world work" derived from that experience. Let's get to work on this world! Seen in this light, the question is not presumptuous at all, but simply self-confident composure.
Let's get to work on this world.

 Translation: Steven Lindberg

MAXION
ONOFILE
KMINSTER FULLER
895 1915
VOL. I
DYMAXION
CHRONOFILE
R BUCKMINSTER FULLER
1915 - 1916
VOL. II
DYMAXION
CHRONOFILE
R BUCKMINSTER FULLER
1916
VOL. III
DYMAXION
CHRONOFI
R BUCKMINSTER FUL
1916
VOL. IV

DYMAXION
CHRONOFILE
R BUCKMINSTER FULLER
1916
VOL. III

DYMAXION
CHRONOFILE
R BUCKMINSTER FULLER
1916
VOL. IV

DYMAXION
CHRONOFILE
R BUCKMINSTER FULLER
1917
VOL. V

DYMAXION
CHRONOFILE
R BUCKMINSTER FULLER
1917
VOL. VII

DYMAXION
CHRONOFILE
R BUCKMINSTER FULLER
1917
VOL. VIII

DYMAXION
CHRONOFILE
R BUCKMINSTER FULLER
1917
VOL. IX

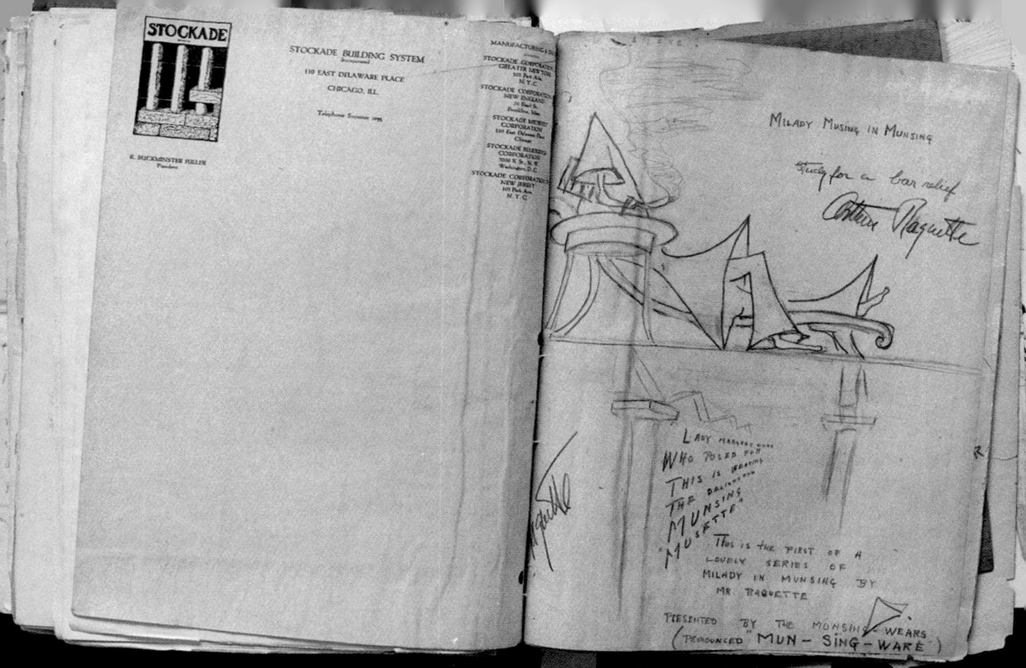

STOCKADE

STOCKADE BUILDING SYSTEM
Incorporated
110 EAST DELAWARE PLACE
CHICAGO, ILL.

R. BUCKMINSTER FULLER
President

MILADY MUSING IN MUNSING

Study for a bar relief

Arthur Paquette

LADY ... WHO POSED FOR THIS IS WEARING THE ... MUNSING "MUSETTE"

This is the first of a lovely series of Milady in Munsing by Mr Paquette

PRESENTED BY THE MUNSING WEARS
(PRONOUNCED "MUN-SING-WARE")

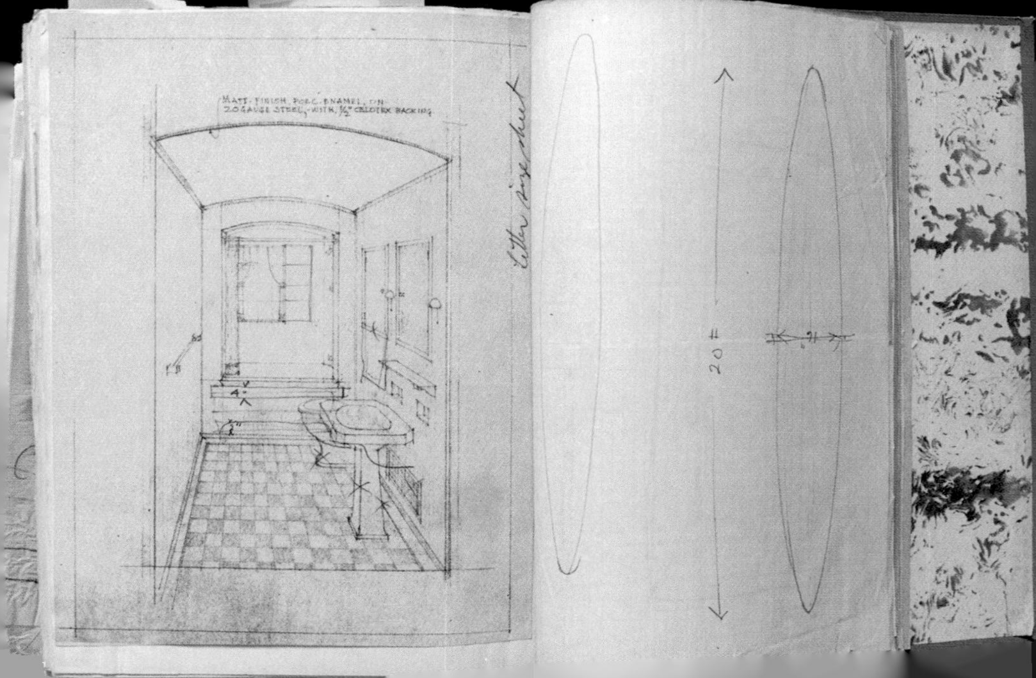

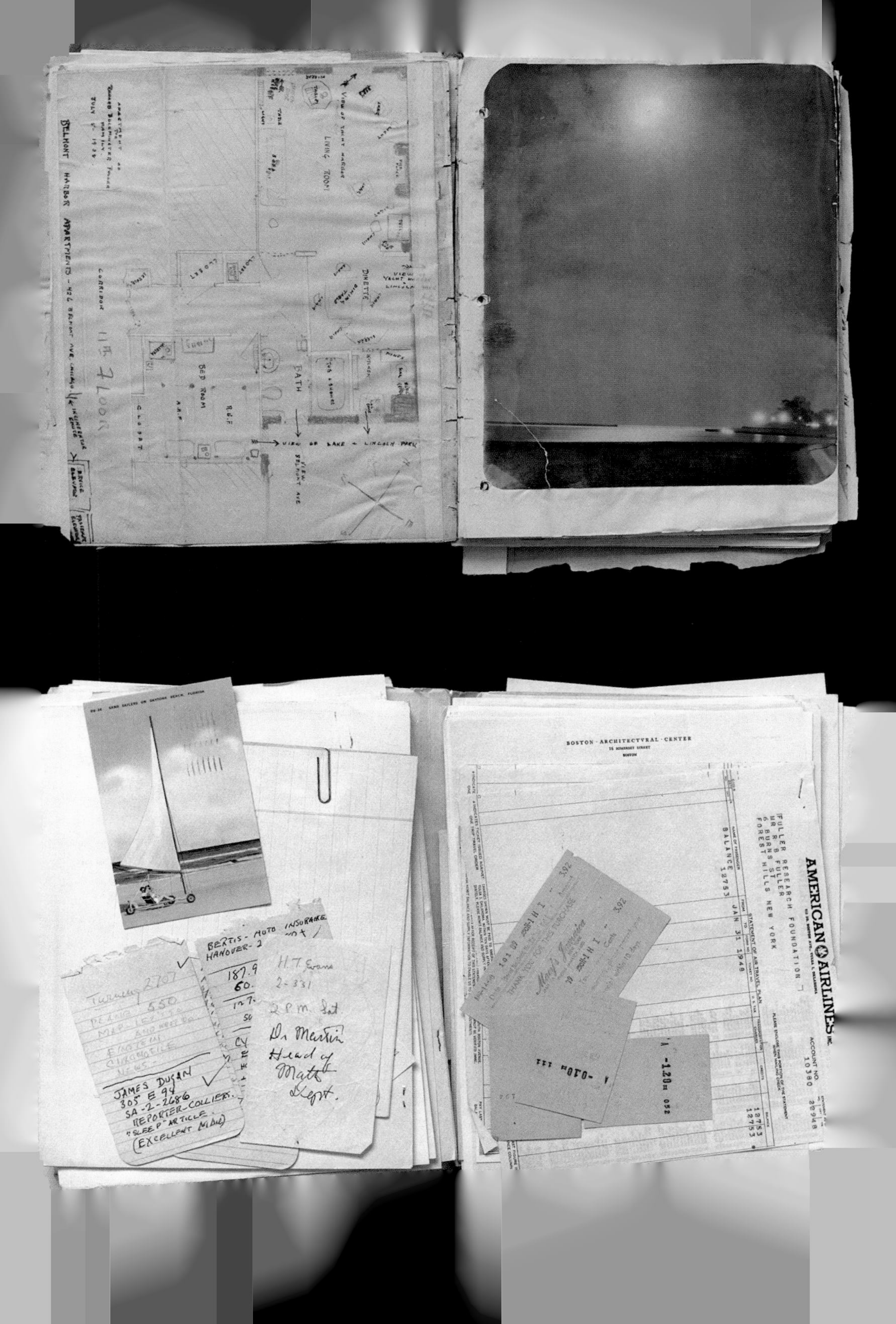

R. Buckminster Fuller

12 July 1895

Born in Milton, Massachusetts. Son of Caroline Wolcott and Richard Fuller, owner of an import company in Boston and descendent of a respected family, rich in traditions and active in politics, that had been resident in New England since 1632. Great nephew of Margaret Fuller (1810–50), writer and founder of the American women's rights movement. Second of four children: older sister Leslie, younger brother Wolcott, youngest sister Rosamund. RBF suffers from extreme farsightedness from birth.

1895 Discovery of X-rays by Conrad Wilhelm Roentgen.
1897 Discovery of electron.

1899

Attended Froebel Kindergarten.

1900 Founding of quantum theory by Max Planck.
1903 First flight in motor airplane, by Wright brothers, Kitty Hawk, N.C.

1904–13

Attended school at the Milton Academy. Grandmother Matilda Walcott Andrews purchases Bear Island, Compass and Little Sprucehead Islands, Penobscot Bay (Maine), as summer residence. · Bear Island, several miles off the mainland, becomes an important site of relationships and experiences for the family. · The young RBF builds many model boats and begins to develop his strong affection for sea travel.

1905 Special Theory of Relativity by Albert Einstein.
1906 Radio tube and crystal detector.
1907 Einstein's $E = mc^2$; equivalence of mass and energy.

1908

Death of father, Richard Fuller. · RBF decides to keep all of the letters written to him and the drafts of letter by him. Over the course of his youth, *Chronofile* develops out of it: a chronological collection that is intended to document the entire life of RBF in model fashion.

1909 North Pole reached, Peary.
1911 South Pole reached, Amundsen.
1912 Sinking of the Titanic. Emergency signal SOS.
1913 Atomic model by Nils Bohr. · Ford's Model T; first assembly-line production.

1913–15

RBF finishes school at Milton Academy. · Studies at Harvard University; is expelled twice and considered a failure.

1914 Radio: Armstrong's Regenerative Circuit. Development of chrome-nickel steel.

1914

Apprenticeship as machine fitter of imported weaving machines in his cousin's cotton factory in "Sherbrook," Québec (Canada). · RBF keeps a technical sketchbook, in which he reflects on his technical and mechanical learning.

1915 General Theory of Relativity by Albert Einstein. · Woolworth Building, highest building, 792 feet. Determination of electron distribution with closest packing of sphere, W.H. Bragg.

 RBF as youth, Bear Island, Maine

RBF's younger brother Wolcott, Equitable Building, NYC

RBF (middle) as machine fitter, Canada, 1914

1915–17
Transportation work, later assistant cashier in canned meat company Armour & Co., New York and New Jersey. · Important insights into the relationship of production, distribution, and consumption; interested in logistics. · Meets his future wife, Anne Hewlett. Engagement.

1917 October revolution in Russia.

1917
Volunteers with the family boat, *Wego,* for wartime service with the Navy. · Patrols off the coast of Maine. · 12 July 1917 On twenty-second birthday, marries Anne Hewlett, the daughter of the well-known and respected New York architect and painter James M. Hewlett.

1918–19
Cadet at the Navy Academy in Annapolis; becomes lieutenant.

1918
Birth of daughter Alexandra. · Communications officer in units of the U.S. Atlantic fleet. · Editor of the Navy newspaper *Transport,* which is published at sea. · Participates in the radio telephone experiments of Lee de Forest, at sea. · Commander in the gunboat *Eagle 15,* built by the Ford Company in mass production. · Accompanies Admiral Gleaves to Europe for the signing of the Versailles Treaty. Subsequently resigns from active service, but remains a reserve lieutenant of the Navy. · Serious illness of daughter Alexandra.

1919–21
Export manager at Armour & Company. · Until company is liquidated, sales manager at Kelley-Springfield Truck Company.

1920 Discovery of neutron. · Beginning of commercial radio broadcasts.
1922 Zeiss Planetarium in Jena, Germany: network cupola as reinforcement for concrete shell (W. Bauersfeld).

1922
RBF as pilot: flies with Anne in aquatic monoplane by Vincent Astor from N.Y.C. to Bar Harbor, Maine, record time. Journey continues to Bear Island and Boston. · Death of daughter Alexander by polio. · RBF reads Christopher Morley's poem "To a Child", which he will cite frequently later.

1922–23
Courses at Graduate School of Business Administration.

1922–26
Cofounder (together with his father-in-law, the artist and architect J.M. Hewlett) and director of the Stockade Building System, producer of filling material and a patented system of light-construction elements. By 1927, 240 houses have been constructed along the entire east coast.

1923 Hubble discovers another galaxy beyond the Milky Way.

 As ensign, around 1918

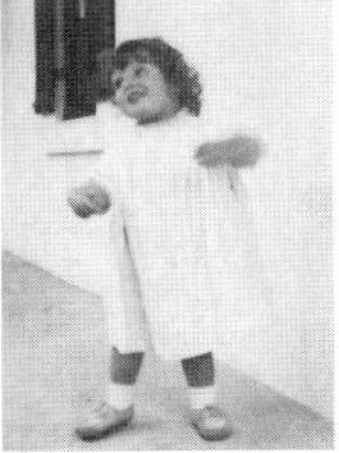

Alexandra Fuller, 1918–22

Stockade Building System: Load trial, pressure test

1924

Moves into an apartment on East 94th St., N.Y.C.

1925 A. N. Whitehead, *Science and the Modern World.*
1927 Lindbergh crosses the Atlantic nonstop in *The Spirit of St. Louis;* photoelectric cell; Le Corbusier's book *Vers une architecture* appears in an English translation by Etchells; Weissenhof housing estate in Stuttgart, Germany. · Heisenberg's uncertainty principle. · Ford introduces the Model A.

1927–28

Stockade Building Co. is bought by Cellotex, and RBF loses his position (resigns as president on 10 February, leaves November 1927). · Anne comes to Chicago, Belmont Ave., 7 August 1927 and keeps a diary. · Birth of second daughter, Allegra, 28 August 1927. · Serious personal crisis. Concentrates on beginning a new job. · Discovers the importance of his great-aunt Margaret Fuller and the other New England Transcendentalists (Emerson, Thoreau). · Reads Le Corbusier's *Vers une architecture* in English translation *(Towards a New Architecture, 1927).* · Anne notes in diary: conversations about "Fuller Houses," December 1927. · *Designs and manuscripts for Lightful Houses* (January to March 1928). (Vol. 2) · Develops 4D Philosophy (architecture as part of a four-dimensional world); published privately under the title 4D, *Timelock.* Distributes the work at a meeting of the American Institute of Architects (AIA) in St. Louis, Missouri, on 17 May 1938; the work pronounces against industrial construction. RBF calls for a *Spirit of St. Louis* in architecture. · Sends the 4D book to friends, acquaintances, and influential people, beginning on 28 May 1928; later editions beginning with 28 July include mimeographed sketches. · Patent for the 4D House is submitted, but not pursued to completion. · Series of sketches for the 4D Concept (4D Houses, 4D Transportation).

1928 Henry Ford publishes My Philosophy of Industry (March–April). · Amelia Earhart, a friend of RBF, crosses the Atlantic. · Airship Graf Zeppelin, Atlantic flight to America. · Five-year plan for the industrialization of the USSR.

1928

September: First presentation of the 4D sketch, called "Hexagonal House," Le Petit Gourmet Restaurant, Chicago. · December: First publication of Fuller's 4D sketch in *The Chicago Evening Post.*

1929

Work with design students in Chicago. · Construction of second *4D House* model. · April: Two-week exhibition of the 4D House in the Marshall Field Department Store, Chicago. · Creation of the *Dymaxion* concept by the Marshall Field advertising staff. RBF holds six half-hour lectures daily. · May: Exhibition at the Harvard Society of Contemporary Art, Cambridge, Mass., then at several locations in Chicago and New York. · Spends the summer on Bear Island and buys the motorboat *Lady Anne.* · Publication of the *Dymaxion House* in *Architecture* (published by Charles Scribner, N.Y.C.). The publisher plans a book on the Dymaxion House, for which a Fuller lectures is scheduled in July and recorded by a stenographer (the book was not published) [see *Your Private Sky: Discourse*]. October: Fox Movie Tone, news reel shows RBF with models of the Dymaxion House. · Fall: The family moves to Long Island, N.Y. RBF remains in New York City. · Meets the Japanese-American sculptor Isamu Noguchi, a student of

 House using Stockade building techniques, around 1926

The author Margaret Fuller, 1810–1850, RBF's great aunt

RBF with structural model of 4D House, 1928

Brancusi. Noguchi becomes one of RBF's closest friends. Noguchi makes a portrait bust of RBF of chromed bronze. · Active in the Greenwich Village scene of artists and intellectuals. · November: Series of lectures in Romany Marie's Tavern, N.Y.C., Greenwich Village; RBF designs the interior for Romany Marie. · December: Exhibition of the Dymaxion House model in the studio of Isamu Noguchi. Exhibition and lecture by RBF in the studio of Dorothy Moise, N.Y.C.

1929 British airship R 100 with geodesic construction (Barns Wallis). · A. N. Whitehead, *Process and Reality.*

1930

In addition to the family apartment in Woodmere, Long Island, Fuller rents a space on the roof of the Lehigh-Starret Building in N.Y.C. Designs a *Tension Room* for Eli Jacques Kahn as a gallery space for Noguchi's sculptures. · Numerous one-man shows with the house model and lectures in New York, Chicago, Cambridge, Washington, D.C., etc. · Hugo Junkers takes notice of RBF and evaluates the construction of the Dymaxion House. Notices on the industrialization of building. · Contacts with a group of engineers and scientists at Columbia University (Technocracy, Inc.); distances himself from Howard Scott and his technocracy movement. Contacts with Diego Rivera and the John Reed Club; distances himself from communism.

1930 Construction of the Empire State Building, at 415 meters to the tip of its mast, it is the tallest building in the world. · Founding of *Fortune* magazine (editor: Henry Luce). · Albert Einstein, "Religion and Science" *The New York Times* (9 November 1930).

1931

Designs a bathroom unit that can be industrially produced for American Radiator's Pierce Foundation. · RBF reads Albert Einstein's essay "Religion and Science" in *The New York Times Magazine,* and it impresses him deeply.

1931 Gödel's theorem of the undecidability of propositions. · Piccard's balloon flight into the stratosphere. · Three-engine aluminum airplane by Ford (W. B. Stout, engineer).

1931–32

RBF cashes out his life insurance and purchases the architecture magazine *T-Square;* renames it *Shelter;* publishes his own contributions (e.g., the series of articles "Universal Architecture"). · Isamu Noguchi creates the sculpture *Miss Expanding Universe* (December 1931), which is used on the cover of the November 1932 issue of Shelter. · After Franklin D. Roosevelt is elected president and the New Deal is introduced, the magazine ceases publication.

1932

Foundation of the Structural Studies Associates (SSA), which RBF along with, among others: Archibald MacLeish, Knud Lönberg-Holm, Frederick Kiesler, Simon Breines. · Criticism of *International Style.* Lecture series for the SSA. Museum of Modern Art rejects and exhibition of the Dymaxion Housemodel. Shows in N.Y.C. und Philadelphia. · RBF participates in a project with Simon Breines, in which an office building abandoned during the Depression is converted to shelter for the homeless.

1932 Isolation of the ninety-second chemical element, completing the periodic system. · *The International Style,* book and exhibition by Hitchcock and Johnson at the Museum of Modern Art.

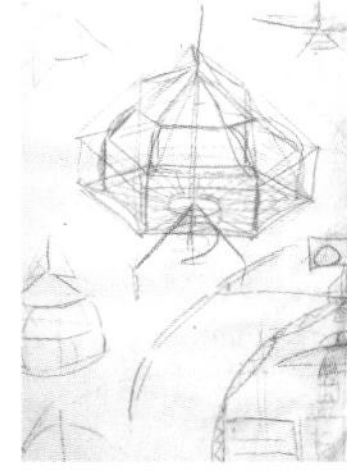

 RBF with second model of Dymaxion House, 1929

Mobile Shelter for the Soviet Union, sketch, around 1932

1933

February: Planning commission of the USSR is interested in the Dymaxion House

1933 Hitler becomes chancellor of the Reich. · Franklin Delano Roosevelt becomes president of the USA: New Deal. · Henry Luce founds and publishes the illustrated magazine Life. · Roosevelt introduces a moratorium on banks, closes all the banks in the USA (March). · *Century of Progress* World's Fair in Chicago.

1933–34

With $ 5000 in cash from the stock broker Philip Pearson and the enthusiastic support of Pearson's wife, Temple, RBF is able to build the Streamline car published in *Shelter.* · Founding of the *4D Company.* · March: Construction of the prototype for the Dymaxion Car begins in Bridgeport, Conn., with the ship designer Starling Burgess and a group of former Rolls-Royce mechanics. · In addition to Fuller, the renowned yacht builder Starling Burgess decisively influences the design. · First trip in the completed car in July 1933. First presentation at the Automobile Show in New York. · Involvement in a serious accident results in negative headlines for the Dymaxion Car. Dymaxion Car No. 3 is introduced with much publicity at the World's Fair in Chicago (1933–35). RBF meets the writers H.G. Wells and Christopher Morley, who will become one of RBF's best friends. · Morley writes the essay *Streamline: Thoughts in a Dymaxion* (1934).

1934 Electrification project of the U.S. Tennessee Valley Authority.

1934

Death of RBF's mother. · Friendship with Martha Graham, for whose dance performance *Frontier* Isamu Noguchi conceived the original stage design (1935).

1935–36

Collaboration in the construction of the research and development division of *Phelps Dodge Corporation,* the third largest copper company in the world. · Works out the research program and completes statistical and historical analyses and maps on extracting and working strategic metals. · The results are used by *Fortune* magazine and the Bureau of Standards in Washington, D.C. · Conceptual work on a prefabricated wet cell; production of twelve prototypes of the Dymaxion Bathroom; applied for patent in May 1938. · Exhibition of the Dymaxion Bathroom at the U.S. Bureau of Standards (1936–38), Museum of Modern Art, etc.

1936 The term *automation* is introduced at General Motors.

1936

Participates in experimental television broadcasts at the CBS experimental studio, under the direction of Gilbert Seldes. · RBF explains to his Noguchi the equation $E = mc^2$ in a telegram. · Work on the manuscript of his first book, *Nine Chains to the Moon* (written summer 1936), published in 1938. · Meets with Albert Einstein to discuss the manuscript of his book.

1937 Splitting of the atom, Hahn and Stressmann. · Jet gas turbine (Whittle); nylon.

1938–40

Technical advisor on the editorial board of *Fortune* magazine. Collaborates on a series on the topic "The New U.S. Frontier" and on the anniversary edition celebrating ten years of *Fortune,* with "USA" as the theme and "U.S. Industrialization" as the lead article. · Advisor to *Life* magazine.

 RBF with the Dymaxion Car, 1933

RBF with Gilbert Seldes, TV transmission test

Publicity poster for book *Nine Chains to the Moon,* 1938

1939

Unpublished book manuscript: *Ballistics of Civilization.* · Dymaxion Bathroom is exhibited in a newly opened building for the Museum of Modern Art.

1939 O. Neurath, Modern Man in the Making. · Long-range bomber of the R.A.F. *Wellington* uses geodesic construction (B. Wallis) in serial production.

1940

***Untitled Epic Poem of Industrialization:* extensive manuscript, first published in 1962. · Model of the Dymaxion House is destroyed while being transported to Rockefeller Center. · Studies commissioned by the industrialist Henry J. Kaiser. · Reengineers metal grain bins as shelters: Dymaxion Deployment Unit.**

1940–42

Request from the British War Relief Organisation to design emergency shelters for war homeless. A prototype is built by the Butler Manufacturing Company. · The design for the Dymaxion Deployment Units is patented. · Because of wartime restrictions on metals, the design is not mass produced. · DDU exhibited in Washington, D.C. (1941). Production of a small series for the Air and Signal Corps; employed in Alaska and Iran.

1941 Japanese attack on Pearl Harbor; USA enters the war (December).

1942 Halford Mackinder, *Democratic Ideals and Reality.* · Uranium atom split at the University of Chicago, the *Manhattan Project.* V2 rockets. · New edition of D'Arcy Thompson, *On Growth and Form.*

1942–44

Leading engineer on the Board of Economic Warfare in Washington, later Foreign Economic Administration. · Studies on wartime economy, for example, an investigation of the industrialization of Brazil. Twice weekly lectures on world energy and geography for coworkers in the departments and administrations in Washington. · Works out the *Dymaxion World Map;* in 1940, the world energy map appears in *Fortune* (February issue); in 1943, first version of the Dymaxion World Map as a tear-out sheet in *Life* magazine. · Geometric investigations, or preliminary studies to energetic geometry. · Sleep experiments (1943) [see *Your Private Sky: Discourse*]. · Lectures in the Cosmos Club, Washington, D.C., for scientists and engineers, on the history of discoveries and inventions (1943). Contact to Vannevar Bush (Office of Scientific Research and Development) and the scientists of the Manhattan Project. · Display panel *Profile of the Industrial Revolution* (1943). · Publication of an essay on the World Map, "Fluid Geography" (1944) [see Your Private Sky: Discourse]. · First manuscript on energetic geometry (1944) [see *Your Private Sky: Discourse*].

1943 Sikorsky Helicopter. · Wendell L. Willkie, One World. · Liberty Ships produced on assembly line (Henry Kaiser).

1944 E. Schrödinger, *What Is Life?,* which predicts DNA.

1944–47

Planning and development of the *Wichita Dwelling Machine* in Wichita, Kansas. · Uses primarily light metal. Production of prototypes in a special program of the Beach Aircraft works in Wichita. · Founding of the Dymaxion Dwelling Machine Company, for the purposes of patenting,

 Advertising prospectus for DDU: Better Homes for Lower Incomes, 1940

Test of acoustics in Wichita House, 1946

licensing, coordinating with government offices and the defense industry, as well as building a network for sales and distribution. · Despite diverse efforts, serial production is never started. · The members of the development team for Dymaxion Dwelling Machine, Inc., include, among others, Cynthia Lacey, Herman Wolf, Edward Applewhite. In 1945, renamed Fuller Houses, Inc.

1945 Death of F. D. Roosevelt, succeeded by H. Truman; Vannevar Bush, *As We May Think.* (computer concept) · Destruction of Hiroshima and Nagasaki by atom bombs. · End of World War II. · Truman's concept *Intellectual reparations.*

1946 *Eniac* computer. · Regular trans-Atlantic service of passenger planes (DC-4).

1946

Obtains patent for the map projection of the Dymaxion World Map. · Founding of Fuller Research Foundation with the goals of basic research in geometry, Fuller documentation, and the preparation of patents. · Private printing: *Designing a New Industry* (1946). · Private printing: *Earth Inc.* (1947) [see *Your Private Sky: Discourse*].

1947–48

Return to Forest Hills, New York. · Two-year period of work on Energetic-Synergetic Geometry. · Discovery of Jitterbug Transformation (March–April 1948). · Work on the concept of geodesic domes. · Application: true planetarium *Your Private Sky* (July 1948) · First great-circle model with four-foot diameter. · Summer 1948: Failed attempt to construct a prototype with a diameter of fifty feet, together with students at Black Mountain College, North Carolina. · First lecture on energetic geometry. · RBF befriends Ruth Asawa, John Cage, Merce Cunningham, Josef and Anni Albers, and others . · Inspired by RBF's lectures, Kenneth Snelson discovers the principle of mulitpolar tensegrity structures (1948–49). In the "magic summer" of 1948, the core of a genuinely American form of art develops at Black Mountain College. · Meets Konrad Wachsmann, Serge Chermayeff, and Peter Blake in Chicago. Construction of the first Necklace Dome there (collapsible structure, i.e., flexible nodes) with a diameter of fourteen feet. · Student projects *Autonomous Package, Autonomous Dwelling.* · Teaches regularly at the Institute of Design, Chicago, beginning in September 1948.

1947 Commercial television broadcasts begin in the USA.

1948 Linus Pauling, *The Nature of Chemical Bond.* · N. Wiener, *Cybernetics; or, Control and Communication in the Animal and the Machine.* Sigfried Giedion, *Mechanization Takes Command.* 1948

1949

Teaching at Institute of Design, Chicago, and the University of Illinois, Champaign. Visiting lecturer at Illinois Institute of Technology: energetic geometry. · First seminar at North Carolina State College. Construction of Necklace Dome in the Pentagon Garden, Washington, D.C. (February). · RBF is dean of the Summer Institute, Black Mountain College; construction of the Skybreak Dome with transparent plastic skin. Meets Robert Rauschenberg. · Essay: "Total Thinking". · Visiting Critic at M.I.T. Seminars and visiting lectures at Bennington College, University of Michigan, Harvard University, University of Toronto.

1949 Mainframe computer for war aims: *Cold War.*

 Early study model of geodesic domes

Construction of the dome from top to bottom, assembly mast, around 1948

1949–53
Extended lecture tours and teaching in the USA, including North Carolina State University, School of Design. · Study project: Automatic cotton factory. · Under the name of "anticipatory design science," he subsumes a mathematical-structural model theory with a preventative planning and design theory. Anticipation of global environmental problems. · Meets Duncan Stewart (1949). · Other important collaborators during the years 1949 to 1955 include T. C. Howard, James Fitzgibbon (Synergetics, Inc.), Don Richter, Kenneth Snelson, Jeffrey Lindsay, Shoji Sadao.

1950 L. von Bertalanffy, *Theory of Open Systems in Physics and Biology.*

1950
December: Construction of the first full-size geodesic structure: a semispherical dome in Montreal with a diameter of forty-nine feet, made of aluminum tubes, each of which weighs one pound. Developed by the *Fuller Research Foundation, Canadian Division* (Jeffrey Lindsay). Aluminum was still rationed in the USA, and thus was not available to RBF, so he turned to Canada. Unpublished manuscript *Noah's Ark 2* [see *Your Private Sky: Discourse*].

1951
Fuller's first use of the metaphor "Spaceship Earth." · Press reports on geodesic domes. · Museum of Modern Art exhibits a geodesic dome. Patent application (December 1951).

1952
Study project *Geoscope,* Cornell University, Ithaca, N.Y. · Paperboard Dome, Princeton University. · Elaine de Kooning, *Dymaxion Artist* [see *Your Private Sky: Discourse*]. · RBF meets Shoji Sadao at Cornell University, Ithaca, during the construction of the geodesic globe "Geoscope"; they collaborate closely from that point on.

1953–54
Fulfillment of a commission from Ford Motor Company to build a dome over the corporate headquarters in Dearborn, Michigan; first commercial application of the principle of the geodesic dome on a lightweight metal framework; duration of project: four months from receipt of commission to completion. · Hyperbolic-parabolic wood construction for the *Geodesic Restaurant* in Woods Hole, Massachusetts (G. Peterson). · Extensive teaching activity at many universities in the USA, as in subsequent years. · First *Tensegrity* Sphere (1953).

1953 Herbert Bayer, *World Geographic Atlas.*
1954 DNA double helix; discovery of the genetic code by Watson and Crick. · Linus Pauling, *Modern Structural Chemistry* (Nobel Prize for Chemistry).

1954
Icosahedron projection of the Dymaxion World Map. · Patent awarded for the construction of geodesic domes. · Test and acquisition of geodesic domes by the U.S. Marine Corps and Air Force. · Development of a dome type without metal for radar stations at the Lincoln Laboratory

 Fuller's studio at a university

RBF with model of Skybreak, 1949

Testing load with students, around 1950

(in parallel with computer research there). · The polyester-fiberglass construction is superior to tires. First Radome installation on Mt. Washington. · Production and assembly of Radomes for the DEW Line *(Distant Early Warning).* · Grand Prize at the Triennale di Milano for the two Cardboard Domes (corrugated paperboard). · Prizes and honors from, among others, the American Institute of Architects, the Marine Corps, Michigan State University, as well as honorary academic degrees; Visiting Professor at the University of Michigan. Lectures on Synergetic Geometry.

1955

Founding of Synergetics, Inc., Geodesics, Inc., with the goal of exploiting the patent, licensing, and carrying out commissions for industry and the state.

1956 Introduction of transistor devices on the market.

1956 on

State purchases domes for U.S. exhibition pavilions at industrial and trade fairs. First geodesic pavilion for the Miel Continent Jubilee, St. Louis, Missouri.

1957 October: First artificial satellite (Sputnik). "Sputnik Shock" in the USA; founding of NASA. Radar early warning system, DEW Line installed along the polar circle; first use of computers. · International Geophysical Year. · First civilian atomic power plant.

1957–59

Construction of large geodesic domes for, among others, Kaiser Aluminum and Union Tank Car Company, Baton Rouge, Louisiana. · By late 1959, more than one hundred licenses to construct geodesic domes have been awarded on the basis of Fuller's patent rights (5% of sales). · *Play Domes,* climbing frames for playgrounds, are mass produced. *Plydomes,* domes with plywood panels, are produced commercially. · Dome built for Synergetics Inc.'s Society of Metals. · Construction of the U.S. pavilion of the American exhibition in Moscow, Sokolniki Park, 1959: geodesic dome, inside the first multivision with film and slide projections by Ray and Charles Eames. · Lecture tour through South Africa, Japan, India, England, among other places, and visiting lecturer. · Exhibition in the Museum of Modern Art in New York (1959).

1958 First American satellite. · Invention of laser. · Atomic submarine *Nautilus* crosses under North Pole.

1958

Invited to England by the Independent Group, Institute of Contemporary Art (Richard Hamilton, John McHale, Alison and Peter Smithson, Reyner Banham. and others). · Lecture R.I.B.A., London.

1959 Worldwide passenger service with jets; Boeing 707; Luna 3 transmits photos of the hidden side of the moon; integrated circuit (Kilby/Noyce).

1959

Professorship at the Southern Illinois University, Carbondale; research and teaching at the Department of Design. · Participates in first Delos Symposium, organized by Doxiadis. · Memorial speech for Frank Lloyd Wright in Taliesin, Wisconsin. · Correspondence with the virus researchers A. Klug and T. Fink on the icosa-geodesic structure of the polio virus. Tensegrity mast is exhibited in the Museum of Modern Art.

 Preparing for the Triennale di Milano, 1954

American Exhibition, Moscow 1959

American Exhibition, Moscow 1959

1960

Project *Dome over Manhattan* two miles in diameter. · RBF and Anne occupy their home in Carbondale: Home Dome, a geodesic dome of wood, a series model by the Pease Company of Ohio, design by Al Miller. · Receives several honorary doctorates. · Publication of *The Dymaxion World of Buckminster Fuller* by Robert Marks.

1960 John F. Kennedy elected U.S. President. · J.C.R. Licklider, *Man-Computer-Symbiosis.*

1961

Patent for Octet truss granted.

1961 Genetic code deciphered. · Yuri Gagarin first man in space.

1962

Meets Marshall McLuhan at Delos Symposium.

1963

RBF is Charles-Eliot-Norton visiting professor for poetry at Harvard University.

1962 John Glenn first American astronaut in space. · Photographs of viruses with electron microscope are compared to geodesic domes.

1962–67

Initiative for a ten-year student program World Design Science Decade with the support of the Union of International Architects (UIA). · Construction of the research, planning, and coordination center Inventory of World Resources, Human Trends and Needs at Southern Illinois University, in which the motifs of global projects are intended to run together. · Executive Director John Meltale. His book *R. Buckminster Fuller* appears in 1962. Tensegrity patent awarded in 1962. *Education Automation* published (1962).

1963 Telstar communications satellite. · John F. Kennedy assassinated. He is succeeded by Lyndon B. Johnson. · Vietnam War. · World Congress of Virologists in Cold Spring Harbor, N.Y. Announcement of the discovery of the protein shell structure of viruses; accords with RBF's formula $10\ F^2 + 2$.

1963

Lecture at the Sixth World Congress of the UIA in Mexico City (Mexico Lecture). · RBF publishes autobiographical writings and essays under the title *Ideas and Integrities;* publication of *No More Secondhand God.* · Research project: Advanced Structures Research for NASA. Scholarship for RBF's student Joe Clinton. · 1963–68 RBF advisor to NASA.

1964

***Time* magazine dedicates a cover story to RBF; cover by Artsybashev. Series of articles, "Prospects of Humanity" in *Saturday Review.* · BBC television broadcast with RBF, on science program of Channel Two. · Participation in the Dartmouth Conference in Leningrad, a meeting between leading figures of the USA and USSR. Founding of the architectural office Fuller & Sadao. · Project "Geoscope" with John Meltale.**

1964 Free Speech movement, Berkeley, California.

RBF with Charles and Ray Eames and Konrad Wachsmann, around 1960

RBF with George Nelson, around 1950

Exhibition at Museum of Modern Art, NYC, 1959

1965–75
Together with John McHale, editor of *World Design Science Decade* (six volumes), with lectures by Fuller as well as statistical material, maps, diagrams for an inventory of the planet, its population, sources of raw materials and energy, its industrial use and circulation, and so on. · Patent awarded for geodesic Monohex structures *(Fly's Eye project)* and Laminar Domes with folding skin.

1965 First commercial news satellite in orbit. · Space walks by cosmonauts and astronauts. · Photos from Mars. · California becomes the most populous state in the USA · U.S. carpet bombings in Vietnam; napalm. · Marshall McLuhan, *Understanding Media: The Extensions of Man.*
1966 Soft landing of one Russian and one American moon vehicle. · Television images of the far side of the moon. · Barbara Ward, *Spaceship Earth.*

1965
At Southern Illinois University Fuller inaugurates the project World Game, a computerized display of the world whose concept is modified in a series of variant projects and then later expanded to include project courses and workshops.

1966
Lectures in Cape Kennedy on transformation and use of the fallout of technology from space travel for an industrialized production of dwellings. · Project Yomiuri Tower, tetrahedral tower that is two miles high (not realized). · RBF profile by Calvin Tomkins appears in *Saturday Review.*

1965–67
RBF proposes *World Game* as the U.S. contribution to the World's Fair in Montreal; rejected by the U.S. Information Agency. · Fuller & Sadao, Inc., obtains commission to plan and design the U.S. pavilion for the World's Fair, Expo '67 in Montreal, Canada. Motto: "Creative America". · Construction of a geodesic dome as a three-quarter sphere with a diameter of seventy-six meters and a height of sixty-one meters.

1967 Cultural Revolution in China. · Heart transplant (Barnard). · Record number of visitors at Expo '67 in Montreal, Canada: more than fifty million in six months.

1967
The Expo Dome makes Fuller and his geodesic domes world-famous. · Jasper John's painting of the Dymaxion World Map is exhibited in the Expo Dome · Special issue *2000+* of *Architectural Design.* · RBF and Anne celebrate their Golden Anniversary. · Purchases sailing yacht *Intuition.* · Fuller interviews printed in American and English underground magazines. · Student groups and youth communes build their own geodesic domes. Project *Tetra City* as floating city with growing tetrahedral structure. · Founding of *Whole Earth Catalog* by Stuart Brand, inspired by RBF. "The insights of Buckminster Fuller initiated the *Whole Earth Catalog.*" · John Cage visits RBF to discuss Cage's manuscript for A *Year from Monday.* · Following a lecture by RBF, the idea to found Drop City (drop-out community) is developed; domes and zomes of junk parts.

RBF with Marshall McLuhan, 1962

Time Magazine, cover by Artsybashew, 1964

World Game Seminar, 1969

1968

Honorary degrees in USA and England; received by Queen Elizabeth. · Christening of ship *Intuition.* · RBF and Norman Foster meet; begin collaborating on many projects. Design for the Samuel Beckett Theater, Oxford.

1968 Serious student unrest at American universities. · Paris – May 68 · *A Year from Monday* by John Cage is published; it makes reference to the ideas of RBF. · Apollo 8 orbits the moon. · Jet offensive of the Vietcong; resignation of U.S. Secretary of State Mc Namara. · Assassination of Martin Luther King. · Hippie movement in San Francisco. · L. von Bertalanffy, *General System Theory.* · Development of ARPA-Net as the precursor of the Internet. Computer networks. · First edition of the *Whole Earth Catalog* appears, later becomes the most successful publication of the American alternative scene. · *2001: A Space Odyssey* by Stanley Kubrick, screenplay by Arthur C. Clarke (final sequence with the star child), RBF: "The new life."

1969

Pilot project studies for World Game at the New York Studio School of Painting and Sculpture with an interdisciplinary group of students. · RBF develops the concept of a global electrical circuit. · Project with homeless youth and slum tenants in New York and St. Louis. · June–July: First World Game Seminar (with Ed Schlossberg) at the New York Studio School. Preliminary stage for a computer patent. Consultation with a patent lawyer. · Project World Man Center, conference building for Cyprus. · Fuller & Sadao (not realized). · October: *Operating Manual for Spaceship Earth* appears as a book, as does *Utopia or Oblivion.* · Nehru Memorial Lecture, New Delhi. Friendship with Indira Ghandi. Fuller nominated for the Nobel Peace Prize. · Construction and opening of ecumenical Religious Center in Edwardsville, near St. Louis, Missouri, a building with a concrete pedestal with a geodesic dome, in whose plastic outer surface the outlines of the continents appear; design by Fuller & Sadao, Inc.

1969 Theory of black holes. · M. Gell-Mann receives Nobel Prize for discovery of quarks. · Rock festival in Woodstock (400,000 participants). · First moon landing (Apollo 11 Mission, 20 July, 1 billion television viewers of a world population of 3.5 billion). · Design for distributive networks: Problems with networking computers are solved (Paul Baran).

1970

The CIA searches RBF offices in Carbondale on a "fact-finding mission." Fuller is suspected of sympathizing with the rebelious students. · RBF's research budget at Southern Illinois University is cut. · Research report for NASA prepared by Joseph Clinton: *Advanced Structural Design Concepts for Future Space Missions.* · Book on John Cage by Richard Kostelanetz is dedicated to RBF. · October: Meets Ezra Pound in Venice. · December: *Dome Book* (1) appears, edited by Lloyd Kahn. · Contains examples of and instructions for geodesic domes; Joseph Clinton compiles the data and calculations for the rod measurements. · Patent awarded for the catamaran construction *Rowing Needles.*

1970 Police activity at several American universities. · Southern Illinois University is temporarily closed because of violent confrontations.

1971

NBC television program *Buckminster Fuller on Spaceship Earth* (R. Snyder). · Climatoffice project together with Norman Foster. · Old Man River's City project for the renewal of a slum in East St. Louis. · Concept: "An Umbrellaed Town" 1970–74. · Book *Drop City* appears (Peter Rabbit,

 RBF on Bear Island, around 1967

RBF with Isamu Noguchi and Ezra Pound, Venice 1970

pseudonym). · RBF's finances at Southern Illinois University cut back sharply. RBF finances World Game from his own pocket; Medard Gabel is his collaborator. · Special editions of *Architectural Forum* and *Architectural Design* on RBF.

1971 Twin towers of the World Trade Center in N.Y.C. now highest building in the world. · Book by Viktor Papanek, *Design for the Real World.*

1972

Planning of *Spaceship Earth* at Disney World, geodesic full sphere, main attraction of the Epcot Center. · Do-it-yourself dome construction of wood slats and plywood, design Fuller & Sadao, published in *Popular Science,* a success with the public. · *Book Intuition* is published. · Hans Meyer develops with RBF a windmill construction to be mass produced at low cost. · RBF is editor of *World Magazine.* · He is offered a position in Philadelphia as World Fellow in Residence. · Moves residence and research office to Philadelphia.

1972 Zero population growth in USA. · Bay Area Rapid Transit in San Francisco, first new subway system since 1907. · U.S. research station at South Pole with geodesic dome. · *Life* ceases publication.

1973

RBF criticizes *Limits of Growth* as Malthusian. · Hugh Kenner publishes the book *Bucky: A Guided Tour of Buckminster Fuller.* · Patent for tensegrity dome granted.

1973 Energy crisis: oil shortage and the first recession in Western industrial nations since 1945. · E. F. Schumacher, *Small Is Beautiful.* · The book *Limits of Growth* by Dennis and Donella Meadows appears as a report to the *Club of Rome.*

1974 Watergate scandal: President Nixon resigns. · Manipulation of genes is debated. · Search for intelligent life in space. · Theory of fluid crystals (P.-G. De Gennes) as a transitional phase in complex physical systems.

1975

January: Two-week-long lecture: *Everything I Know* is recorded on video: playing time is forty-two hours. Patent awarded for asymmetrical tensegrity construction. · RBF receives his first license as an architect, from the State of New York (at the age of seventy-nine). · Fuller's magnum opus *Synergetics* appears as a book (vol. 1), collaborator: Edward Applewhite. · RBF professor emeritus at Southern Illinois University and University of Pennsylvania. · Project Fly's Eye, Fuller's last design for an industrially produced dweling at low cost; geodesic dome of standard shell elements, round openings are left out, monohex principle.

1976

Hugh Kenner, *Geodesic Math. and How to Use It.* · Synergetics exhibition for the opening of the Smithsonian / Cooper-Hewitt Museum of Design. · RBF publishes book *And Came to Pass – Not to Stay.* · *Tetrascroll* published as artist's book in limited edition by Tatyana Grosman's U.L.A.E. · August: Unpublished manuscript: *Sense of Significance: A Memoir of Friendship,* dedicated to the friendship between the poet Christopher Morley and RBF, written by Louise Cochran, Morley's daughter.

1977 Star Wars. · Alaskan pipeline put in service. · Military research: cruise missiles, neutron bombs, laser cannons. · Fiber optics for data transfer, tests conducted by Bell Laboratories.

RBF drawing in wet cement, Bali, around 1975

John Cage, RBF, Merce Cunningham, around 1978

1979

Second volume of *Synergetics* is published (collaborator: E.J. Applewhite). · Project RBF/ N. Foster: International Energy Exhibition, Knoxville, USA. · Harold Loeb teaches geometry to design students at Harvard University on the basis of RBF's *Synergetics.* · World Game Laboratory founded. · Medard Gabel publishes the books *Ho-Ping: Food for Everyone,* 1979, and *Energy, Earth, and Everyone,* 1980.

1979 Atomic reactor accident near Harrisburg, Pa., leads to a reevaluation of the atomic energy program. · The engineer Paul MacCready crosses the English Channel with muscle-powered airplane *Gossamer Albatross* and dedicates a photo documentation to RBF. · Gaia thesis of James Lovelock: Earth as living creature, biologistical parallels to Fuller. · First Personal Computer: Rank Xerox (Menlo Park).

1980

New edition of the Dymaxion World Map by R. Grip, C. Kitrick. · Moves from Philadelphia to Pacific Palisades, California.

1981

RBF: *Critical Path* (St. Martin's Press, N.Y.C.)

1982

Fuller/Foster: Project *Autonomous House,* a project for Fuller himself. Double shell, the idea of the Garden of Eden. · Disney World Epcot Center with geodesic sphere *Spaceship Earth* opens. · 23 February: Fuller is awarded the *US Medal of Freedom* by President Reagan.

1982 Nobel Prize for Chemistry to Aaron Klug for research on the structure of viral protein shells. Benoit Mandelbrot, *The Fractal Geometry of Nature.*

1983 Construction of the "Spruce Goose" geodesic dome, with a diameter of 415 feet: the largest self-supporting cupola in the world (Don Richter, Temcor). President Reagan initiates the Star Wars defense program SDI (Strategic Defense Initiative).

1 July 1983

Death of Buckminster Fuller from a heart attack. His wife, Anne Hewlett, dies just two days later, on 3 July.

1984 Founding of Buckminster Fuller Institute, Philadelphia, later Los Angeles. Discovery of quasi crystals by Dan Shechtman: for the first time crystals also have five-point symmetry. · Boom in chip production in Silicon Valley, California. Personal computer becomes widespread.

1985 Edward J. Applewhite, *Cosmic Fishing.*

1985 Adrien Fainsilber, "La Géode," Parc de La Villette, Paris: geodesic dome with a polished metal outer surface that curves continuously, diameter twenty-six meters. · Founding of the first public online service, *The Well,* started by the *Whole Earth* group around Stuart Brand and Larry Brilliant: the first virtual community in the world. · Discovery of the carbon C-60 atom buckminsterfullerene by the British-American research team of Harold Kroto, Richard Smalley, and Robert Curl. · In recognition of the inspiration they received from Fuller's Expo Dome, they called the *C-60* molecule *buckminsterfullerene* and the molecular family *fullerene.*

1986 K. Eric Drexler introduces the concept of *nanotechnology* for technical operations on the molecular level.

1990 Kretschmer, Stephanoupoulos, Houffmann: simple methods to produce fullerenes for research. Feverish fullerene research worldwide. Applications are conjectured in the field of half-ladder and supraladder technique and in carbon chemistry generally. Popularization of the Internet with WWW software. · Fall of the Iron Curtain and the collapse of the USSR. · Discovery of the fullerite (carbon-metal compounds).

1991 Experiment Biosphere II in Arizona: Four women and four men live in a closed ecosystem; the experiment lasts for several years. The construction of the large glass house in the form of a stepped pyramid made use of the octet truss.

1992 RBFs book *Cosmography* is published posthumously.

1994 Buckminster Fuller Institute moves to Santa Barbara, California.

1997 Nobel Prize for Kroto, Smalley, and Cure for the discovery of the buckminsterfullerene.

1998 Mathematical proof of Kepler's assumption that cubic-planar-centered packing of spheres is the closest possible one in space (Thomas Hale).

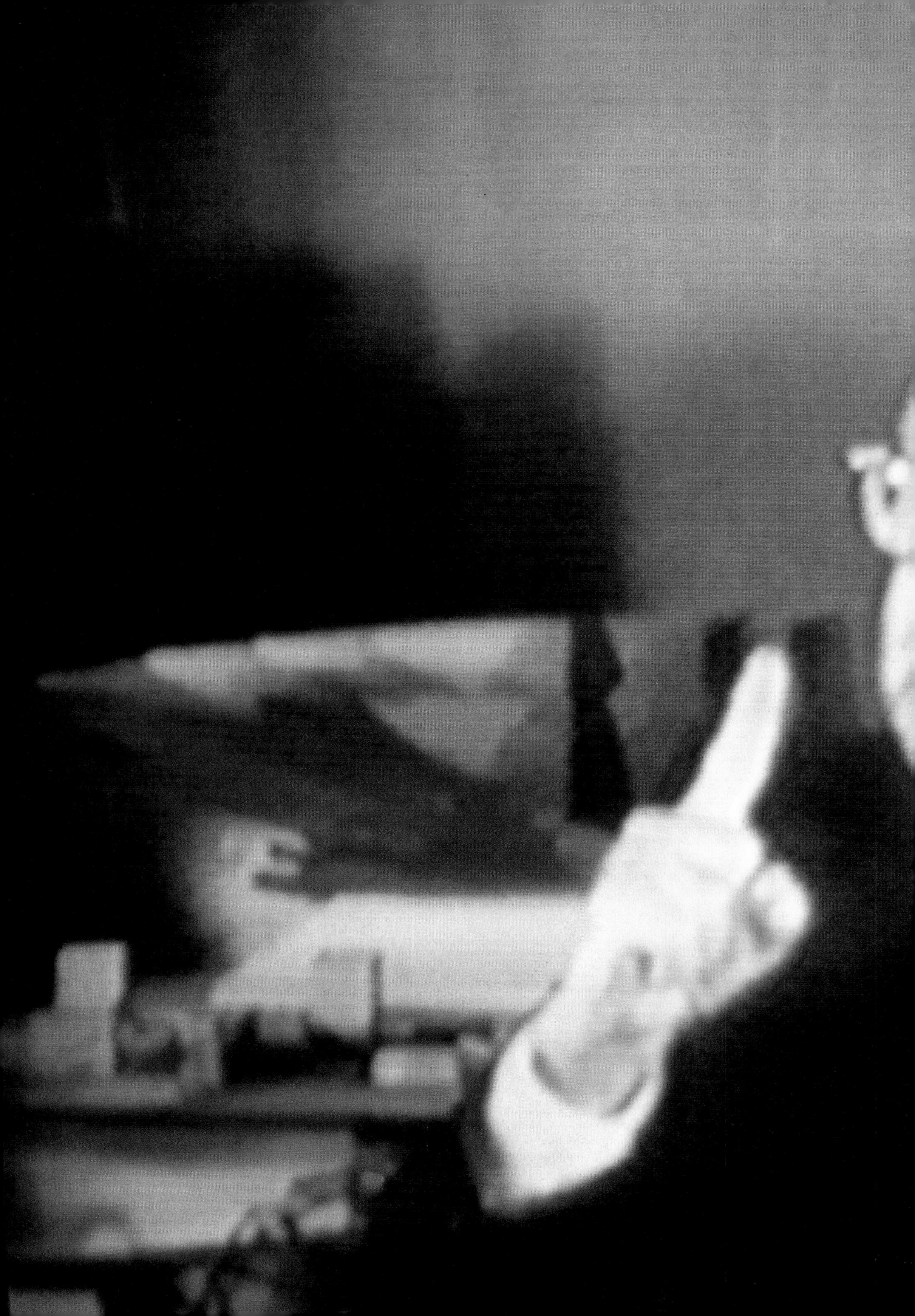

For Buckminster Fuller, the origin of civilization lay in the oceans. Sea travelers faced different challenges than the home-bound "landlubbers" (as he polemically called them); they endured the risks posed by the environment: the endless sea, strong waves, darkness and cold, storms or lulls, shallows and currents. The technologies associated with the rise of science include, for example, navigation by the stars, tacking against the wind, exploiting the trade winds. For millennia, technology took the form of the evolution of effective ships and their "cybernetic" instrumentation. Fuller's term of service as a naval officer in World War I was a decisive influence on his life and work.

NAVIGATOR

2269
2269

My First Invention

My teleologic stimulation first grew out of boyhood experiences on a small island eleven miles off the mainland, in Penobscot Bay of the state of Maine. There, floatable at will, in and out of nature's tidal dry docks, with a fifteen-foot flood rise twice a day, boat-building was the parent technology, and the devices for its original design and fabrication, together with its subsequent sparring, rigging, beaching-out, wintering, cardling, rebuilding, launching, and upkeep in general were so broadly effective as to govern spontaneously almost any technical tasks to be effected on the land, whether this was building of dwellings, barns, well houses, or water-course controls (for water conservation on the island was as essential to survival as was our ability to pass successfully over the waters around the island, away to other islands, or to the mainland). Fishing was the primary local industry, and such tension systems as seines, trawls, weirs, scallop drags, lobster pot heads, and traps, together with all their respective drag and buoy gear, insured an ever-present abundance of stout cordage and light lines as well as experience in net-weaving, tying, splicing and serving. Here men "passed a line" and "took a turn" in deft tension techniques as spontaneous as those of spiders.

This boyhood experience on an island-farm included those first turn-of-the-century days of individual, or family, small tonnage water transportation almost exclusively by sail or rowboat, leading to the experimental inclusion of the newly-invented internal combustion engines. We had in our sloop one of the earliest auxiliary gasoline engines within many miles, and this induced a whole line of inventiveness, along with gallons of sweat, relevant to priming the engine, testing the spark, and rolling over a flywheel. But the rowboat had to serve its complementary tasks, and as I had to row each day on a four-mile round trip to another island for the mail, my first teleologic design invention was a mechanical "jelly fish," or teepee-like, folding, web-and-

sprit cone which was mounted like an inside-out umbrella on the submerged end of a pole. This pole could be hand-pulled through a ring over the stern, drawing the self-folding cone on the pole's water end through the water with little resistance. When pushed by the pole, the cone opened and gave inertial advantage, almost as though touching bottom, to push-pole the boat along far more swiftly and easily than by sculling or rowing.

These trips were frequently rowed in the fog and across strong tidal currents which involved complete dependence upon calculations and compass. The push-pole made it possible to see ahead, having been frustrated in back-towards-bow rowing.

Influences on My Work, in: *Ideas and Integrities* (1963), 9

 ·43: Buckminster Fuller's Yacht *Intuition* off Bear Island, ·RBF in self-built canoe (lath-wood frame, cotton covering), 1915

"But chiefly the seashore has been the point of departure to knowledge, as to commerce. The most advanced nations are always those who navigate the most. The power which the sea requires in the sailor makes a man of him very fast, and the change of shores and population clears his head of much nonsense of his wigwam."

Ralph Waldo Emerson: *Society and Solitude* (1870)

 ·Sketch on the backside of a photograph; of an old ship

LADY ANNE
BEAR ISLAND
MAINE

S.P. 1196

"I had learned in the Navy that when we wanted to light a cigarette on board ship, you would go up in front of the nearest housing to discover almost still air."

RBF, *Designing a New Industry* (1946)

·Rescue of airplanes downed at sea (photo: RBF)
·RBF ▲ with rescue team

·RBF piloting the patrol boat *U.S.S. Ketch Argo*, 1917

Letter to Vincent Astor

I have always liked you so sincerely well, and admired your unostentatious love of the sea, marked ability as a sailor, and mastery of mechanics, which I recall as absolutely natural in you. Indeed, your interest in these affairs was the fundamental background of our friendship.

Of this real significance of yourself I find the public has little or no knowledge. If the public were aware of such significance, it would be unappreciative. The public is still so enveloped in bunkum, as to relative values, that it vaguely prizes you more for the value of a name, long connected with wealth, than for your own character, ability, and individual identity; more as an institution than the damned fine person you are.

For instance, I know none who recall that you commanded on its post-war voyage to this country, the German submarine sea raider, U-34, demanding more than ordinary ability, even amongst professional seafaring men. There are few of the public who, thinking of submarines at all, other than to picture them as a form of hell trap, realize the accidents that have come upon them, were of personal equation origin, usually external, and no more the fault of the machine than of a Rolls-Royce in a street smash, the submarine machinery having functioned, internally, far beyond expectation in such accidents. The public fails to perceive the "hell trap" idea to be undeserved by the submarine, as a mechanical unit itself, "the elder" portion of the public being mechanically incompetent mentally. They do not know that the submarine is the greatest single concentrated grouping of mechanically perfected units, devised to date, and, whether eventually useful as a war weapon, or not, at least it has been the proving ground of many of the most beneficial of mankind's mechanical servants. Unaware of the intricacy of a submarine they are naturally unappreciative of your ability in commanding one.

page from the Na\ … l by Ford Motor Co

Again, few know that you were one of the first to patronize flying, as an every day accessory to life, by purchase and practical commuting use of a plane; to wit: – An Hearst paper editorial here, in Chicago, a short while ago, which queried, scathingly, as to why some of the rich men of today were not taking up aeroplaning as a good example to the public, after the fashion of Mr. Vanderbilt with automobiling in its early days.

I remember discussing with you one evening at your house in Newport, with pleasurable revelation of each other's fund of thought on the subject, our mutual interest in everything to do with boats and mechanical developments. Commonplace as this may seem now, it was actually significant then as being considered slightly moronish, and quite flavoring of the "garage man," for people to be really interested in mechanical developments.

That you were honestly interested in naval mechanical organization and perfection, not in just "dressing up in uniform to play officer" as is the customary unthinking accusation, to the extent of sincerely taking part in, and living aboard the Eagle boats, not having your quite available, and far more comfortable, yacht trail us; that, with me, you were as interested in the perfection of cooking and feeding, in mechanical solutions, in maintenance ability of sanitation in living quarters, and in mental organization able to subordinate an intricacy of mechanics to the useful and easy service of the mind; plus your ability to play, unhypocritically, exhibiting a profound yet happy understanding of a life, was a great, corroborating, inspiration to me, which I have always carried with me, with great elation in reflection upon it, for there was then, and for that matter, still is, little comprehension and appreciation of these things by our "socially smart" acquaintances.

Incidentally your attitude of responsibility, pride in ships, sure, easy, cooperative command that made infinite detail of orderly compliance an industrious pleasure to your ship commanders and crews upon the occasion of minor, though none-the-less of historical interest in navy

·54/55: Vincent Astor's airplane, on loan to RBF for two weeks in 1922. Astor was a Navy comrade of RBF's and one of the [illegible] private pilots in America

·56/57: RBF ▲, U.S. Naval Academy, July 1918
·58/59: *U.S.S. George Washington,* RBF accompanying Admiral Gleaves to the signing of the Versailles Treaty, 1919

annals: to wit the first inspection by Admiral Sims of the Eagle boats, then under your squadron command, which he had ordered to supplant the destroyer forces; which was as well, the occasion of his last inspection, afloat, before his retirement from the naval service. This event was typical of your subordination of extra prejudicial conditions to the best interests of all, your identity being solely that of squadron commander, your ships and their personnel thereby revealing themselves to the best advantage of the service. I have the Admiral's flag, which was flown at that time and which we also flew on our memorably pleasant "cruise," with Admiral Gleaves as our guest, in the Eagle Nr. 15.

Sailing boats were, at that time, naturally "quite alright" as something to play with personally. They had historic precedent, but anything mechanical seemed to smack of poor sportsmanship to the dictators of "what is being done." Steam and mechanical yachts were something to have operated by paid hands. Navy activity outside of war time, to say the least, was termed "naive." In inferring their error I am laying no stress on motor boat racing as being alluring, though a means of mechanical test, for the same thoughtless element are interested in horse racing, not for the love of the racing unit itself, but for the lazy men's hope of getting something for nothing, and the inherent fascination of speed in any form. Speed is the progressive essence of temporality. That our sympathy with these and other interests have been so similar, I have always born with me. Once a really concordant friendship is created, upon such a reasonable basis, I doubt if it may become broken. For this reason I write you at such length on the subject which I perceive to be the most important that will ever confront either of us, that you may serve as you best see fit.

It was actually suggested to me, not long after I had given up command of the 15, that you were a little queer to be "fooling around" with the Navy.

Compare such supercilious attitude, as that, to the words of so distinguished a publicist as Richard Washburn Child, in an article written during the world war, commenting upon the congressional investigation of the War Department, a quotation from which was used in the Foreword of our Annapolis Naval Academy Memory Book: "Are all the good things they are saying about the Navy true? Plenty of Americans are asking. The answer is an emphatic 'yes.' If it were all as good as the Navy, as clean, as simple, as free from buncombe, boasts, and botches, then we would have a war machine indeed; ten minutes with the Navy is enough to convince any observer that the Navy organization bears to the War Department organization the same relation that a Greek Temple bears to a dish of scrambled eggs... Armies are sensational. But it was and is the British and American Navies which can and will save the world... The Navy is the key to our self-respect and our self-defense. So it will be until the sea is solid... the Navy – praise be – has it."

While written under far different world conditions than those of today – Mr. Child was inadvertently forecasting the events which I now perceive to be close at hand, wherein the naval officers will, with their training, far superior to any other which I know of (and which incidentally showed many of our Harvard men to disadvantage under the grueling requirements of the Naval Academy), be called into executive activity of which the run of business executives are incapable.

Relative to the same thought the entrance of Lindberg and Lanpierre into commercial aviation direction, is significant of the step-up of the caliber of world industry management from bullying management without technical knowledge, in hit-or-miss-with-a-bang, to wide scientific and practical knowledge (not worn on the surface, any more than on a neatly uniformed officer, of vast training).

15730-1

Returning to the crux of my argument, Naval officers, capable of handling fleets made up of a myriad of battleships, destroyers, submarines, airplanes, supply and hospital ships, with crews of thousands of men, who must be fed, rested, clothed, employed, entertained and trained, moving in orderly fashion in fair weather or foul, divested of all reference to so called "land advantages," beyond an orderly delivery and equally orderly distribution of supplies; these naval officers, trained to ably handle any and all mechanism of this service, through the whole range from submarine signal and torpedo mechanism, etc., to the most delicate of range-keeping instruments, trained in instantaneous logic as to "what do," in an infinity of emergencies, from "man over board" to "enemy submarine reported on the port bow," will be called into the handling of centrally manufactured, decentralized-hook-up housing, which none of our "smart" businessmen, who primarily cannot even drive their own automobiles, let alone conceive of such a housing industry, have naturally missed. The naval officers who have been "admitted" to business life, to which they have tried to adjust themselves, truly "settling down to business," have had their senses so dulled and befuddled by the disorderly and colossal ramification of business, which they had been led to "believe" were so business-like, in comparison to the Navy, by its blustering thick-headed feudal leaders whom they honored, and they had learned to honor their seniors in experience, as to have been forced to consider themselves "stupid" for, being unable to detect the "system," being unaware of the colossal bunkum that hid from them the fact that there is little or no system in business in comparison to the orderliness and efficiency of the Navy, outside of the money collecting systems of the bankers.

Letter to Vincent Astor, 1928, in *4D Timelock* (1928/1972), 88

"The inventor has natural and immediate access to all the potentials of the universe. Edison, Bell, Marconi and the Wrights needed no licences from anyone to light the night, to shrink the earth and interlink all of humanity."

RBF: *Utopia or oblivion.* 1969

Communications Officer at Sea

"For the landsman, a trip around the world means the conversion of a bank account or an Irish Sweepstakes prize into a procession of hotels; to the sailorman, it is the logical fulfillment of his work, punctuated by his visits to the beach." Or: "For the land dweller, 'the East' and 'the West' are places; for the sea traveler, they are directions in which he can move." Fuller expressed himself with this kind of acerbity in his important text "Fluid Geography" (1946, see *Your Private Sky,* vol. 2: "Discourse"). For Fuller, there is an irrefutable difference between those accustomed to the sea, who have to deal with the constantly moving element of water, and those who live within the dull certainty of the mainland. Everything man has produced in terms of progress, according to Fuller, either came directly from sea travel or from a hunger for scientific knowledge that is an extended sense of the desire to break away from the known to reach new shores. For Fuller himself, the stimuli derived from his years with the navy were decisive for his life: experience with navigation, with anticipating turning or breaking maneuvers; planning supplies of food, fuel, and water, plotting routes, schedules, and the daily routine; sensing the time of day or the season, of phases of the moon, of tides; and not least the personal experience of crossing the ocean in day and night travel, slowly but steadily. Not only did he know what the map of the world represented, he was familiar with it as well.

4 5
11 12
18 19
25 26

My Navy experiences ranged all the way from those small commmands to subsidiary functions within much larger command patterns. The assignments and commissions brought, for instance, incidental participation in the first development of ship-to-plane radio-telephony. They also brought subsequent participation in the conversion of radio-telephony from seventy-mile-range spark sets to long-distance, arc-type equipment. The first instance of the above electronic experiences occurred when a small ship I commanded was selected for the experimental work of Dr. Lee de Forest. While aboard, he established the first successful voice communication ever heard between a ship and an airplane.

I was later fortunate to be Force Radio Officer for the Cruiser and Transport Force of the United States Atlantic Fleet which carried and escorted the American troops across the Atlantic in World War I, and, as personal aide for secret information to its commanding admiral, had intimate experience in large teamwork maintenance of colossal dynamic pattern: for instance, that of maintaining the high-velocity trans-Atlantic turn-around of 130 major troop-carrying ships, together with their cruiser and destroyer escorts. Here was the pattern of secret high-speed exact communication and the strategy of sea group formation, all logistically maintained by expertly scheduled supply and maintenance men at their inspired best, accomplishing the impossible under war-induced spontaneous cooperation. After the war I compiled the official Navy statistics of that operation. During the war I had been editor of its sea-printed publication *Transport*, which had monthly circulation in our fleet of 130,000 copies.

Immediately after the Armistice of World War I, the *U.S.S. George Washington*, one of our ships, was selected for President Wilson's trip to France to attend the Versailles conferences. In the *George Washington* we installed the first long-distance wireless arc telephony which dramatically graduated the ship-to-ship telephone from a seventy-mile-range spark set squawk to an effective two-way trans-oceanic voice communication. On President Wilson's second trip to France, the human voice was transmitted trans-oceanically for the first time in history, as a man was heard through the receiving instrument in Arlington, Virginia, speaking over the transmitter in the radio shack atop the *U.S.S. George Washington* at anchor in Brest, France.

Influences on My Work, in: *Ideas and Integrities* (1963), 15

64 ·RBF with his patented invention, *Rowing Needle,* a 1947-designed catamaran skiff, Bear Island, ca. 1975

“A sailing ship opens up the sea, but unlike a bulldozer it doesn’t hurt the sea.”

An Autobiographical Monologue Scenario (1980)

Though not a state unto itself, it is nevertheless the heart of America's self-image: New England, the coastal region of the northeast. The place where the Puritan immigrants from Britain settled. Where Europe is still palpable, but also palpably on the other side of the water, part of the past. New England, the place where America became aware of itself, where the new is held up to the old and consciously chosen over it – or consciously rejected. Buckminster Fuller, the son of a respected family, saw himself as the enterpris-ing heir to the affairs of the Transcendentalists of Concord, Massachusetts, in which his great-aunt Margaret Fuller-Ossoli had played an essential role alongside Emerson and Thoreau.

NEW ENGLAND

R. Buckminster Fuller came from a respected family that had lived in New England for many generations. One of his ancestors, Thomas Fuller, came to America in 1632 as a British officer, a member of the first generation of colonists. The Fuller family distinguished itself especially as part of the abolitionist movement. Fuller's paternal grandfather, Timothy Fuller, died as a high-ranking officer in the Civil War between the Yankees and the secessionist states of the South. The Fullers occupied political offices as members of government or ministers; several of them were respected jurists. Generations of the talented male members of the family studied at Harvard. Fuller's failure at this renowned university was an especially painful break in this family tradition. It is remarkable, though, that even female members of the family managed to develop within the confines of proud Puritanism. Fuller's great-aunt Margaret Fuller (1810–1850) was extremely important as an advocate of women's rights, as a literary critic for the "New York Herald Tribune", and as a member of the Transcendentalists of Concord, Massachusetts, around Ralph Waldo Emerson, Henry David Thoreau, and others. Within this circle, she was surely the thinker who was most unbiased in confronting the question of keeping current with respect to dynamic technical and social progress. During the late twenties Fuller learned about the significance of his great-aunt by examining her writings. He wrote about that discovery in a letter to his younger sister Rosamund.

· RBF with his grandmother and sister Leslie, Summer 1897
· RBF's puritanical ancestors, photograph 1846. Margaret Crane Fuller (▲ Margaret's mother, RBF's great-grandmother), surrounded by her daughter Ellen (Margaret's sister) and three sons.

Letter from RBF to his Sister, Rosamund Fuller

"What I wanted to say to you is that I have been absolutely astounded at the significance of Margaret Fuller which none of us (present generation of our family) have realized, and upon which I have more or less stumbled at this distant point from home. It is exciting to find Women's clubs 'in these parts', and in many parts of the world, named Ossoli Club, after her. The Library here has many books about her, being published from time to time by Philosophy professors, popular writers, biographers, etc., as well as the earlier biographies written by Emerson, Channing, Thomas Wentworth Higginson, Olcott, Howe, and Grandfather Fuller. It is because of the generations of the Fuller family have been so drawn out, and because Grandfather Fuller, whom she brought up and affected so deeply, was killed in the Civil War, when Father was but a little boy, and because Father died while we were so young, that, with the usual casualness of people to that which is closest to them, we have missed her importance entirely.

She was, it would seem to me, almost as important to civilization as such men as Lincoln, but her efforts were so much more abstract and indirect, being applied primarily to the small coterie of thinkers who formed the original nucleus of an American culture, freed from the feudalistic oppressiveness that characterized its literary work up to that time, under her directive effort, known as the Transcendentalists. These people translated and wrote down many of her truths which are now materially benefitting the whole world, through their brain child, the "Natural Recreative Industry" of present-day America. Her genius was as a rule revealed only in conversation. This method while by far the most unselfish form of creation, thinking nought of capitalizing new truths for self, but seeking untiringly to give off and thus ever increasing the river bed through which the truths may flow with natural facility, has by far the greatest influence, but is often temporarily halted by those seeking to redistribute the truths, as of their own making, there being no material testimony contrarywise.

In fact, those doing the latter may not do so consciously, but are rather of the great army of mental kleptomaniacs, who bundle off truths, and later, by virtue of the fact that the truth has been stored in their brain cells, perceive them to be their own. Margaret Fuller's one great term-poral fault, in my opinion, was her intolerance of her father. Had she in addition to her other great powers the ability to see the raft of advantage heaped upon her by her father's tremendous if well suppressed love (in the best New England fashion), she would be on everyone's lips today. Tolerance is a great virtue and will eventually be greatly rewarded by even keener perception, by virtue of the impassionate reasoning which it makes possible. The most perfect example of what I am pointing out is that of Christ, whose unselfish existence, mythical, symbolical, or real (inasmuch as material temporality is only of secondary or temporary consideration, it does not matter) has for its greatest significance the fact that he left behind not so much as one written word, not one material token. His significance lies in his complete faith in love to perpetuate its recreation to eternity whithout resort to any temporality whatsoever. It is my opinion that the testaments of his teachings and existence have done more to retard the truth of Christ than to aid it. It is like the hypocritical nurse maid testifying to the goodness of the little child. If Christ had wanted his word written he would have written it himself. If he had done so it would have been done in one word, I am sure – Love. The actual writing of his teachings in the form of scriptures, arbitrarily selected for publication, and certainly exquisitely beautiful as word composition, as are the carven images of other eras handed down from generation to generation, ever and anon, acquiring an additional beautifying artistic touch, and always the beauty of patina, are to me much the same, when reasoned out, (with full faith in love) as the correspondence school courses on how to be as successful as Washington or Ford. Everyone of the Saints who wrote the scriptures was temporal and as such committed many sins. While the motive in their writing of his word was well covered by the beauty of the truth of which they wrote, it would have been far more lasting and beautiful if handed on by them in the form of word and deed alone, instead of being confined to a cliquish group of politicians, as it eventually was, who alone are capable of reading the written language, and were able to mete it out to their self-aggrandizement. We find the action of the Saints continually portrayed in the action of contemporarios, who no sooner than a person dies heap laudation upon that person, at the same time discreetly revealing their own intimate relations with that person (usually exaggerated or fictitious), that the dead person, unable to deny the allegations, the merits heaped upon them may redound to the 'fine' influence of their speaker. I feel that this is the fundamental basis of the majority of public funerals. Lord help me I credit the grief of the intimate minority of these occasions. All the scriptures or correspondence school courses, no matter how well done, are but one-track formulas for reestablishing the complete faith in love with which children are born, and live in "unconsciously, and therefore perfectly, until it is broken by their elders; who drudge worn, by the every-day immediate living conditions, suddenly blaspheme that perfection of faith and love, which, once broken, cannot become perfect again in the realm of temporality. All the drudgery can be mechanically removed from our living, with the realization of the stylistic bunk that "aesthetically" controls its design in archaic form, in the tailor-made practice still employed in house building, self-consciously, which is selfishly, thought to indicate distinctive personality to its owner.

All this is the basis of our 4D house philosophy.

Emerson, who is ever more often acknowledged the greatest of philosophers to date, said to Margaret Fuller, who, as first editor of the Dial and literary critic of the New York Tribune, (in which her articles were always featured on the front page), had verbally lashed him into exquisitely unselfish thinking, 'Of those for whom I would wish immortality most, there is none for whom I would wish it more than Margaret (Fuller), for it is she whom we can least be without.'

71 · RBF and his father, Summer 1897
· RBF and his youngest sister Rosamund in self-built play-yard, Bear Island 1906

· RBF▲ and his brother Wolcott, 1904

“I thought I was going to be a great athlete, and then I busted my leg playing football…”

RIBA discourse 1958

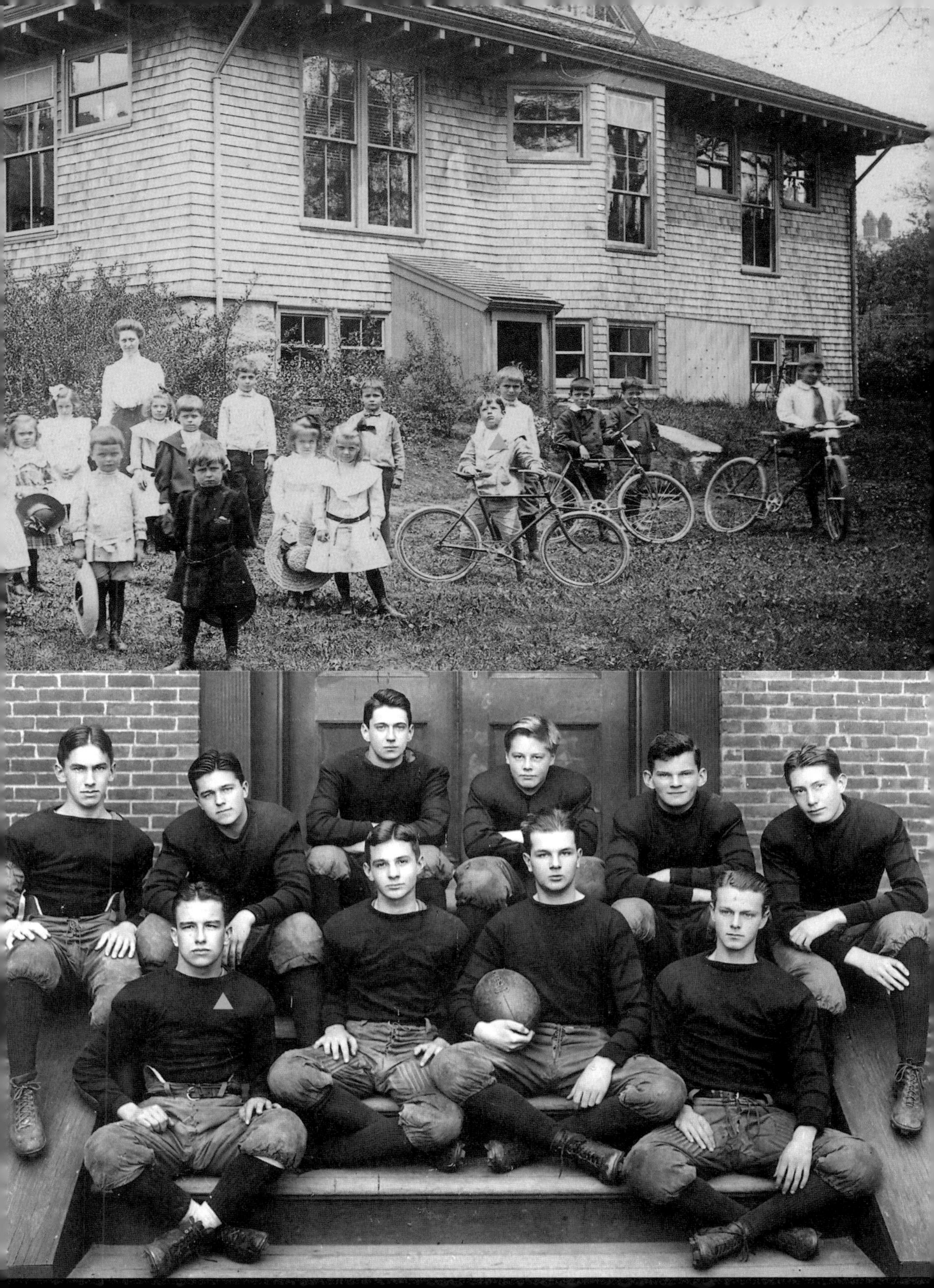

It seems to me, Rosy, that you should learn as much as possible about her. She was well known abroad, too, having been received by Carlisle in England, as the head of a deputation from Emerson; and done heroic work in Italy, while her husband Marchese Ossili fought in the civil war there, just before their final return to this country, on the voyage which ended in their being shipwrecked and drowned off Fire Island, L.I., where the Women's clubs of America have erected a memorial to her. There is not self-aggrandizement in recognizing these facts about her, which, by the natural law of inheritance, may be taken up where left off, when the personal-equation and circumstances are sufficiently licked to allow the truths to come forth, and the work taken up where it was left off.

Of the many books about her there are two particularly good and short ones, as a preliminary study of her. The chapter in Gamaliel Bradford's "Portraits of American Women" and "Margaret Fuller and Goethe" by Professor Braun, published by Holt in 1910. Both are in the public library. I tell you all this, before you go abroad, as I think the effort and time expended by you in reading these, before going, will be amply rewarded by your greater appreciation of all that you observe, in your first vivid impression, of foreign travel. It is fortunate that you go without an older guide. There could be nothing sadder than the truth indicated in the picture (living) which I witnessed the other day of a little girl dragging a still littler group of brothers and sisters through the zoo, whirling and wheedling them by the cages to which they were drawn with natural fascination, it being their first trip to the zoo. The reason for her tenacity and interference with them was that she wanted to get them to look at the zebu, which bore greater significance to her because it had been pointed out to her on her first trip by a favorite uncle, who, unknown to her, picked on it as a target for his own 'smart' remarks. This is the way I heard it, and it illustrates the overemphasis of personal equations. Beware Zebu bally-hooists in your travels.

I am sending you a copy of 4D which has been added to since the copies sent to mother and Wolcott. I hope you will find time to read it, and to bring me comments on your return, brought about by your observations abroad, relative to it.

4D-Timelock (1928/1972), 79

Stockade

The stockade construction system was developed and patented during World War I by Fuller's father-in-law, J. Monroe Hewlett, and Fuller continued to develop it. It uses large but light bricks of straw pressed and cemented together. They were made with round holes that would line up in the bond so they could be filled with concrete. The concrete columns that resulted provided the support; the remaining material in the brick served as insulation. The columns remained hidden inside the wall and gave the process its name (a stockade is a structure made from posts). The Stockade Building Corporation conducted tests on materials – weight stress at collapse – and beginning in 1922 constructed 240 buildings under its young director Fuller before it was bought by Cellotex in 1927 and liquidated shortly thereafter. The two main concerns that Fuller had about the process were construction delays caused by weather and the ability to bear compressive loads.

·RBF's father-in-law James Monroe Hewlett, painter and architect
·Stockade lightweight construction technique, developed by J.M. Hewlett and RBF (building blocks of cement-bound straw, joined together by concrete posts cast on-site), title page of the firm's brochure, ca. 1923

"Baby's Record"

In spring 1927, Fuller received the news that his position as president of the Stockade Building Corporation would soon come to an end. His departure followed in December. For three quarters of a year, that departure weighed heavily on Fuller, as was clear from a long correspondence. In August 1927, his daughter Allegra was born. Fuller began to compile statistics, marking a time line with the child's growth, weight, and important firsts like the first smile or the first articu-lat-ed sounds. In the course of this difficult year, Fuller had a kind of mystical experience that showed him the direction for his future life. These statistics have a significance that is typical for Fuller. They contain among other things notations of accompanying temporal events. For example, on one of the three sheets that show the child's growth curve, "Daddy's leg" is drawn in as a basis for comparing size. For Fuller, statistics were not just a matter of numbers, instead he tried to relate them to himself. Fuller was not a positivist. This detail expresses how attentive he had become to the question of coordinates during this period; he sought coordinates for his life and his task, and the newborn child was a guiding star in that effort.

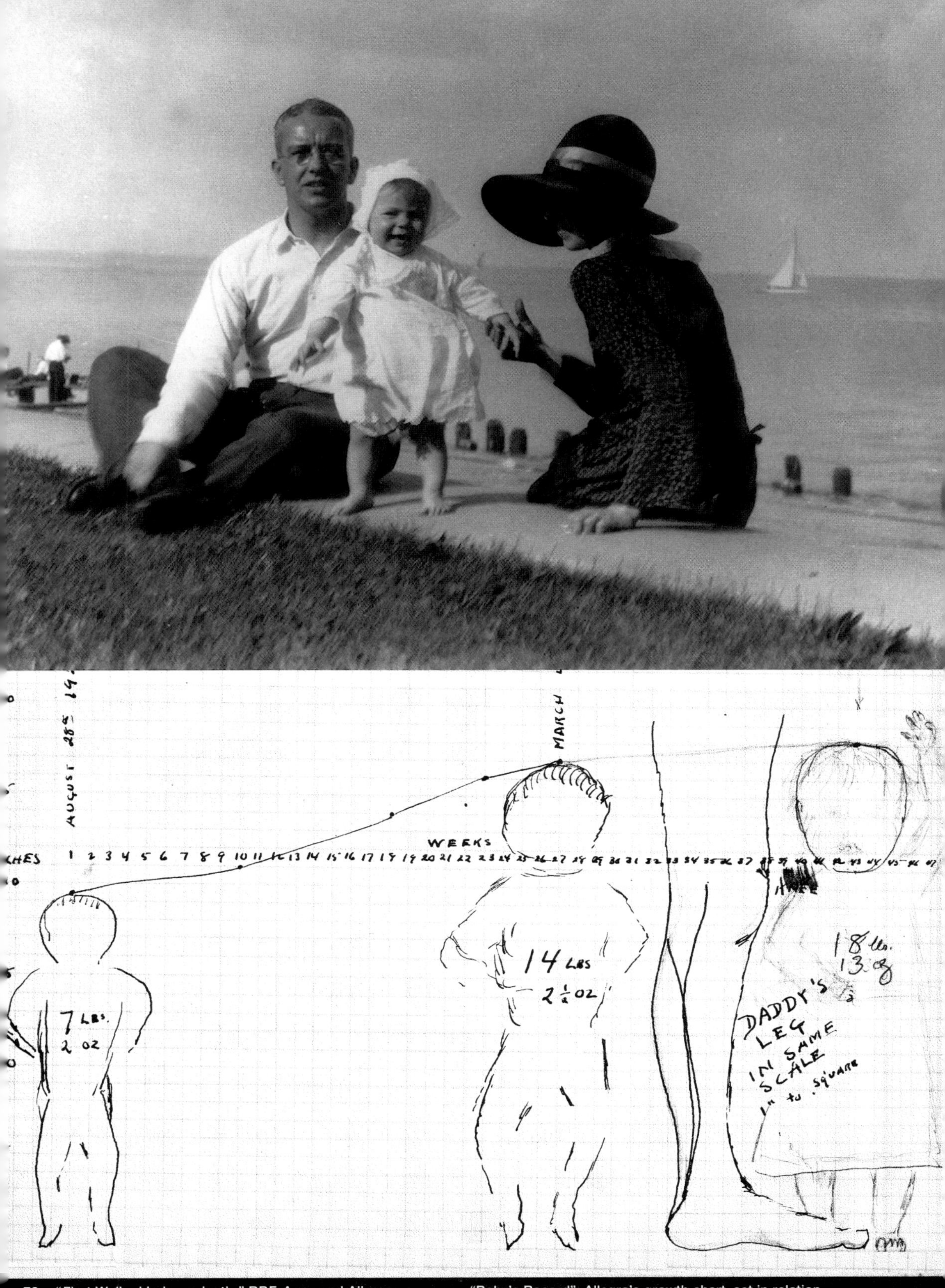

79 · "First Walked Independently," RBF, Anne and Allegra at the shore of Lake Michigan, Chicago, July 1928

· "Baby's Record": Allegra's growth chart, set in relation to 'Daddy's Leg.' Drawing by RBF (*Chronofile* 30/1927)

The circumstances of Fuller's life led him into a deep crisis out of which he found his way by setting himself no less a task than redesigning the house for the world. He called this program to find a new way of housing mankind "Lightful Houses." It stood for "full of light," "lightweight," "de-lightful," "light-Fuller," and so on. From it, he derived the concept of 4D. Fuller wrote his polemic "4D/Time Lock", which he had copied and sent to relatives, friends, and selected exponents of American intellectual life. In it, he sketches a vision of a united world: The continents together form an archipelago beneath the "air ocean." It is made accessible by the most modern means of transportation and telecommunication. People live in 4D light-construction towers that are transported and set down everywhere by airships. Henceforth, Fuller took the name 4D to mean design that was always based on time.

LIGHTFUL HOUSES/ 4D

Time, Light, Transportation
Life, Lightness, Sheen, Strength, Color
Loveliness, Cleanliness,
Godliness, Truth
LIGHTFUL
HOUSES

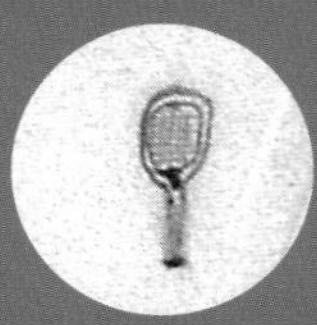

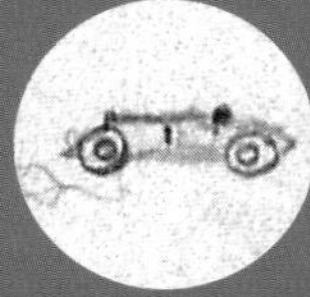

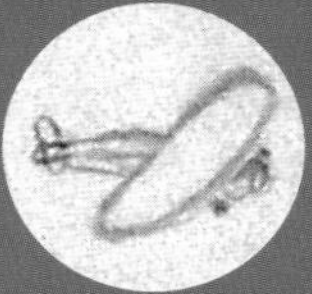
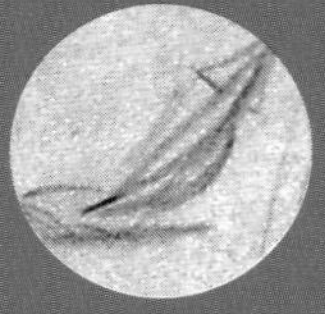
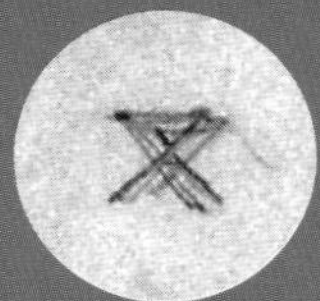
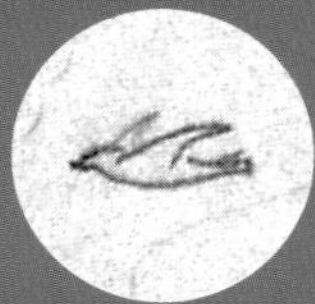

**Programmatic drawing
of Lightful Houses**

**Main coordinates:
Heart / Baby / Sun / Spirituality**

**Inner circle: everyday objects
(foldable, tension-stressed, convertible, movable):**

**tennis racket
umbrella
auto
steamer trunk
airplane
sailboat
folding chair
bird**

**Outer circle:
world-tree
Schenectady experimental high-voltage plant
mooring mast with dirigible air ship
pagoda**

**In between:
skyscraper
obelisk
ship with transmitter tower (Schuchow type)
lighthouse**

TIME EXQUISITE LIGHT

TIME SLOW MATTER

TIME METAL MECHANICS

TIME FELLOWSHIP PRODUCTION

20

24

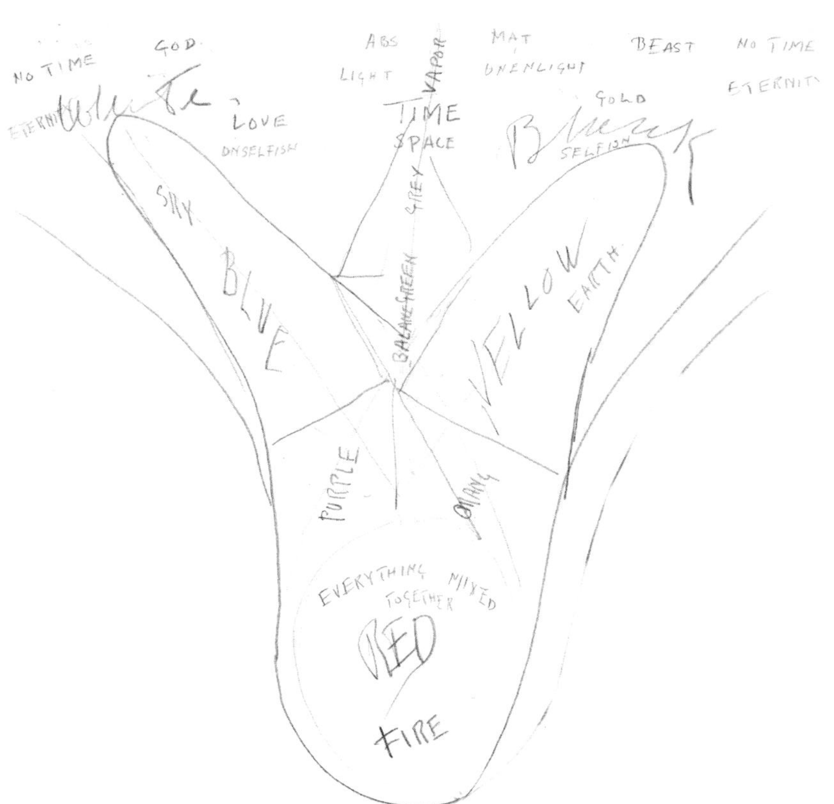

Fountain of life

"... there is in every creature a fountain of life which, if not choked back by stones or other dead rubbish, will create a fresh atmosphere, and bring to life fresh beauty. And it is the same with the nation as with the individual man."

Margaret Fuller, 1842 (Ideas and Integrities, 70)

 ·Conceptual model of origin and growth: the seed

·The other models are: the expanding sphere

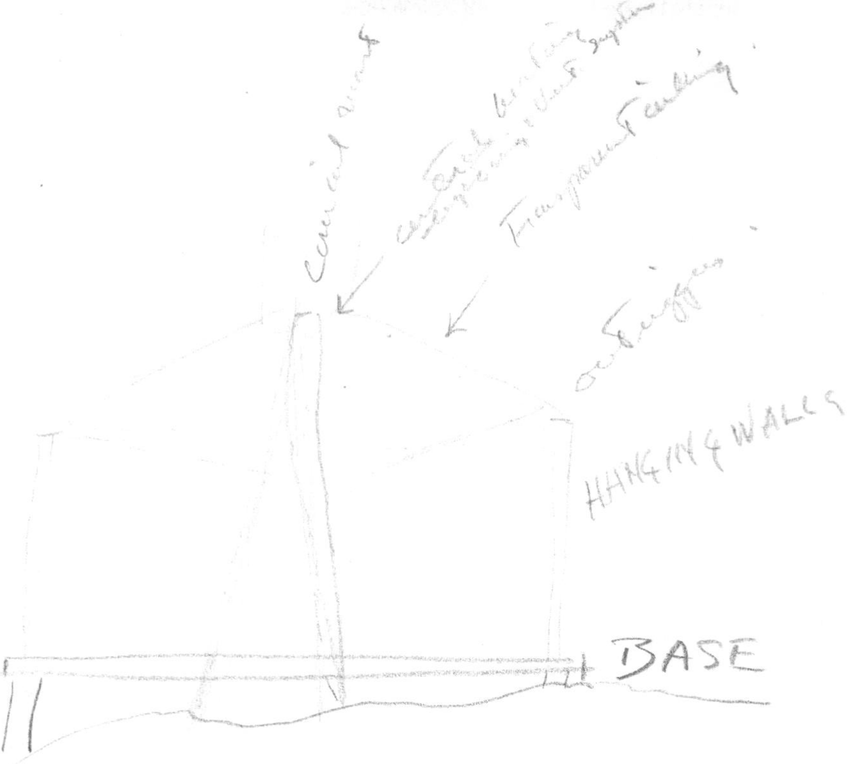

Chassis Construction

In a folder containing manuscripts and a few sketches from the time of the birth of the concepts that led to the 4D house and the 4D book, there are hints about Fuller's beginnings as a designer. It is highly likely that none of Fuller's manuscripts and drawings date from before 1928. All of the dates with the year 1927 were added later and are incorrect.

The first reflections on the planned house – originally called the "Fuller House" within the family, and its subsequent manufacturer was sometimes called the Cosmopolitan Home Corporation and sometimes Fuller Houses – were written down in January 1928. For the trademark and trade name, Fuller used the word "Lightful" he had devised. He called the houses "Lightful Houses" and the furniture, household items, and so on, "Lightful Products." Fuller dictated the script for the first approaches and development of the 4D house. Fuller sees three classes of buildings in contemporary architecture: (1) industry buildings, (2) office buildings and large apartment buildings, (3) small private dwelling. As for the first class, stimulated by reading Le Corbusier's "Vers une architecture" (translated into English as "Towards a New Architecture", 1927), Fuller saw the origins of the new style of architecture in factories, power plants, and so forth. In the second class, the skyscrapers of densely populated cities are constructed using the new materials steel and reinforced concrete and use electricity and modern methods of transportation. The third class includes the small, private dwelling, the "American Home," which is technically the most conservative and is based on local handicraft production. Fuller started with the third class and proposed introducing a new method of construction that the various manufacturers of materials could help to develop. The new construction would have the

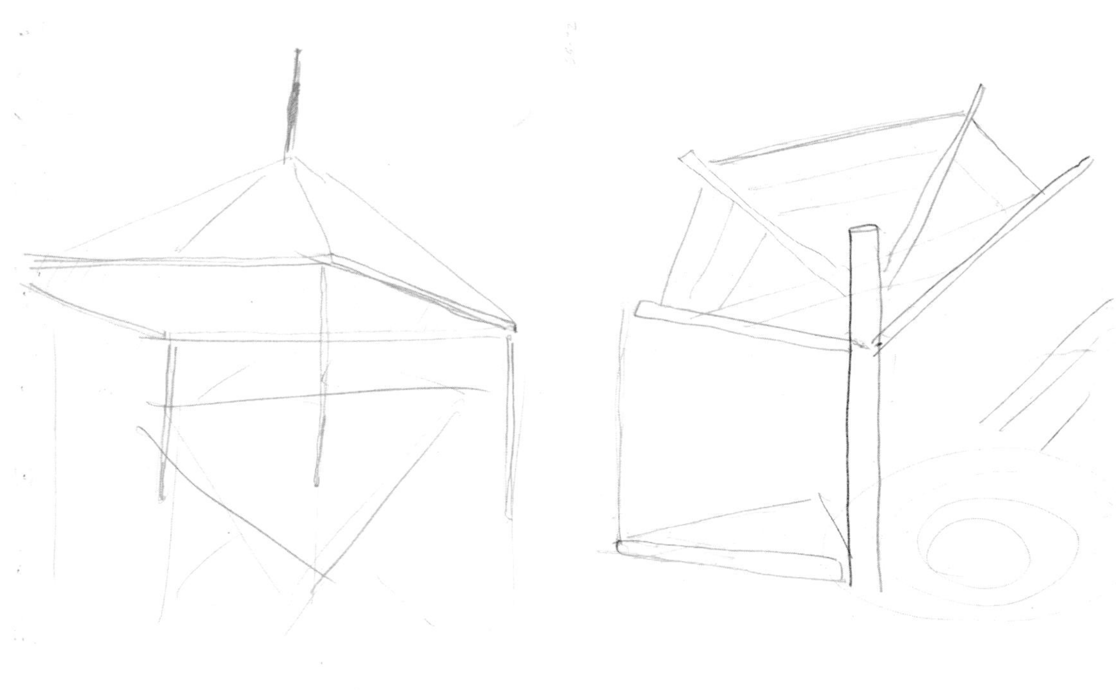

chance “to serve the architect in building the utilitarian portion or the chassis of a house that would correspond to the utilitarian framework of the second class of buildings, i.e., the city skyscrapers.” Fuller had in mind to combine forces for research, development, and marketing. “The final manufacturing of the product for market of mass goods and the preparation of a considerable market for the ‘utility chassis’ for various price classes. The mass production of ‘utility cabinets’ with a market not just for the small one-family house of the third class but also for the second class and for building.”

Fuller began with the technical core for the buildings, the supporting structure and the units of the systems for provisions. Independently of the class of building, he looked to industrial mass production for these things. Elsewhere, he developed the idea of gradations of prefabrication and so he divided the houses into four categories, from A to D. Class A contained only the delivered “chassis”; B also contained additional installations; C is delivered finished; and D also contains the complete fixtures, including furnishings. Class D is intended for the “well-paid worker.” Four months later, the 4D house would develop out of it, and a year later the Dymaxion House.

He concerned himself first with “chassis construction.” The precursors or points of reference for it can be seen in the drawing “Lightful Houses,” from February 1928, in a typology of towers that spans the globe. Fuller’s “chassis” was based especially on the airship landing pylon that appears on the sheet pointing downward. This pylon was first developed by the British airship and airplane designer Barnes Wallis during World War I. Fuller selected its three-legged, tetrahedral form for his chassis construction. He describes its features as follows:

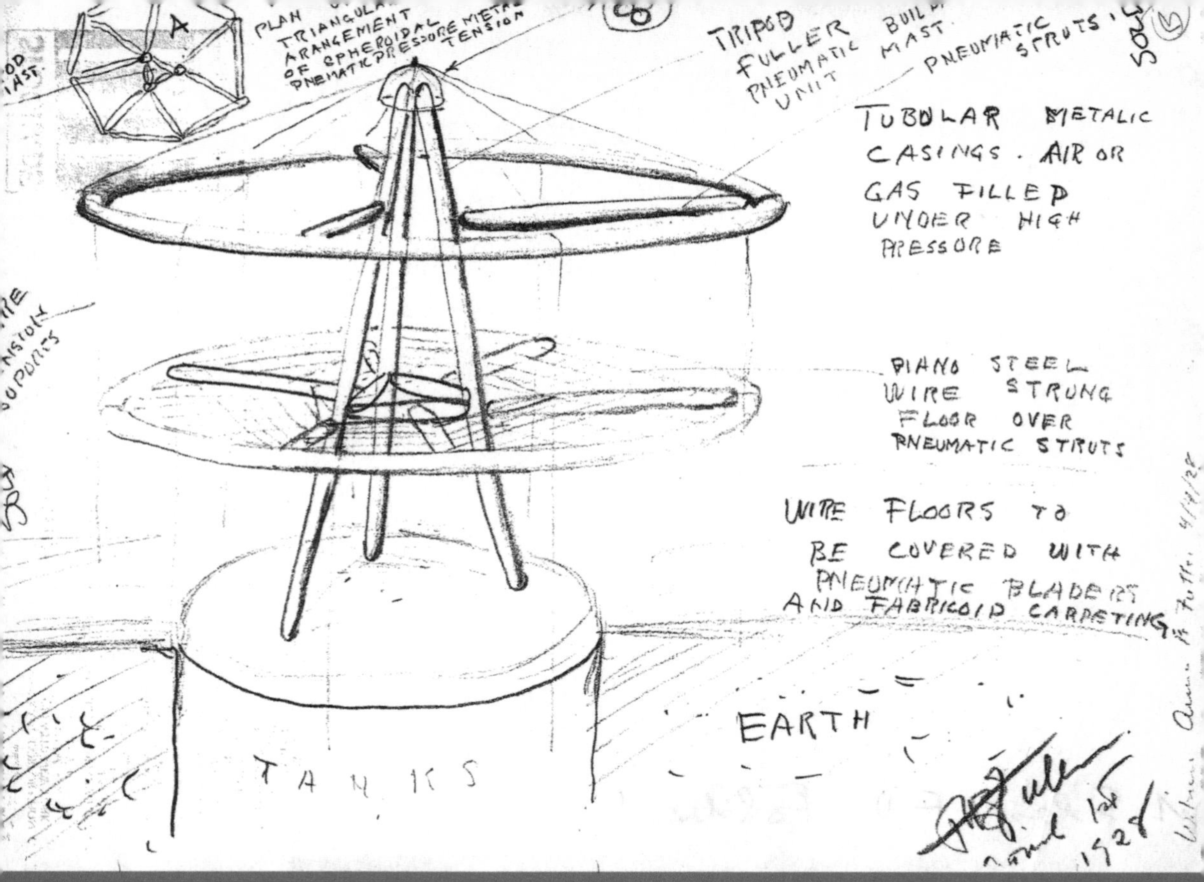

"The Benefits of 'Chassis' Construction": The basic idea of the construction is that all elements shall be suspended from above rather than rest upon support from below.

In a small house a central mast of reinforced concrete or steel will envoy trusses which will intersect at such an angel that their extremities will mark the four corner of the body of the house. The guy cables for this mast will run down through the partitions so as to be invisible, and the mast itself can readily be marked in a similar manner. From mast and trusses will depend walls, floor and all other elements of the construction.

The bearing load of the outer wall will be eliminated. Thick walls of brick or stone will not be needed. The wall of the house, instead of serving the double function of a protection from the natural elements and a support for the upper stories, will become merely a jacket for protection purposes, insulating the living quarters from rain, wind, heat and cold, fire and frost.

This concept was developed while Fuller was already hiring someone to complete the drawings for a patent application.

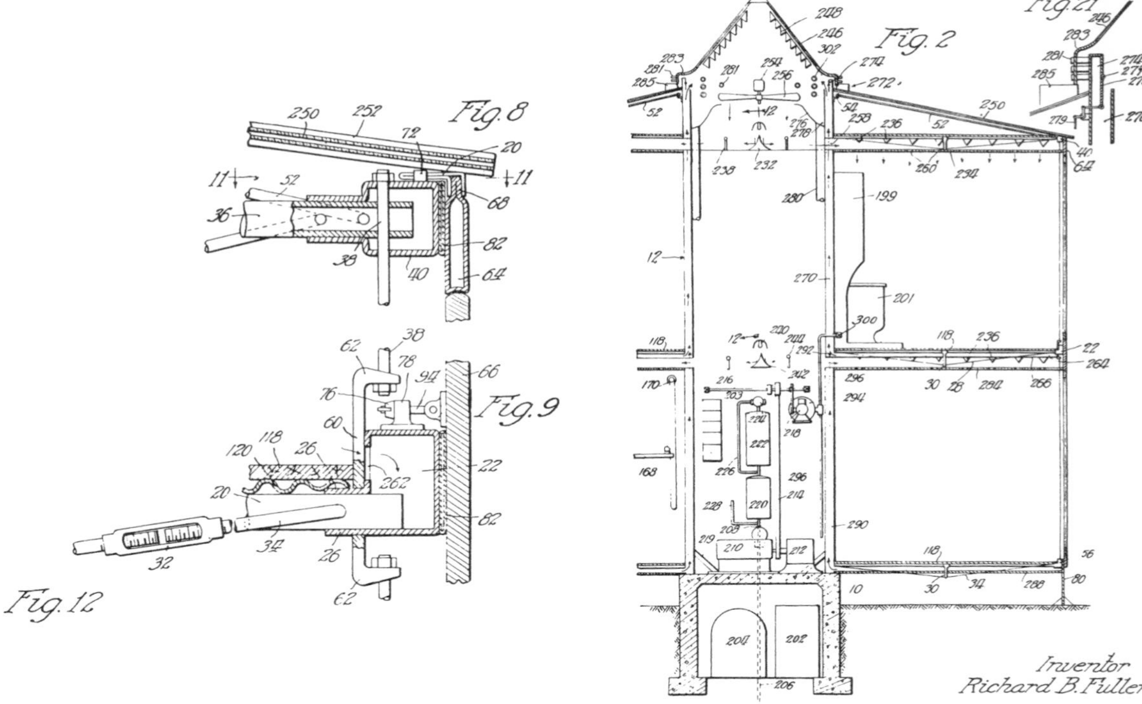

The drawings for the hoped-for patent are from spring 1928. By the time they were completed, in late March or early April, they were already superseded by the development of the chassis concept. Fuller was still thinking of a rectangular floor plan. The central supporting shaft can be seen clearly, but the drawing does not yet represent the system of stays seen in the later version of the concept, but is rather a cantilever construction that was also seen in the early sketches. Fuller later referred to these drawings, especially the rectangular floor plan, as a pragmatic compromise to make the 4D House seem less shockingly new. In reality, they are evidence of the dynamics of the design process in which a series of momentous decisions about the plan followed one another in just a few weeks. The main difference between the rectangular and hexagonal floor plan is that the former does not allow the forces to abate symmetrically, and so it requires cantilever arms in the diagonals of the rectangle, i.e., in two different corners. The center symmetry of the hexagon, by contrast, achieves equality on all sides. There is extensive correspondence on the content of the patent. Fuller was not able to obtain binding patents on the ideas visualized in the drawings, so he withdrew his application. Surely he did this thinking that the other, later innovations also deserved to be included in an updated patent application.

·Patent drawings of 4D House, drawn by his co-worker

·Rectangular floor plan, central support shaft, floor construction

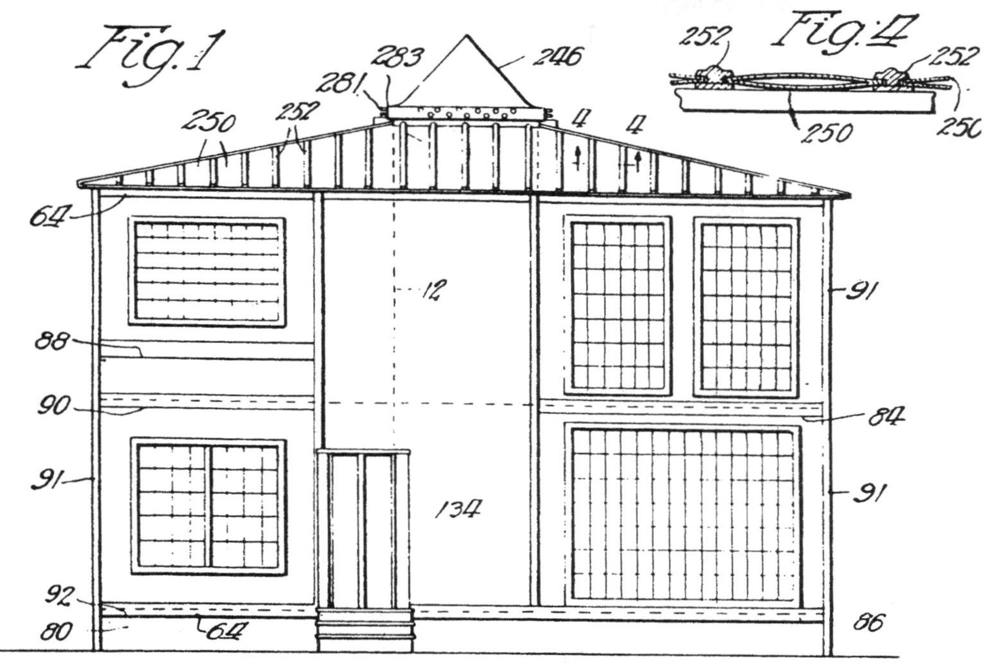
Fig. 1
Fig. 4
252
252
250
250
283
246
281
250
252
4
4
64
12
91
88
90
84
91
91
134
92
80
64
86

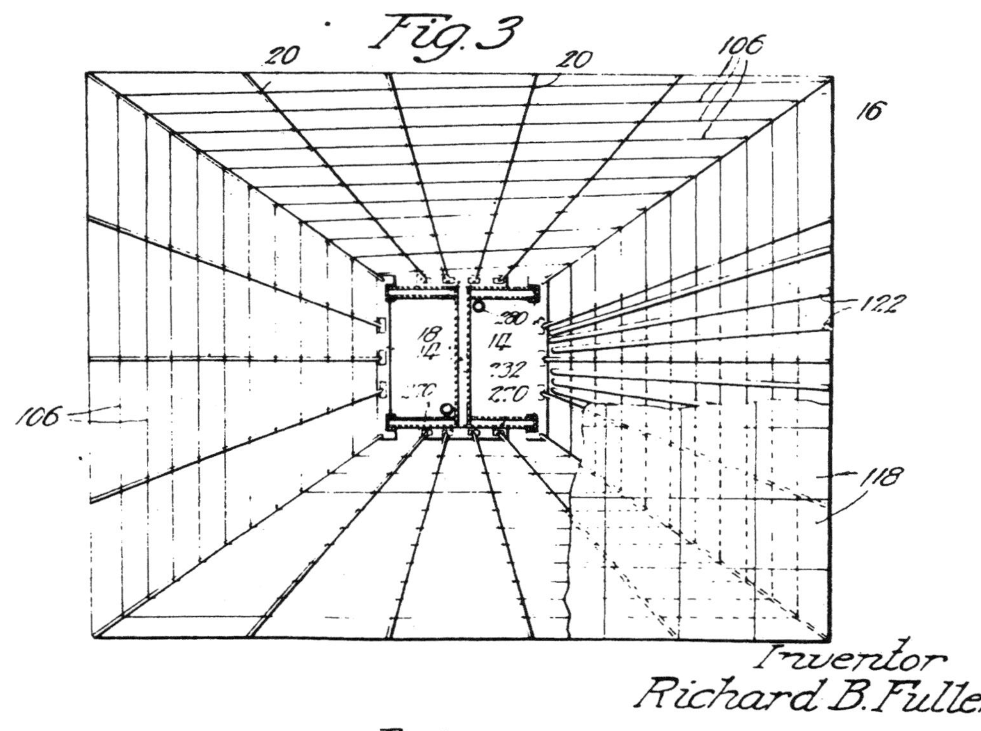
Fig. 3
20
20
106
16
122
18
14
14
232
270
270
106
118
Inventor
Richard B. Fuller
By Emery, Booth, Janney & Varney Attys

On the Evolution of Tower Construction

At first, Fuller did not in any way take it for granted that the supporting structure of the 4D towers had to be a hanging construction. Sketches from the early planning stage show the initially uncertain exploration of hexagonal symmetry (p. 91) – i.e., a property of crystals – and the criterion of a cross-section that tapers off as it moves from the bottom to the top – i.e., the principle of the world of plants. On this side, the sheet documents the attempt to find a synthesis of both these principles: The drawings depict various types of towers, short and tall, slim and compact. It seems more likely that Fuller was thinking of phases of the construction process. The integrated construction crane that sticks out on either side is to install the hexagonal floors of either equal or varying size. The tower results out of this stratification, in which the lowest floor nevertheless lies well above the ground on an enormous construction of props. On the right is a sketch of a variant using equal-size floors. It seems that Fuller viewed the core of the building as a shaft that was continually being driven upward, and upon which the beams of the crane could climb as the floors are built in succession.

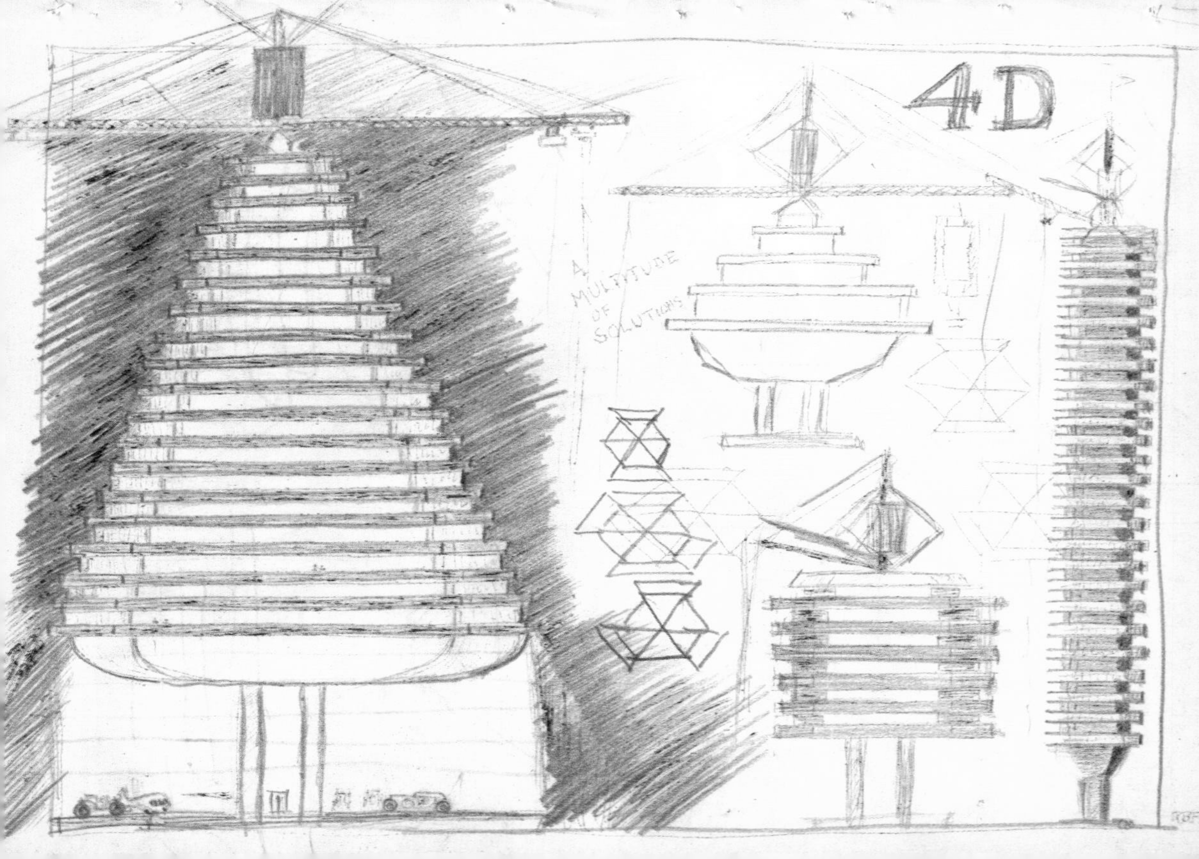

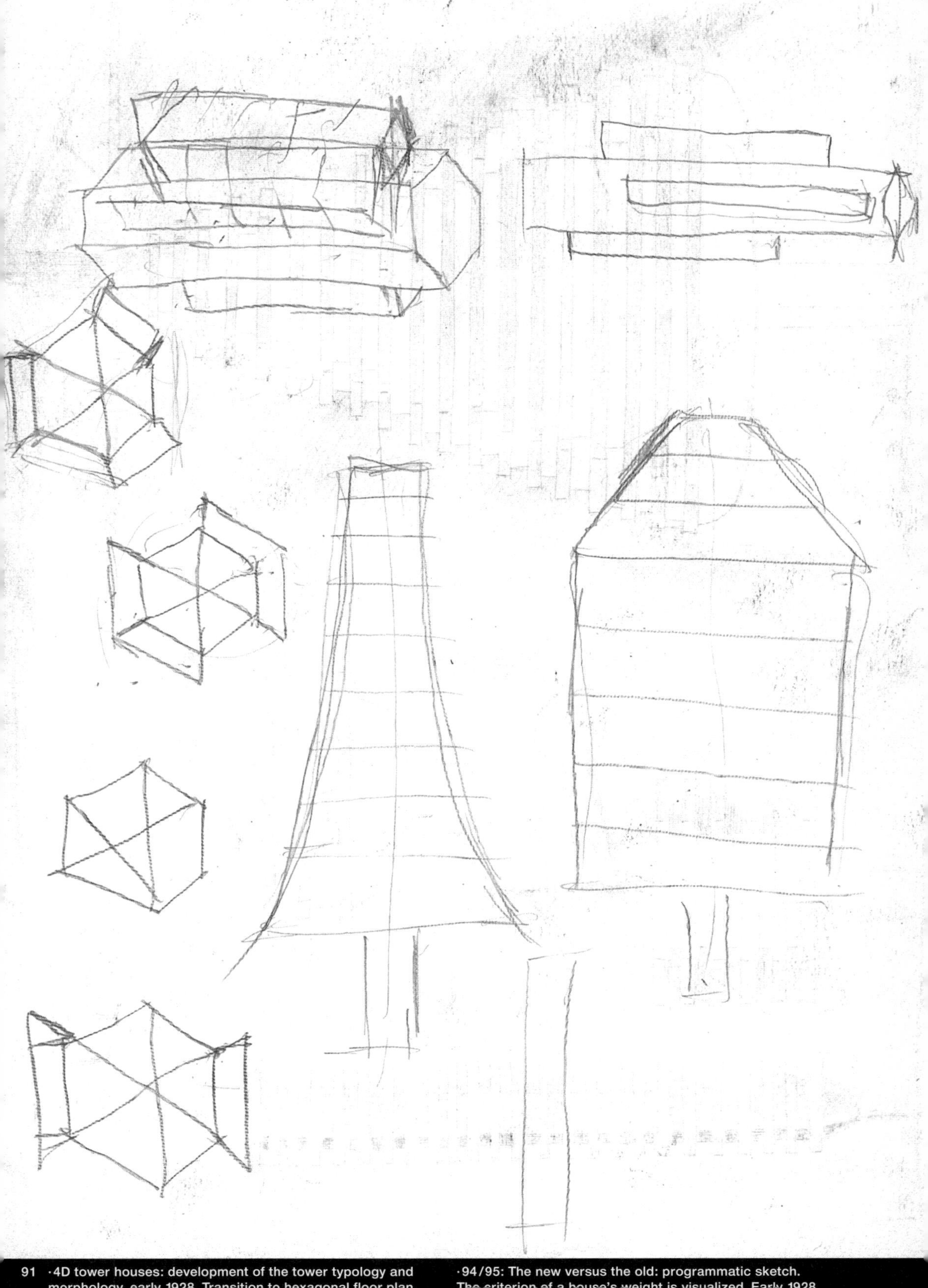

91 ·4D tower houses: development of the tower typology and morphology, early 1928. Transition to hexagonal floor plan

·94/95: The new versus the old: programmatic sketch. The criterion of a house's weight is visualized. Early 1928

“Doing

with the

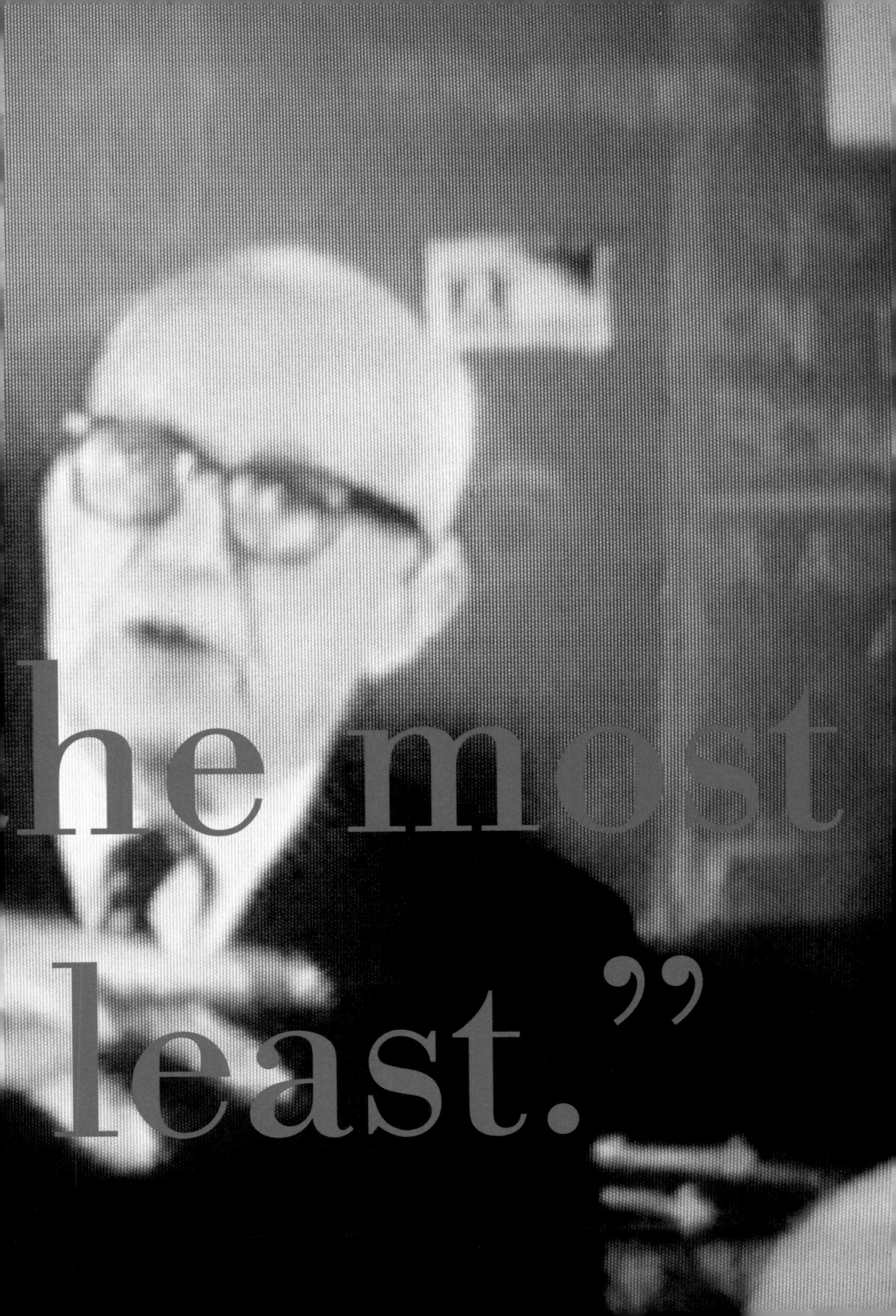

he most
least."

1927 mimeo-sketch
by R.B.Fuller

IT TAKES 10 TAILORS TO MAKE A MAN
BACK IN THE DEAR "OLD" DARK AGES
← THE PRIVY
TWIN MODEL
← ATTIC A GOOD PLACE FOR OLD JUNK FURNITURE AND YESTERDAY'S STYLES
HAVING REMOVED LOTS OF DIRT TONS OF HARD DIRT IS PUT BACK
MA HANGING OUT THE WASH
A GOOD "OLD" MANY SONS HOUSE
← PA GOOSE STEPPING OFF TO AN HONEST DAYS WORK
← DAUGHTER MARY LEAVING "HOME" WITH HER "GREAT SHAME"
ICE
TITLE IN PA'S NAME
MORTGAGE IN "UNCLE'S"
A CUTE LITTLE "HOME AND VEGETABLE" HOUSE

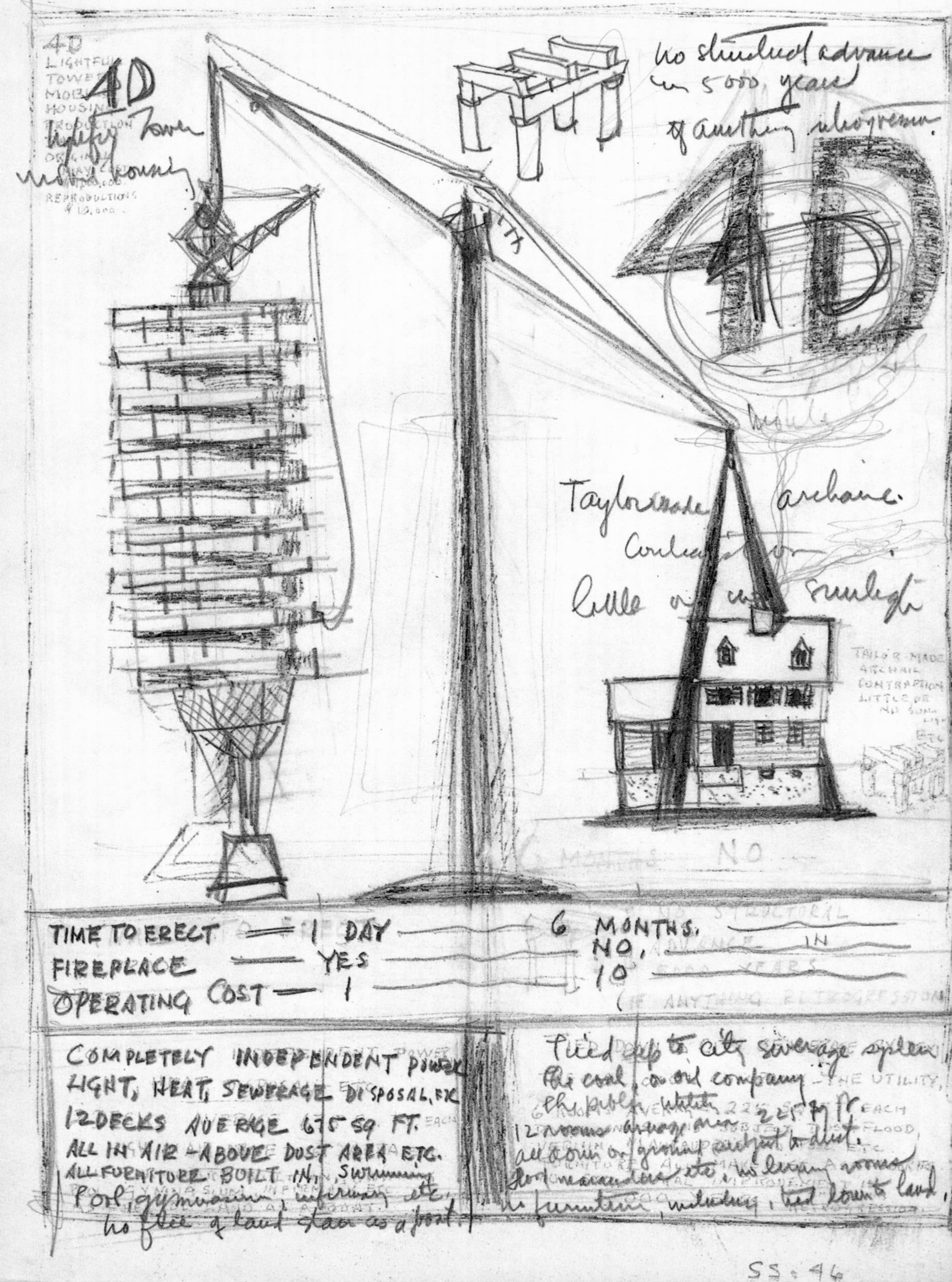

4D
REPRODUCTIONS
no structural advance
in 5000 years
if anything retrogression.
4D
Taylormade archaic
contraption
little or no sunlight
TAILOR-MADE
ARCHAIC
CONTRAPTION
LITTLE OR
NO SUN
ETC.
NO
TIME TO ERECT — 1 DAY — 6 MONTHS.
FIREPLACE — YES — NO,
OPERATING COST — 1 — 10
NO STRUCTURAL
ADVANCE IN
5000 YEARS
(IF ANYTHING RETROGRESSION
COMPLETELY INDEPENDENT POWER
LIGHT, HEAT, SEWERAGE DISPOSAL, ETC
12 DECKS AVERAGE 675 SQ FT.
ALL IN AIR - ABOVE DUST AREA ETC.
ALL FURNITURE BUILT IN, Swimming
Pool, gymnasium infirmary etc.
Tied up to city sewerage system
the coal, or oil company - THE UTILITY,
the public utility
225 sq ft EACH
FLOOD
ETC.
SS: 46

97 ·Cover design for the book 4D – *Time Lock:* globe,

"… In which the great combination is revealed, if thoughtfully

26% of earth's surface is dry land.
85% of all earth's dry land is here shown.
86% of all dry land shown is above equator.
The whole of the human family could stand on Bermuda.
All crowed into England they would have 150 sq. feet each.
"United we stand, divided we fall" is correct mentally and spiritually
but falacious physically or materially.
2 000 000 000 new homes will be required in next 80 years.

cean-World-Town-Plan, 1928

·100–105: The first representation of building as a scenario

ANCHOR OUT

THE BOMB
IS DROPPED

VIEW OF THE SHELL CRATER
AND ANCHORED ZEP. FROM A BU
HOUSE BEING MANOEUVRED IN T
POSITION.

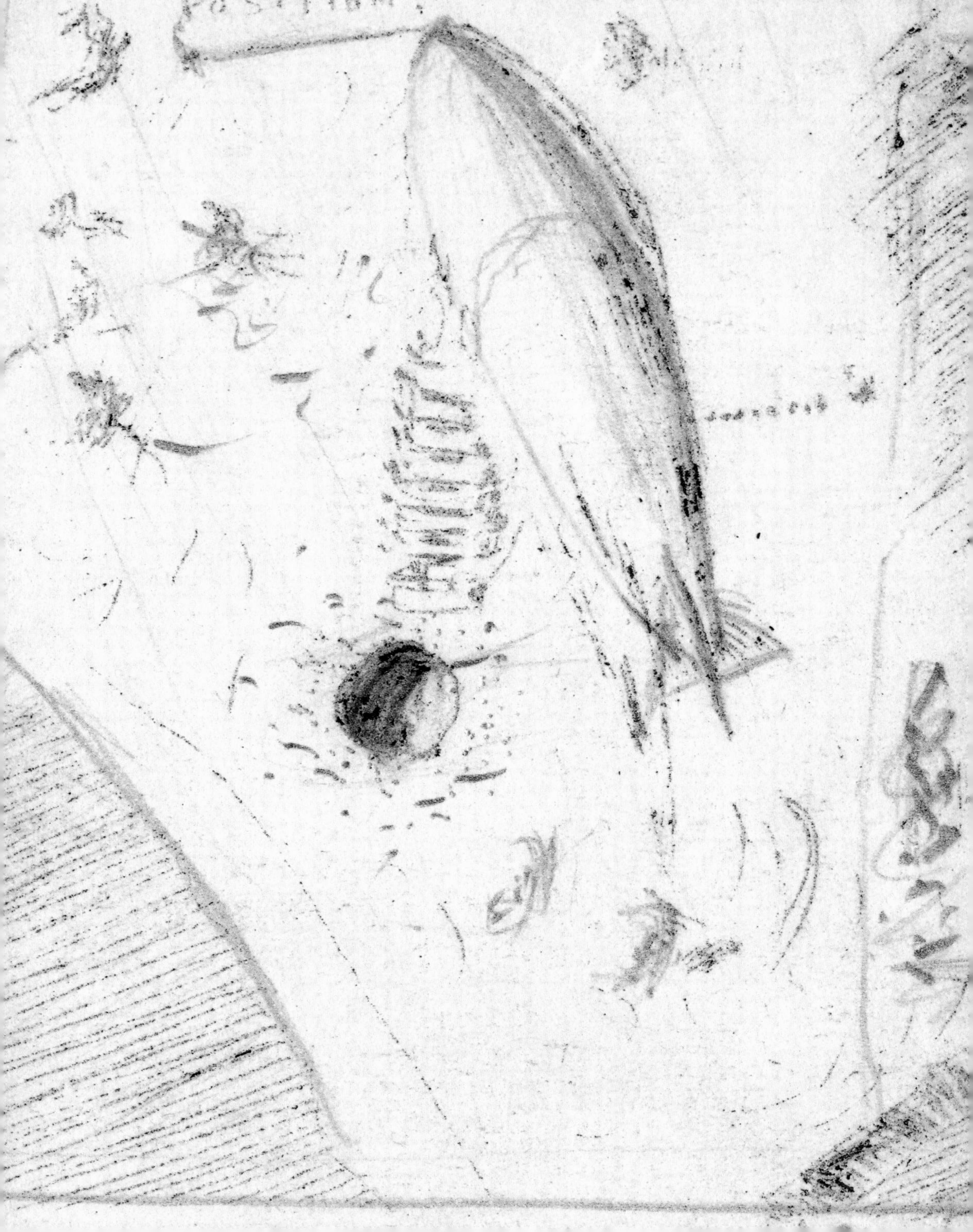

Comics (funnies-Philosophy)

"Undeniably the 'funnies' are at the most generally inspected portions of our daily newspapers, and may be considered the economic frosting that sells the cake. It is more than significant that these funnies have completely lost race of 'slapstick' and have become serials of handy philosophy. The laughs are being turned on self and improve proportions of life, its pleasures and necessities."

4D Time Lock, (1928/1972)

Dirigible

"In 1927 I gave myself a theoretical problem which turned out to be feasible in high performance per pound engineering. I gave myself the problem of delivering large structures by air. Why? We can assemble large structures under preferred environmental and mechanical conditions as for instance we build airplanes at high advantage in the controlled environments within a factory and with a galaxy of efficiently arranged jigs, fixtures, templates and powerful tools. I can deliver an ocean ship from the 'ways' into the sea with ease and without obstructions. There is nothing to obstruct the ship; I can deliver an airplane into the sky, nothing to obstruct it; but if I try to make large buildings in a factory and try to deliver them over highways or railways, they will run into highway or railway tunnels and bridges which are too small for their passage. These holes through mountains occurred because of man's having to transport

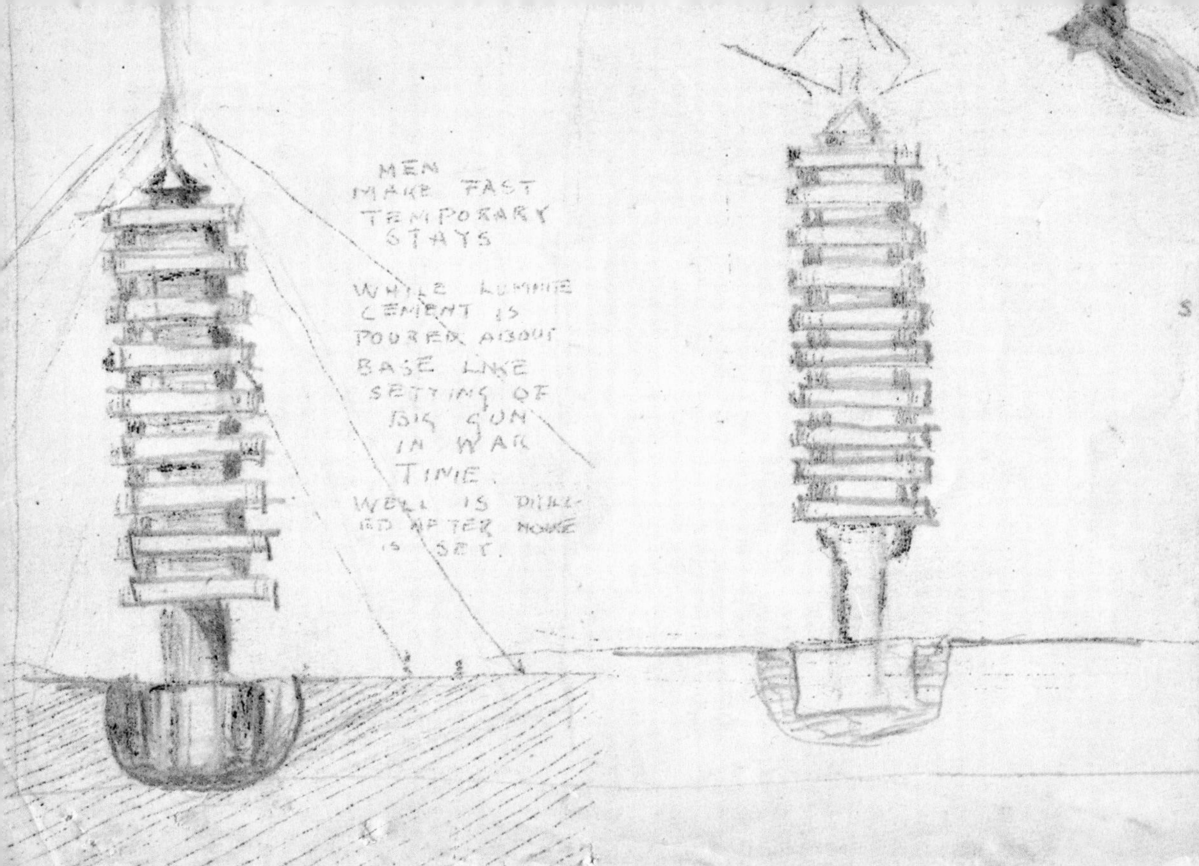

heavy loads overland at almost level grades pulled only by friction-overcoming locomotives. To accommodate the latter man punched holes through mountains etc. Therefore in 1927 I thought of the idea of trying to deliver large structures by air to be above all obstacles.

Therefore I gave myself a problem in 1927 of delivering a ten-deck apartment building to the north pole, during the arctic summer. In designing my ten-deck apartment I used the wire wheel principle. The wire wheel is a tensegrity structure because the island of compression in the hub is completely islanded from the compressional 'atoll', – the rim. These two compressional areas are cohered in stability by the minimum of twelve tensional spokes. I found I could make a ten-deck building so light that it could be carried by the Graf Zeppelin, suspended horizontally under the Zeppelin's belly. The Graf, upon reaching the arctic site, could drop a bomb, make a crater, lower the building into the crater, plant it like a tree and fly back home leaving the building occupied. I saw that all this was feasible even though in reality I would not be enabled to use the Graf Zeppelin and even though I would not have, for another quarter of a century to come. The necessary high weight-strength alloys which I had assumed to be available by 1952 in making my 1927 calculations."

World Design Science Decade: 2 (1965), 63

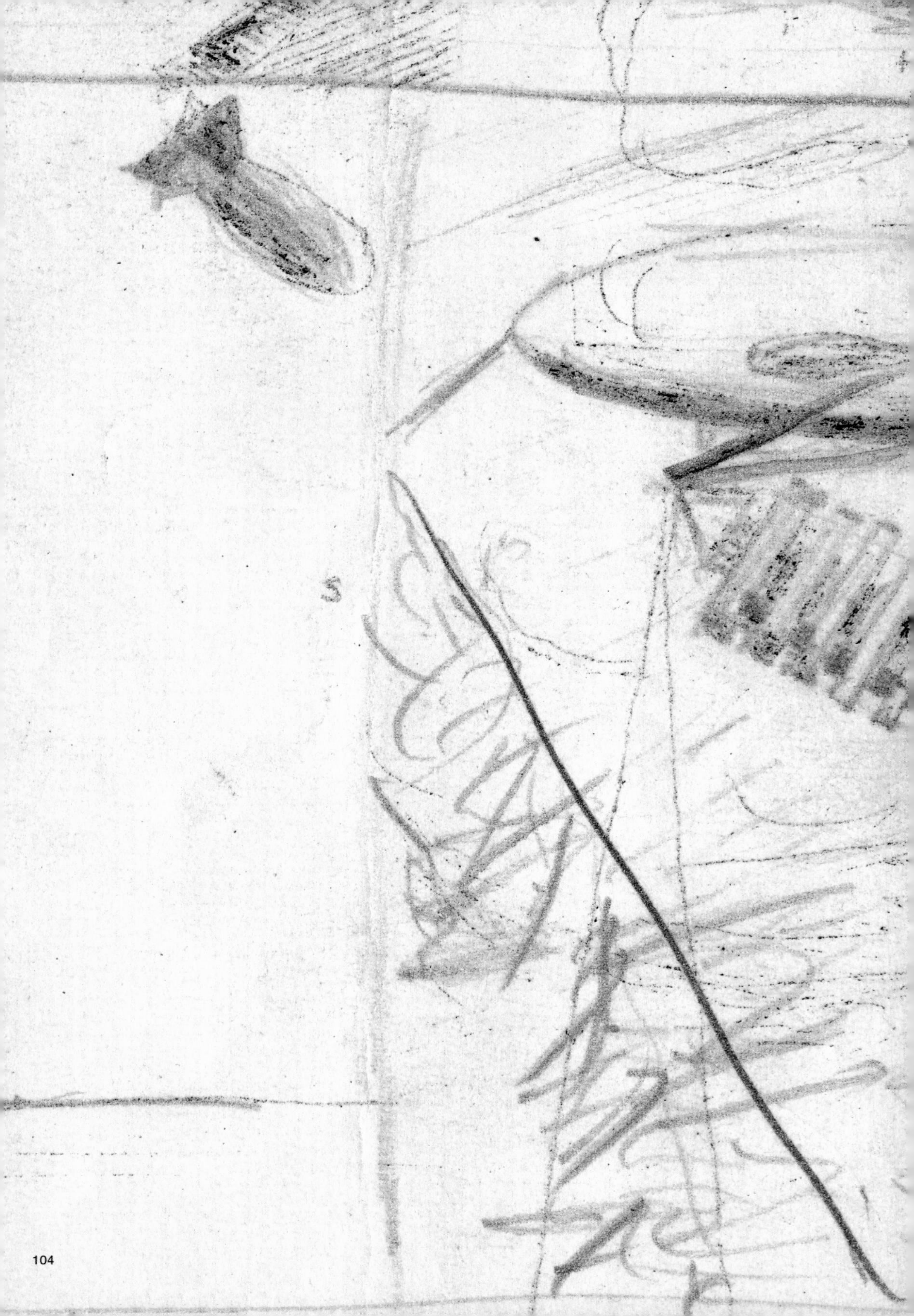

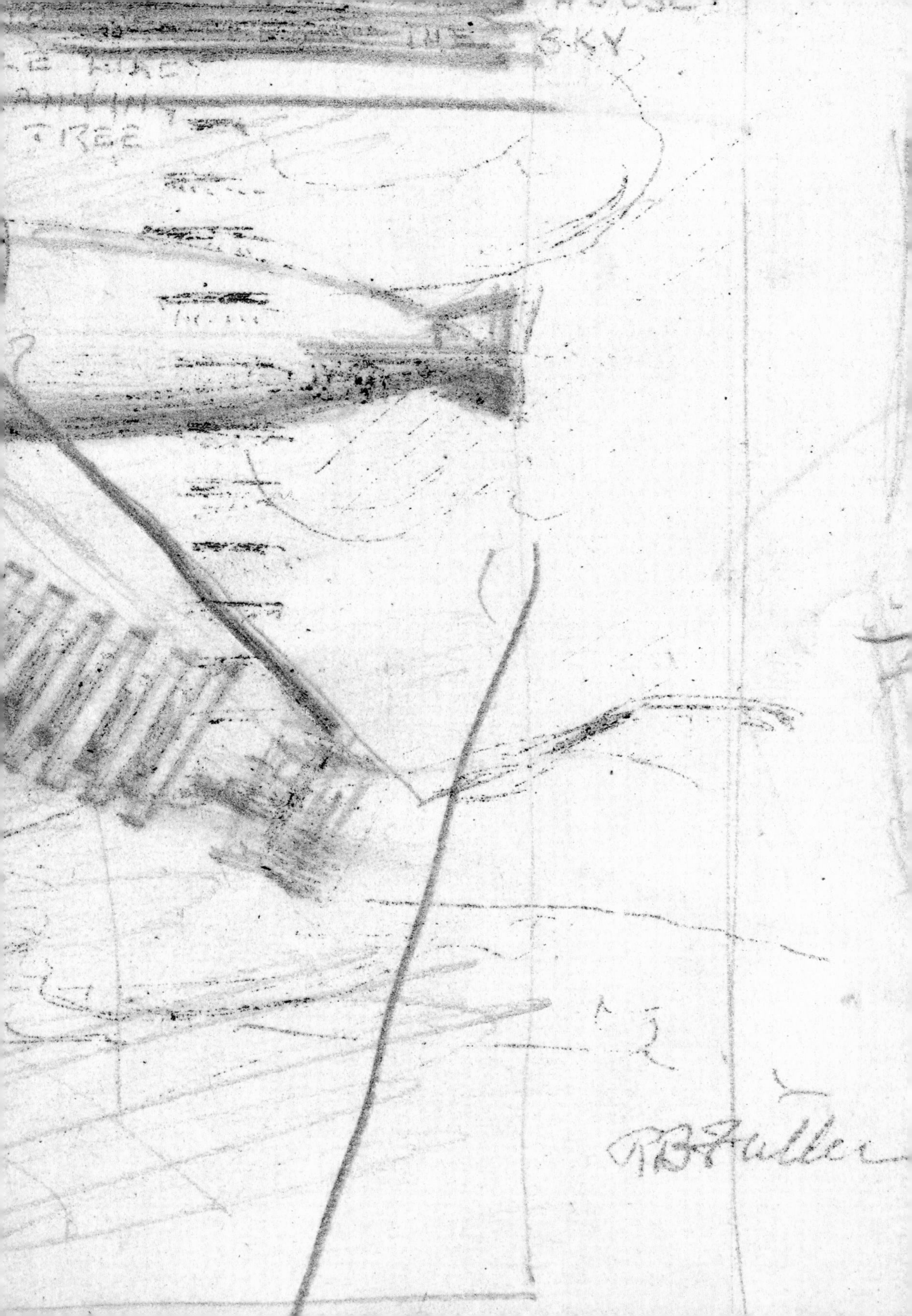
SKY
TREE
RBFuller

Buckminster Fuller's 1928 Prophecy

"There will come a time when in our individualistic harmonious state all work will consist of thinking and crystalizing thought into sound or directionable spheres which will set in motion machinery or controled fourth dimensional design. In our individual homes no matter where we may be we can speak into our combination radio-telephonic-recording-dictator recording our commands graphically (word or picture) to our machines which will involve only real thinking and study of statistics by the directors of industries machines. The rest of the world will be free to do what it wishes so long as it is truthful, harmonious, unselfish. The stock markets will grow vastly and attain more and more machine practice and eventually will be the voting place rather than in politics. People then will survive through not being lazy and through constant consideration of what the rest of the world is doing and by recording their votes by way of purchase or sale of shares of industry. They will have to vote then or they will starve. As long as they play the game they will have most of the time to themselves for harmonic creation in art, philosophy etc. or science which will become an art when we have properly analyzed the fourth dimension which has already been researched. As we save time we increase our speed and we compass the world faster the world being of a fixed fourth dimension and humans with the power to board any dimensional unit by proper control of it. This is relativity."

Lightful Houses (1928)

“Ferro-concrete architecture may be likened unto the plastic cocoon of the archaic worm from which will emerge the 4D butterfly.”

Letter to Rosamund Fuller, August 1928

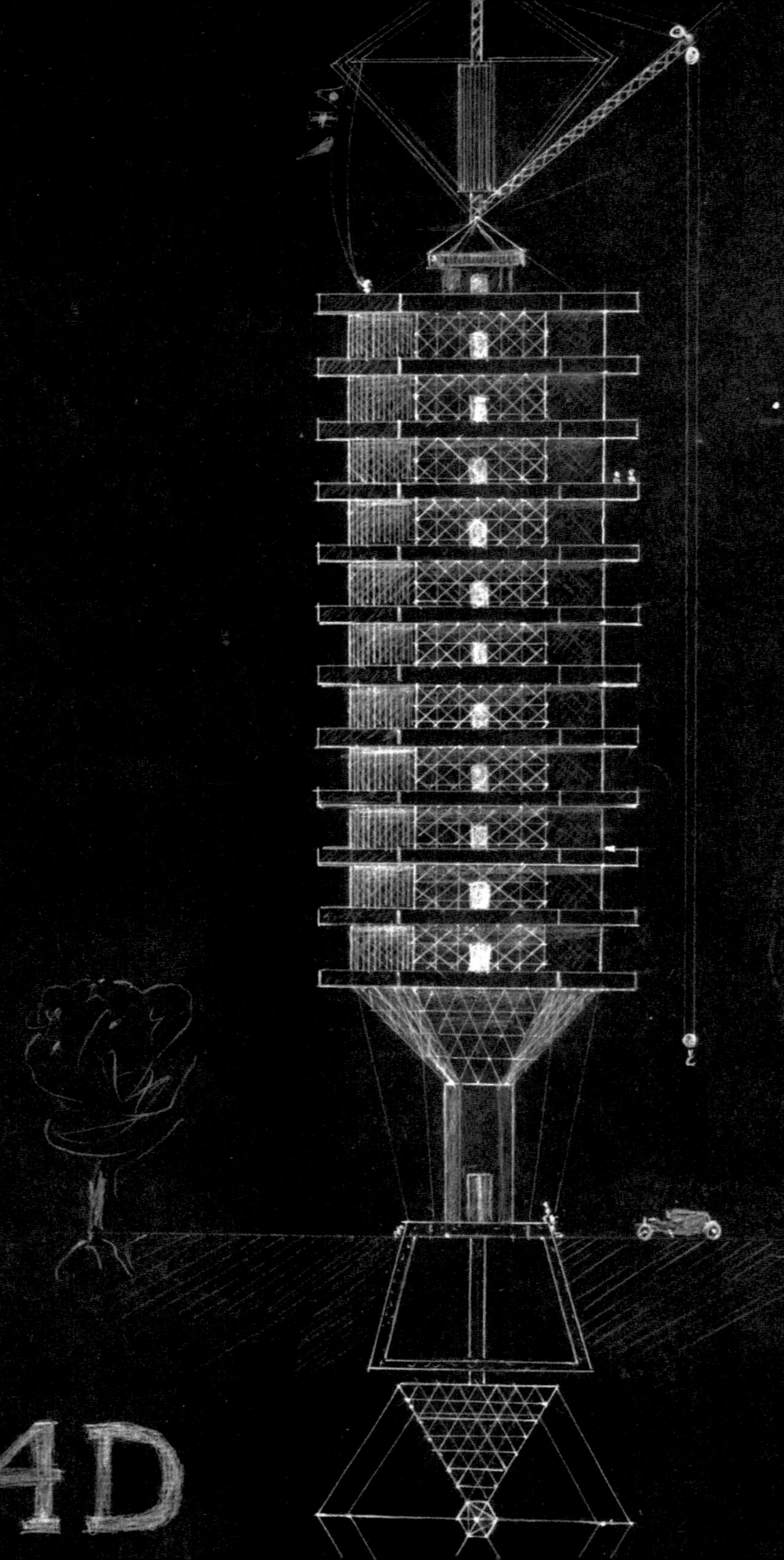
AIRPLANE
RADIO
FLETHER AIR
MILL
HOIST & SIGNAL
MAST
1 SKY-PROMENA
2 POWER-RADI
3 GRILL-BAR-MUS
4 STORAGE
5 LIVING PR
6 "
7 "
8 NURSERY
9 HOSPITAL-SER
10 GYMNASIUM
11 DIVING PLA
12 POOL
4D

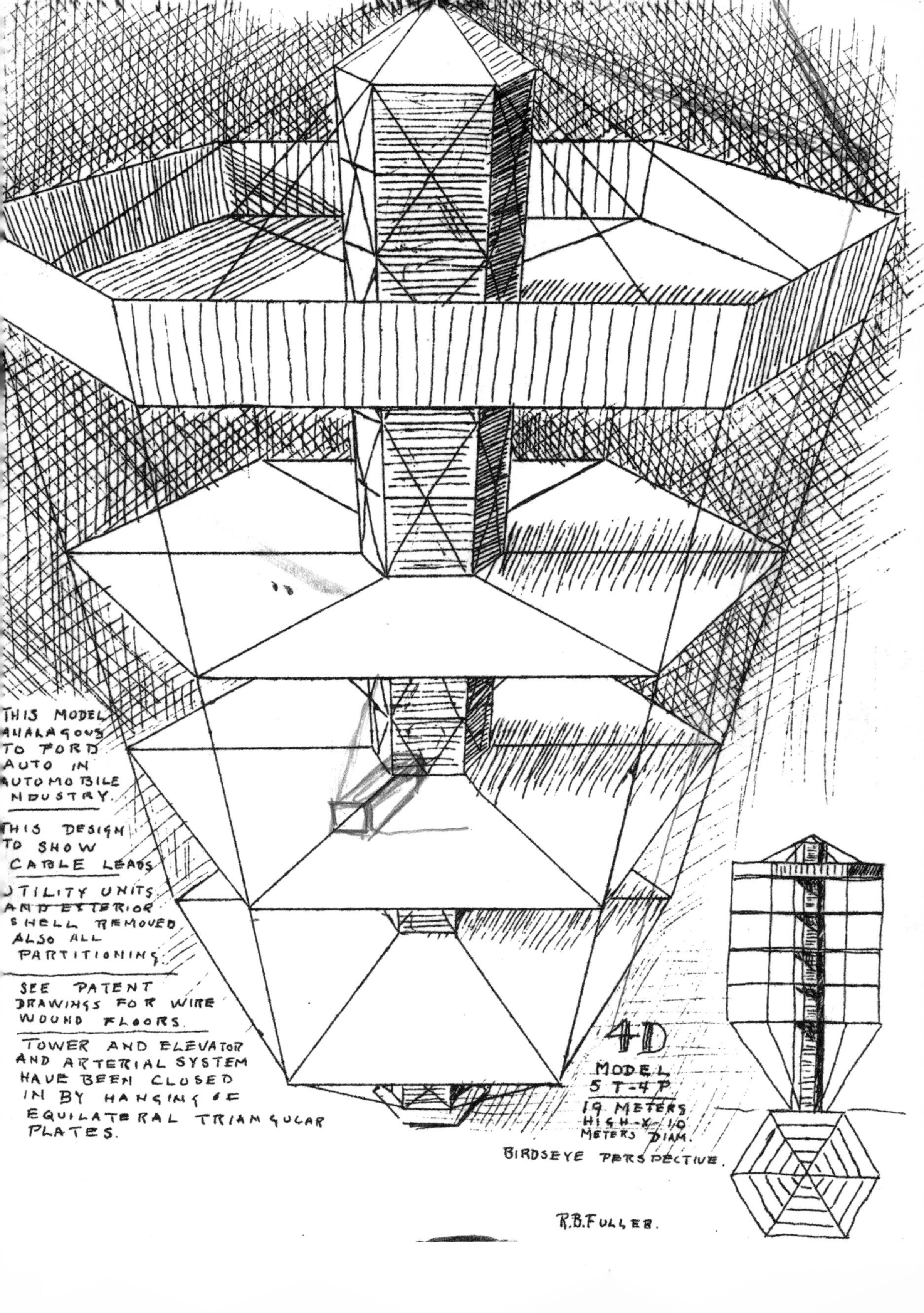

THIS MODEL ANALAGOUS TO FORD AUTO IN AUTOMOBILE INDUSTRY.
THIS DESIGN TO SHOW CABLE LEADS
UTILITY UNITS AND EXTERIOR SHELL REMOVED ALSO ALL PARTITIONING.
SEE PATENT DRAWINGS FOR WIRE WOUND FLOORS.
TOWER AND ELEVATOR AND ARTERIAL SYSTEM HAVE BEEN CLOSED IN BY HANGING OF EQUILATERAL TRIANGULAR PLATES.
4D
MODEL 5T-4P
19 METERS HIGH X 10 METERS DIAM.
BIRDSEYE PERSPECTIVE.
R.B.FULLER.

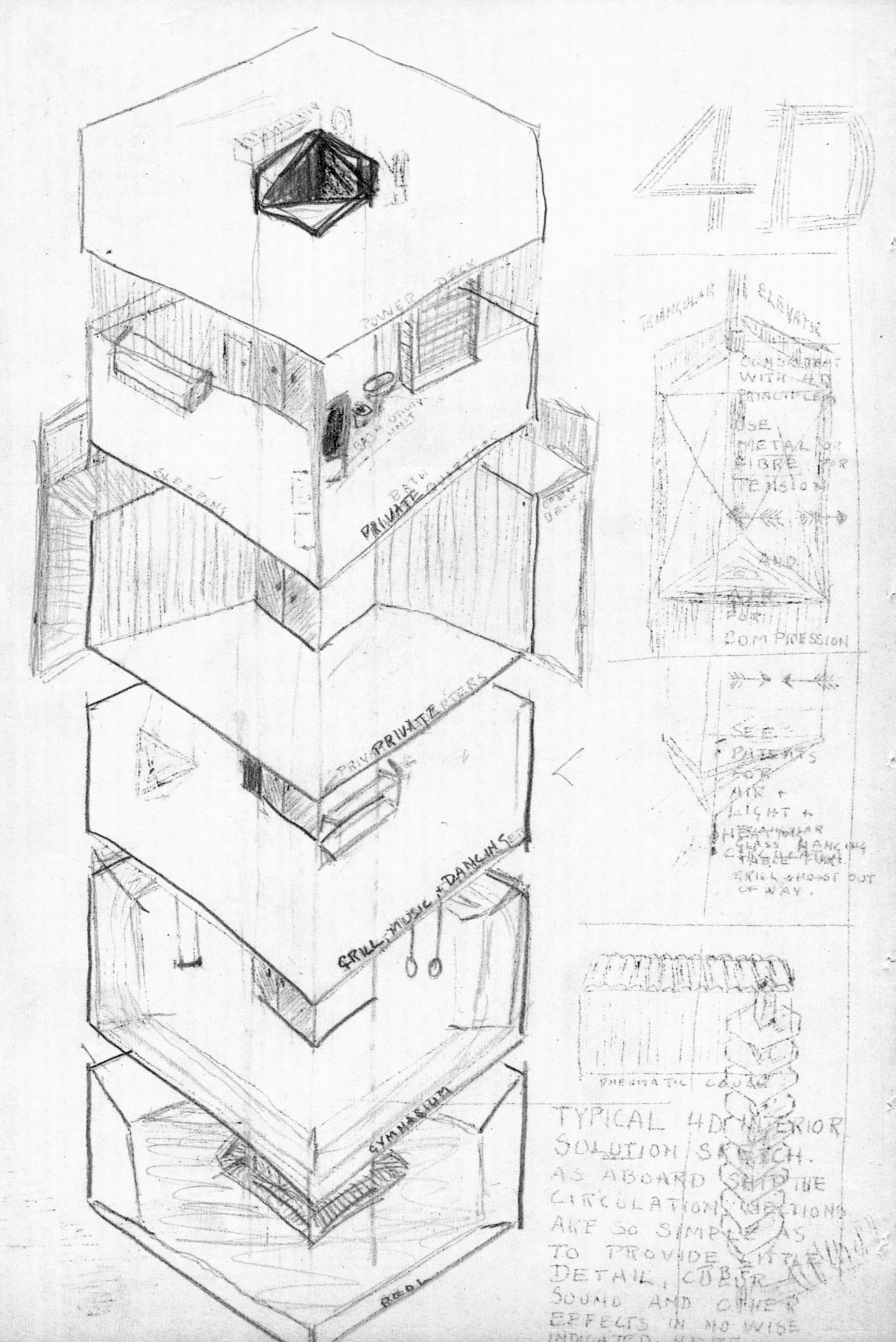
4D
POWER DECK
BATH UTILITY UNIT
SLEEPING
BATH
PRIVATE QUARTERS
OPEN DECK
PRIVATE QUARTERS
GRILL, MUSIC + DANCING
GYMNASIUM
POOL
TRIANGULAR ELEVATOR
CONSISTANT WITH 4D PRINCIPLES
USE METAL OR FIBRE FOR TENSION
AND AIR FOR COMPRESSION
SEE PATENTS FOR AIR + LIGHT + HEATING CIRCULATION
TRIANGULAR GLASS HANGING TABLE FURN.
GRILL & HOIST OUT OF WAY.
PNEUMATIC COLUMN
TYPICAL 4D INTERIOR SOLUTION SKETCH.
AS ABOARD SHIP THE CIRCULATION SECTIONS ARE SO SIMPLE AS TO PROVIDE LITTLE DETAIL, COLOR SOUND AND OTHER EFFECTS IN NO WISE INDICATED.

"A room should not be fixed, should not create a static mood, but should lend itself to change so that its occupants may play upon it as they would upon a piano."

Chronofile 36/1929

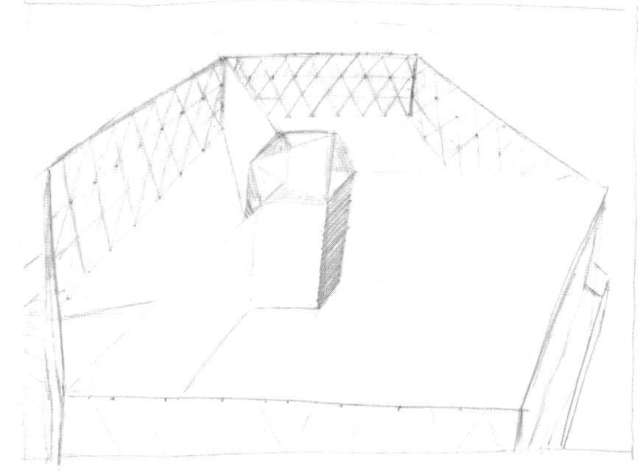

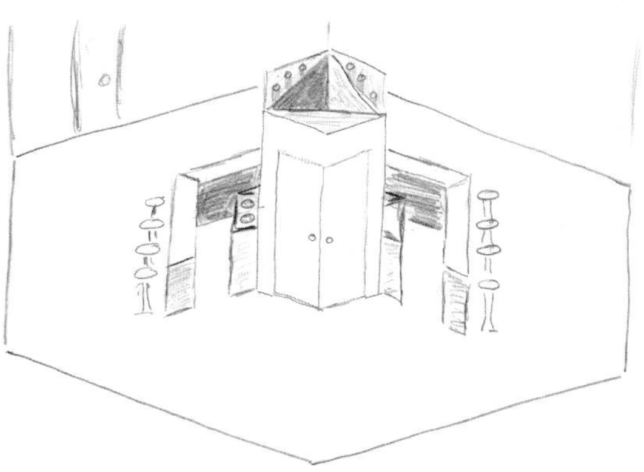

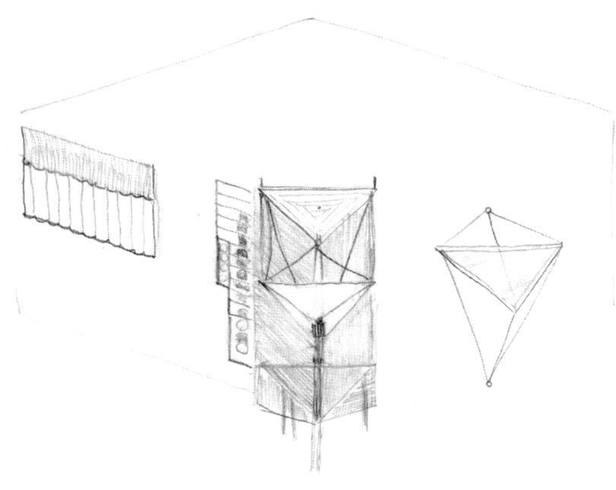

113 ·Proposal for the 1933 World's Fair in Chicago (*A Century of Progress*): 4D Tower as an auto garage.

(Compare to Paul Nelson: Maison suspendue, Paris 1938; Frank Lloyd Wright: among other designs, Guggenheim Museum,

“The basic idea of the construction is that all elements shall be suspended from above rather than rest upon supports from below.”

Lightful Houses, Manuskript (1928), 13

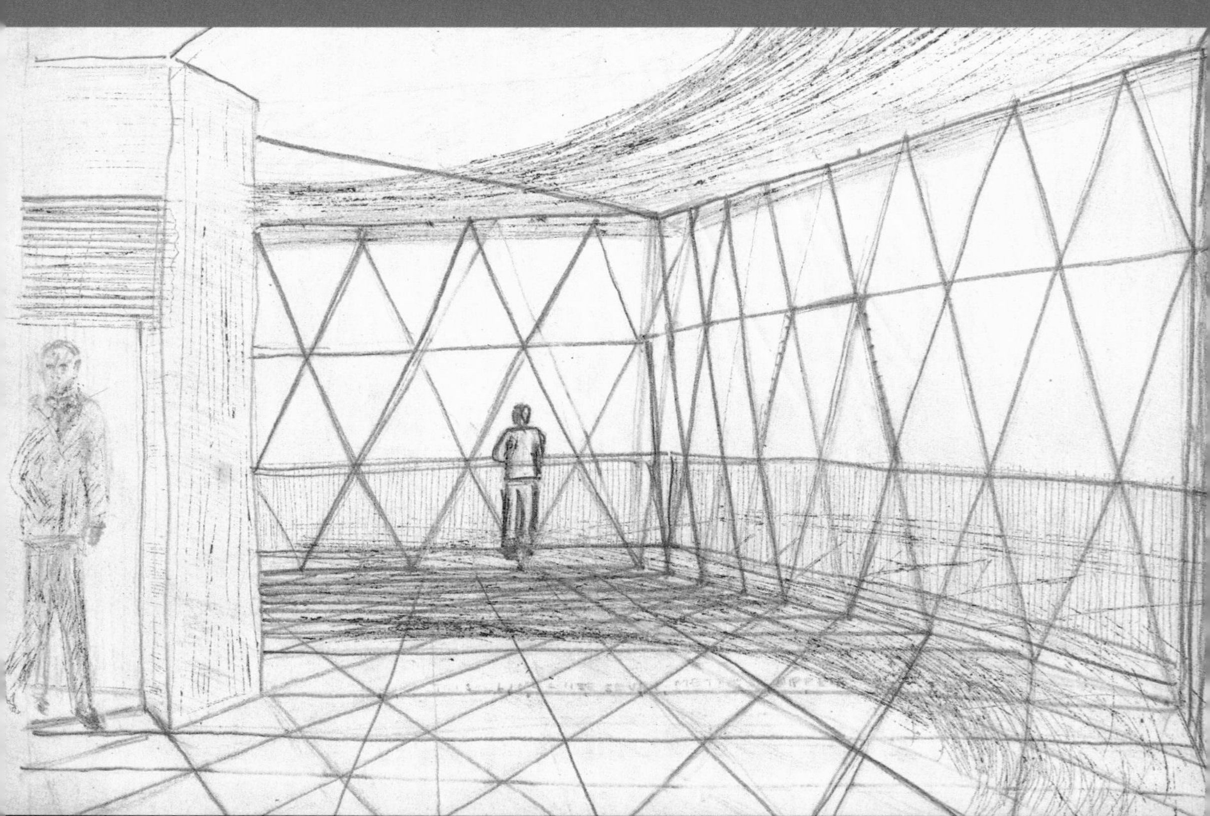

114 · Time-based plan of the 4D House: triangulated floor and facade elements. “Time Interval 1 Meter”: measure of length understood as a measure of time away from the central axis; sketch 1928

4D
TIME INTERVAL 1 METER
Copyright 1928. 4D

Ephemeralization

"There will be of course opposition by such of the interests to whom progress seems to spell disaster. For instance, just as 'one track' or confined transportation of railroad or trolley is being supplanted to trackless bus, auto and airplane; and wire or fixed line communication is being replaced by wireless broadcasting, so will steam pipes, electrical wires, etc. in the home be supplanted in the new house by indirect central broadcasting systems. This will be stubbornly opposed by vested interests in conduit, etc."

4D Time Lock, (1928/1972)

"We are concerned with mooring first class habitation to the earth's surface."

Designing a New Industry (1946), 30

On Mimeographing

Synchronously with the development of his industrial machine-wrought architecture, Buckminster Fuller has developed a new method of mimeographed drawing. To be found in almost any American office of size is the Mimeograph Machine. Its work has, as a rule, been confined to turning out something "cheap," – propaganda for the waste-basket. Mr. Fuller feels this is profoundly unjust to the machine. Instead of "going back" to woodcutting or etching to present an artistically reproducible composition, which is by nature limited in reproduction, he has developed a mimeograph-cut, which is very reasonable of realization. The 14 compositions shown in this exposition are typical of the proper field for mimeographing. It is in reality the field of multiple sketching – reproduction of useful sketches which one could ill afford to duplicate except on the mimeograph.

4D. Buckminster Fuller's Dymaxion House. The Harvard Society for Contemporary Art (text probably by RBF)

rushing out
TEN
THOUSANI
SEASON'S
GREETING
from the
partners of
GREATER
CHICAGO
LETTER
SERVICE
oan
sbaner
1928-1929
Leon
Keller

Media and Publication Strategy

In order to disseminate his design philosophy, Fuller developed in 1928 an original publication strategy and forward-looking ideas about the use of modern media. Lacking money and unknown as a designer and author, Fuller developed a method using mimeographs, the cheapest means of reproduction available, to put his 4D manuscript and his other sketches into circulation in limited editions. Isbaner's Stationery and Copy Shop in Chicago helped him to do this. To express his thanks, he designed a New Year's card for the small shop to give its customers. Fuller sent the printed 4D books to family members, friends, acquaintances, and prominent members of society and their intermediaries. In this way, he hoped to make his multifaceted manuscript familiar to Henry Ford, Bertrand Russell, Le Corbusier, Christopher Morley, among others, but also to publishers, university presidents, manufacturers, and architects. According to American law at that time, copyright was binding when it could be shown that 200 copies had been received, which required proof that the addressee received it. On 30 July 1928, only 128 copies of the 4D book had been sent; only in 1929 did it reach 200 copies, so that Fuller obtained copyright for his book.
This procedure resulted in an extensive correspondence with the recipients, which Fuller published as a supplement to the text in later editions of the 4D book. The recipients themselves often exchanged letters that Fuller included in the growing collection if they related to 4D. The combination of the process of mimeographed copies and the correspondence used as an open letter results in something that in an editorial in the magazine "Shelter" Fuller would later call a "correlative medium." Everyone should be able to get involved with everyone else in intellectual exchange.

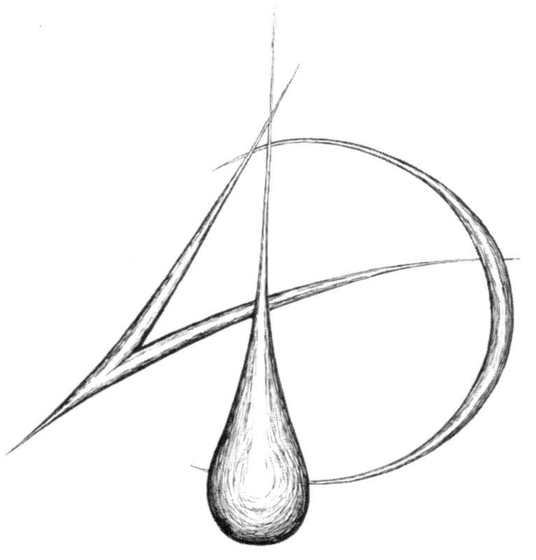

My seat in Turbo-jet to New Orleans to receive Honorary membership in A.I.A. 31 years after trip to St. Louis to present 4D IDEA to A.I.A. June 1959.

4D

IMPORTANT INFORMATION

Subject to Tariff Regulations

CHECK-IN: Passengers departing on flights are requested to check in at the ticket counter for:

DOMESTIC FLIGHTS—at least twenty minutes before scheduled departure.

INTERNATIONAL FLIGHTS—at least thirty minutes before scheduled departure for flights to/from BERMUDA, CANADA or SAN JUAN; at least one hour before scheduled departure for flights to/from MEXICO CITY or to South American destinations on the Eastern-Braniff International Interchange flights.

ALL FLIGHTS—Passengers presenting themselves at the ticket counter less than 10 minutes before scheduled departure will be refused passage, if boarding them will delay the flight.

Free baggage allowance, including checked or unchecked baggage, is 40 pounds except in conjunction with international travel where higher allowances apply. Small parcels, briefcases, hat boxes, portable typewriters, cosmetic kits and other items may be carried in the cabin ONLY after being weighed. Weight in excess of free allowance will be transported only after payment of standard industry charges on file in the tariff. Retain your receipt for any charges paid.

Please retain your passenger receipt. If your flight is interrupted en route, a refund of the incomplete portion may be obtained immediately at the airport, or at any time later, by surrender of your passenger receipt to the Revenue Accounting Office, Miami International Airport, Miami 48, Florida. Unused tickets may be refunded at any Eastern Air Lines Office.

GROUND TRANSPORTATION LEAVES ______ AT ______ AM PM

FLIGHT LEAVES ______ AT ______ AM PM

PLEASE CALL ______ AT ______

FOR RETURN RESERVATIONS OR TO RECONFIRM

EASTERN AIR LINES

EASTERN AIR LINES USES SINCLAIR AIRCRAFT OILS EXCLUSIVELY

The powerful engines of Eastern Air Lines' Great Silver Fleet depend on Sinclair Aircraft Oils exclusively for continuous efficiency. These oils protect Eastern's fleet over 300,000 flight-miles a day.

The same top-flight care and engineering experience behind high-quality Sinclair Aircraft Oils are used to develop superior petroleum products for your car, such as *New* Sinclair Power-X Gasoline with X-Chemical and *New* Sinclair Triple X Multi-Grade Motor Oil. You can rely on them for extra-dependable car performance.

YOUR CAR DESERVES THE SAME HIGH-QUALITY PROTECTION

SINCLAIR

Have a pleasant trip!

SINCLAIR

Beginning in 1928, Fuller always had an eye on the possibilities for a network of communications. His Conning Tower of 1932 was the prototype of a multimedia center. Ever since Lee de Forest's experiments during World War I with radio and radiotelephone, in which Fuller took part as a Navy officer, he knew that the roles of sender and receiver were technically interchangeable. In 1936, Fuller learned about the beginnings of television from Gilbert Seides, who was running the first experimental studio for television in the USA. In the early sixties, Fuller predicted a future with two-way channels; in 1969, he was involved in a patent for a computer. In the same year, he wrote about embryos connected by telecommunications network, which did not want to come out of the womb until peace ruled on earth. It was the first generation that grew into the age of the PC and the Internet. And Fuller is, just as much as McLuhan, one of its mythologizers.

The publication of the Dymaxion House in 1929 made Fuller suddenly famous on the American architectural scene. The house was more uncompromising than anything that had come before it: a turn away from the orthogonal plan, hexagonal symmetry, a tripod supporting pole in the center from which cables stretch out to the decks – like spokes and rim from the hub of a wheel. Fuller conceived the floors as stretched piano wires with a pneumatic layer over them. The outer walls of vacuum casein elements are a non-load-bearing screen – opaque, transparent, or translucent. The rooms are lit indirectly by a system of mirror prisms. Fuller thought of the house not as private property but as something available long-term, a service product like a telephone. "Something New under the Sun" was Fuller's title for the announcement of his lectures. How true!

DYMAXION HOUSE

F. S. LINCOLN PHOTO ©

f the 10-deck *Strea*

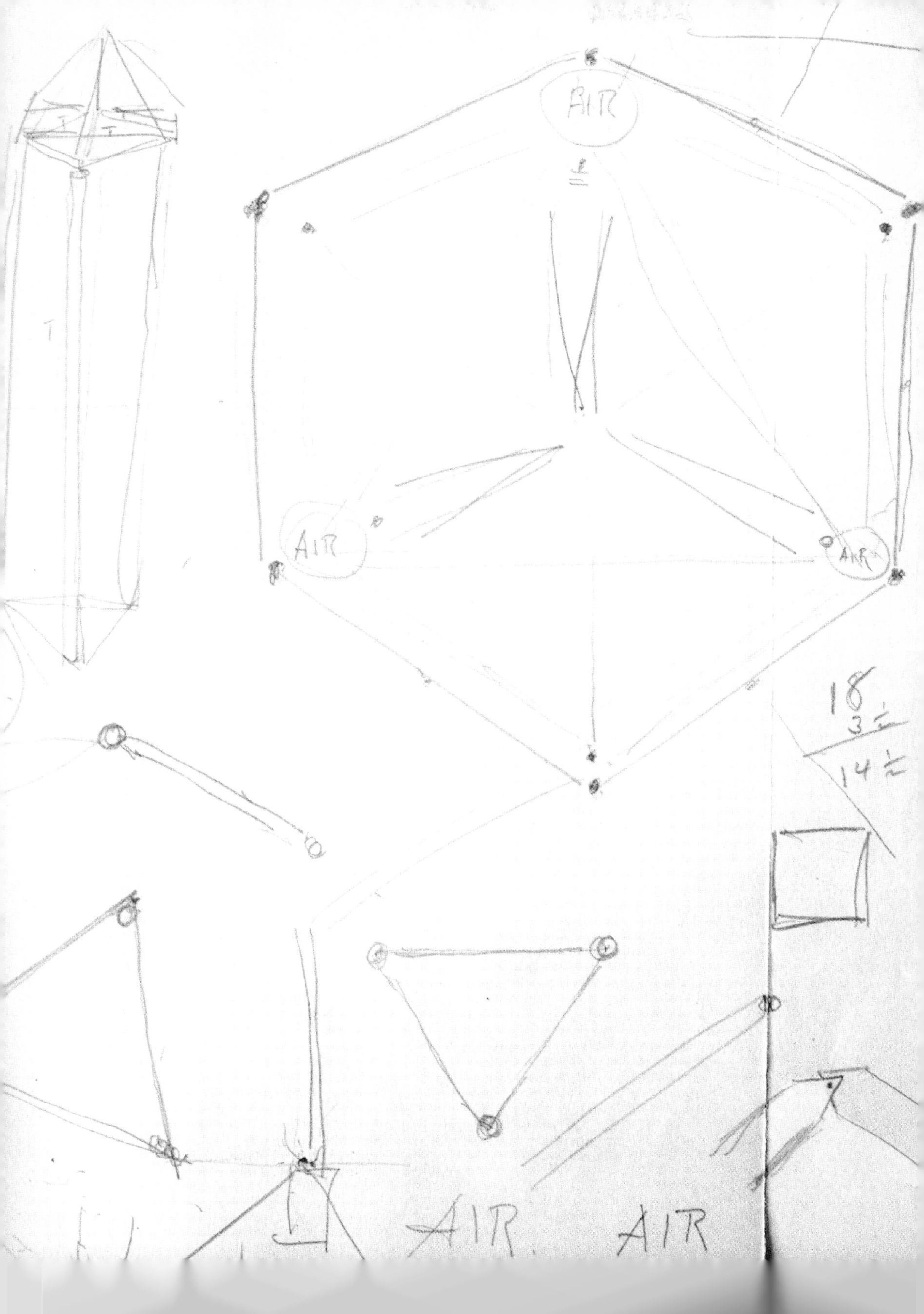
AIR
AIR
AIR
18
3½
14½
AIR
AIR

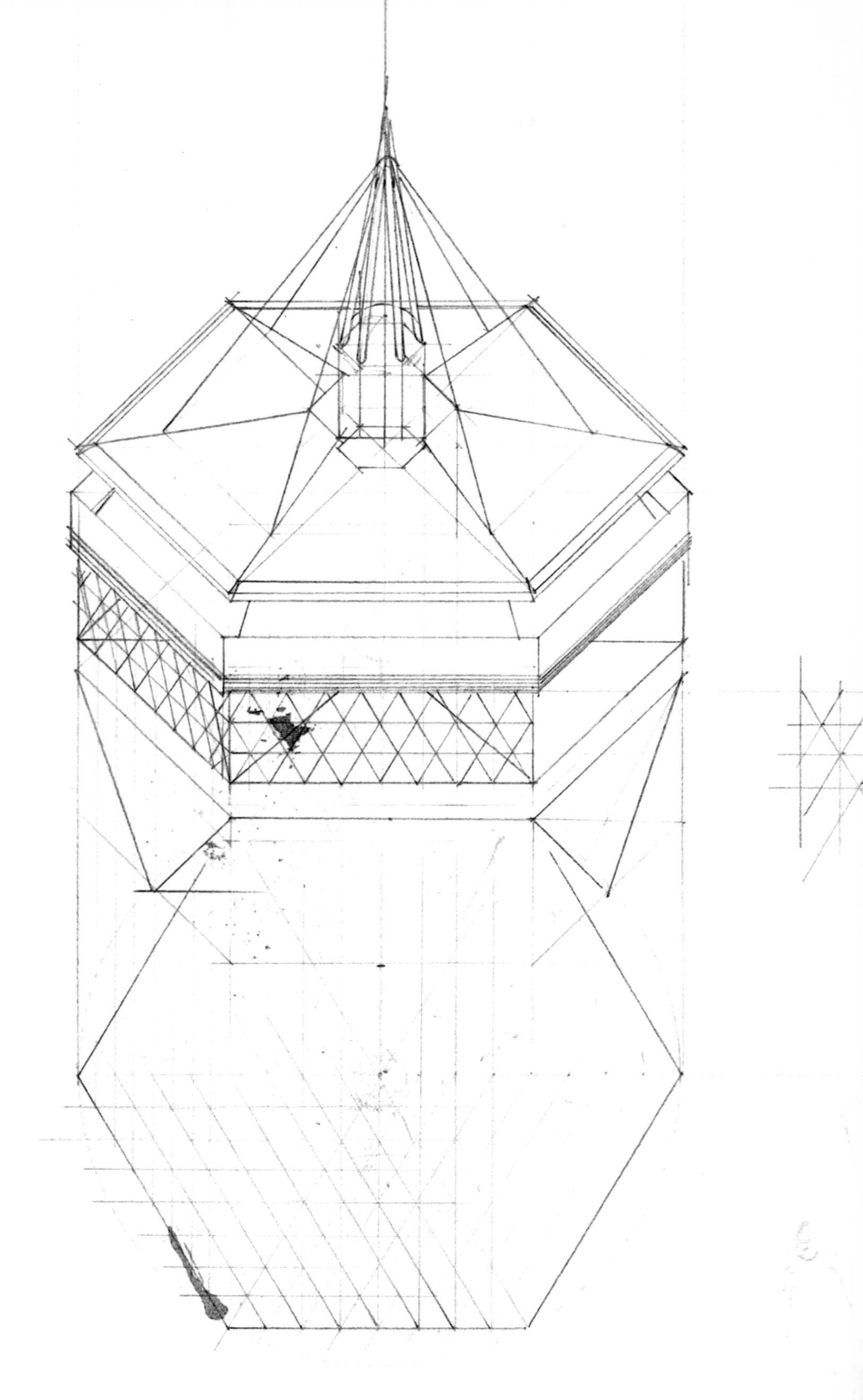

"A home, like a person, must as completely as possible be independent and self supporting, have its own character, dignity, and beauty or harmony."

Lightful Houses, Manuskript (1928), 22

·Analogy: Ship's mast with Schuchow support structure and look-out enclosure. (*Chronofile* 36/1929)

The postcard with the lighthouse and the images on the following double-page spread are documents from the "Chronofile". Fuller glued them in, along with his own sketches, letters to him, by him, overdue-payment notices for the membership dues from his tennis club. Everything is not equally important, but nothing is unimportant. They seem to be merely glued found objects from life in America, without any commentary, but, in fact, the act of collecting them is the commentary. The lighthouse, the lookout onto the lattice pylon or the rotating storage container for tools, nails, and screws – for Fuller, these things were like fingerprints of the civilization process, evidence of practical applications that corresponded in some hidden way to his thinking, an inner correlation that revealed itself to him, and which he was reworking out of this perception and into something else. The lighthouse stands for autonomy; it is an aid to navigation, freestanding and dependent only on itself – an image of proud independence and, in addition to its functional justification, a metaphor for the true – that is, intellectually independent – human being. The Revolvo-Displays are another example for the significance of Fuller's work that almost always transcends the functional. Fuller came across these representatives of a typically American view of reality at a trade fair and then corresponded with the manufacturer. They gave him an idea how the occupants of his house could store their appliances in a space-efficient way. They were certainly not appliances that he intended to use exactly as they were. And yet they were clues that brought him further. The symmetry of the round stands, which require that the contents be distributed in approximate equilibrium, points beyond the simple practical aspect of the installation and gets at the essence of the house as a whole, and that is itself something spiritual. See "Your Private Sky": Discourse "Lightful Houses" and "Dymaxion House" (a lecture at the Architectural League).

gy: Lighthouse (*Chi*

o with display stan

matic
play
BEARING

The
Economy
Way
STABILIZED

Cost **DISPLAY STANDS** They Pay

S MERCHANDISE

ıds are of outstanding value. displayed where customer can REVOLVO" Stands make better space, save valuable time, give l increase sales and profits. Can e ever wanted. Every item is nonstrated practical, useful and ousands of users.

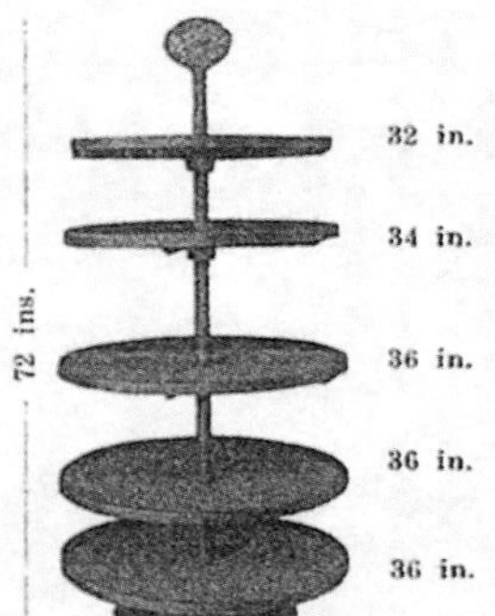

No. 5

CONSTRUCTION

Stands are all steel. Center stems supporting sections are double strength steel molded solidly into stabilized circular, cast iron base. Sections are made of heavy gauge, roller leveled sheet steel, bolted and electric welded and revolve either way, independently, on ball bearings. Patented stabilizing features prevents sections from sagging or getting out of line. Usage has not set a limit on "REVOLVO" serviceability. Will work a businees lifetime.

28 in.
28 in.
28 in.
28 in.
28 in.

S. 88

niversal Stand ns with 1 in. running full Each section ders at given quipped with le bins. Each ide at front, 4 ront, 12¾ in. . deep at back.
ight 275 lbs.
ce $85.00

No. 5—Display stand for paints, oils, aluminum and many other kinds of merchandise. Five sections of display, require circular floor space 36 in. in diameter and are equivalent to 47 ft. of shelving or bin display. Sections are adjustable up or down on center standard at various intervals.

Shipping Weight—295 lbs.

List Price, $90.00

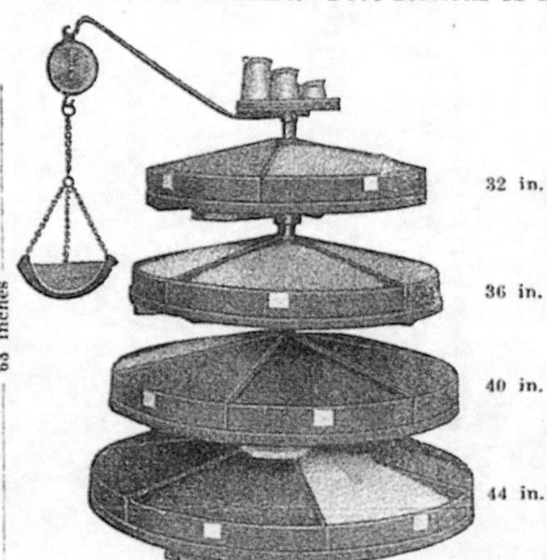

No. RC-4

Removable Bin for No. US-88 and No. RC-4

Net Price 50c each

22 in.
25 in.
72 ins.
28 in.
31 in.
34 in.

No. 534

No. 534, Another display stand of five revolving sections requiring circular floor space 34 in. in diameter and equivalent to 36 ft. of shelving or bin display. Five wings anchor each section to central casting. Each wing is 7 in. high at casting and tapers down to 1½ in. at edge of section. Bound by 1½ in. retaining band. There are 25 compartments.

Shipping Weight 264 lbs.

List Price $90.00

22 in.
25 in.
28 in.
31 in.

No. 40-40, a display stand of four under braced, flat top sections, is equivalent to 27 ft. of shelving or bin display. It requires circular floor space of 2½ ft. in diameter.

Shipping Weight 150 lbs.

List Price $60.00

No. RC-4, "REVOLVO" Combination display stand is used for seeds, bulk merchandise, canned goods, paint and other stock. Comes equipped with four sections and 20 removable bins each with a capacity of ½ bu. Paper bag pockets under each section. Occupies circular floor space 44 in. in diameter and is equivalent to 39 ft. shelving display. Can be bought without removable bins.

Shipping Weight 410 lbs.

List Price Without Scales **$100.00**

No. 534-A has the same construction and specification as No. 534 excepting a ⅜ in. round steel guard, elevated 1¾ in. above sections, runs full circumference.

Shipping Weight 280 lbs.

List Price $90.00

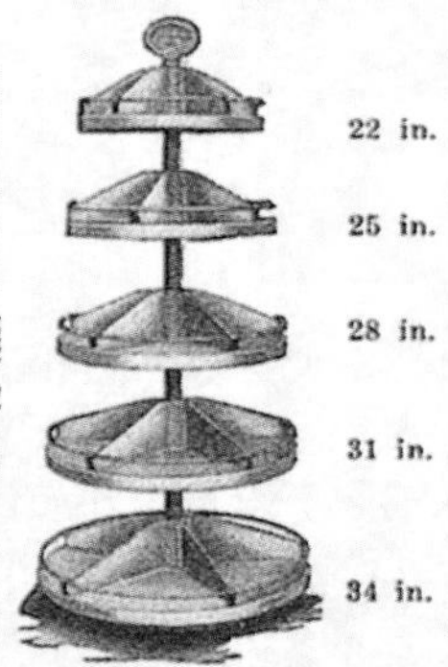

No. 534-A

SEE DISCOUNT SHEET

ALL EQUIPMENT F. O. B. FACTORY

130 · *Dymaxion House.* Second model, following the 1928 structural model. Depicts the below-ground building elements (foundation and tanks). Fuller utilized this model for lectures he gave in 1929.

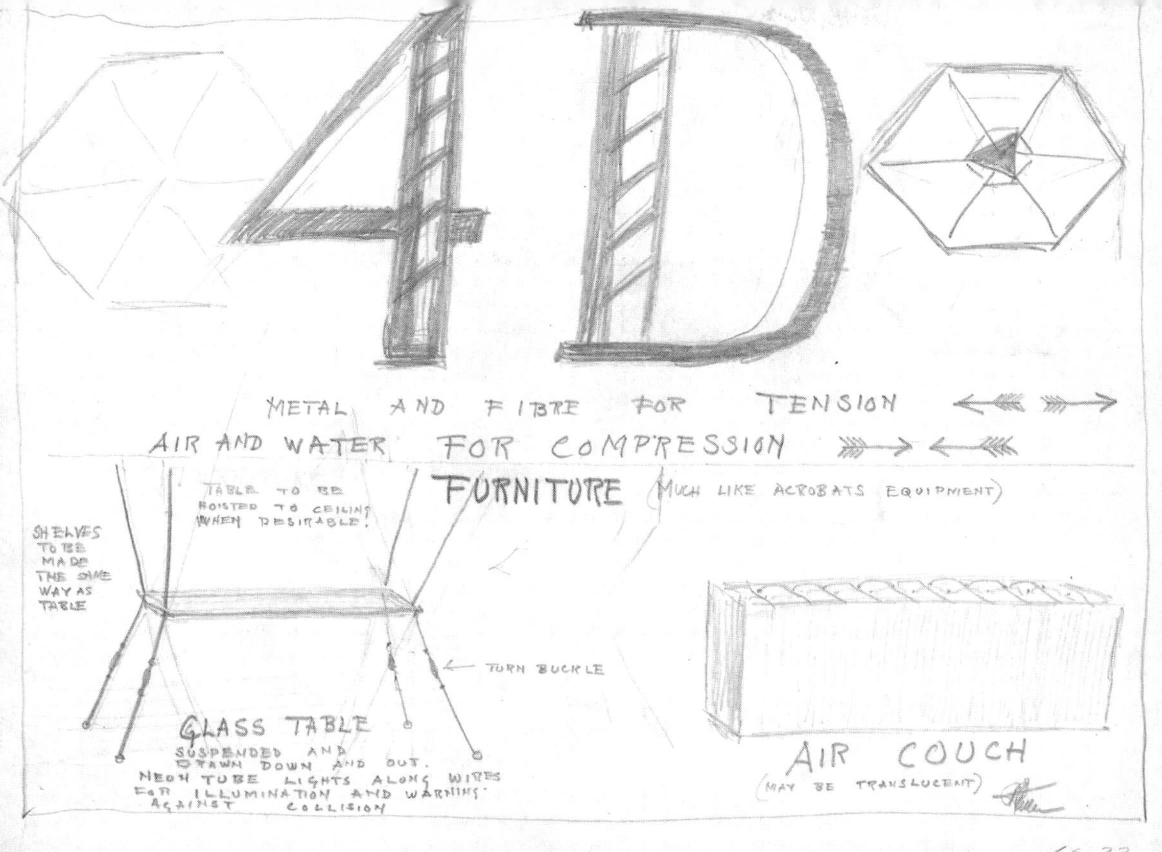

"We will have arrived at our new artistic era of architectural expression when our buildings have lost their last trace of feudalistic depression; when we arise in our buildings in concentrated area of compression in opposition to gravity by means of mast or caisson reach out in space from the vertical by tension and compression, compression diminishing as we fall off from the vertical, until we finally flow downward in direct tension. Then will our exteriors, hanging from the outward flow of the top like a great fountain be full of lithesomeness, light and color."

Lightful Houses (1028), 15

ketches of 4D furni
ouch; metal and fib

DY(NAMIC)M

The name "Dymaxion"

"... It was in Chicago that the name Dymaxion was born. G. Ray Schaeffer, then the Marshall Field promotion manager, asked Fuller to demonstrate his house in the interior-decorating galleries of the store, and Fuller agreed to do so.

Before the house was displayed, however, it had to be given a name. Marshall Field wanted all its merchandise, if possible, to have a name.

Schaeffer called in the late Waldo Warren, a merchandising namesmith who had won considerable renown for coining the word 'radio' to take the place of 'wireless'. There was an instantaneous meeting of minds. Fuller was an amateur namesmith himself, having just coined the word 'debunk' in order to point out the flaws in traditional architecture which his type of housing was intended to correct. Warren agreed that 'debunk' was a very good word indeed, and so did everyone, for, like radio, it quickly became a part of the American language.

Warren spent two days listening to Fuller talk, jotting down such typical Fuller jargon as 'maximize', 'sumtotally', 'earthian', 'dynamic', and 'teleologic'. After he had compiled a list of several hundred words, Warren turned to Fuller and said: 'You're obviously a four-syllable personality, and you'll have to have a four-syllable name for your house.'

He returned a few days later with the word of words. «Dymaxion» was it. Warren had formed the word by making endless combinations of nonsense syllables taken from Fuller's lingo until he had struck the word which seemed to him to dramatize the Fuller personality."

Leigh White: Buck Fuller and the Dymaxion World, The Saturday Evening Post, 14. Okt. 1944

AX(IMUM)ION

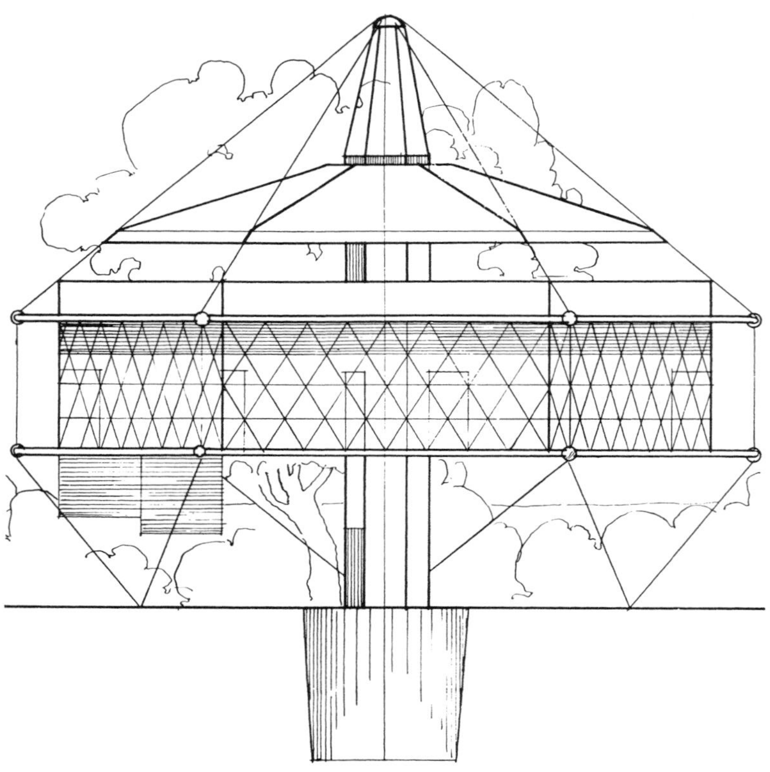

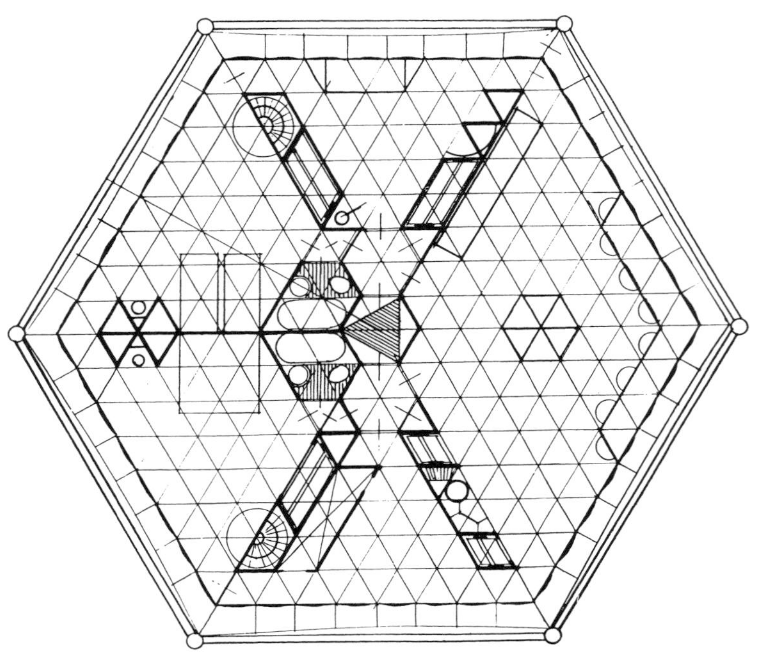

134 ·*Dymaxion House.* Project phase 1929. Elevation and floor plan, roof screen is not weight-bearing but springs back upwards, hung on tension cables; Tension struts with ground-level anchor points set slightly set inward. Floor plan, clockwise from above: library, living room, utility room (kitchen, laundry), two bedrooms. In center, elevator, two bathrooms

The Dymaxion House

(...) Having once freed our minds of the customs and traditions that have bound us since the days of the earliest shelters, we can attack this dwelling problem just as we would attack the problem of building some other device or piece of machinery that had never before been made. The dwelling, after all, should be a machine for the efficient and comfortable conduct of family life under shelter. Whatever will contribute most effectively and economically to that end merits our consideration; whatever does not so contribute, though it be hallowed by centuries of tradition, need not concern us. There is one important point here that we should recognize at the outset; we can ignore all considerations of cost. The old necessity of "cutting our coat to fit our cloth" – the ever-present cloud that hangs over every home-building operation – is gone. The world's greatest tool, quantity production, will make possible for all what would otherwise be possible only for the very few. (...) It is said that Mr. Ford's new model, if but one car had been built, would have cost $ 43,000,000. Reproduction of that first unit costs but $ 500, or approximately twenty-two cents per pound for completely harnessed synthetic and mechanically coordinated materialism. There is no reason to believe that a dwelling should cost more.

This talk of "cost per pound," however, brings us at once to the realization that for our reproducible dwelling we are going to have a minimum of masonry. Our house must have lightness of weight, for easy distribution, combined with strength sufficient for all exigencies. Fire, flood, wind must be resisted with unquestioned superiority of defense.

(...) The Dymaxion house is the product, not of a whimsical dream but of years of concentrated study. The design here shown is by no means offered as the sole solution of the problem. The modern automobile changes year by year, as we discover and develop new ways of improving it; so with the house. As a start, however, the design herewith is developed somewhat along the lines suggested by the natural system of human beings and trees – a central stem or backbone, from which the structure depends and in which are the supply, distribution, and waste systems. The central tower, or mast, is made of duralumin tubes, inflated to high pressure, in triangulation with piano-wire steel – similar to the battleship mast or dirigible mooring. With this employment of steel in its most effective condition – that of tension – we are able to construct the complete mast for a twelve-deck house which, together with its tubular perimeter beams, has a total weight of less than one thousand pounds.

Starting with the central mast principle, we discard at once all rectangularity of plan. The logical development from the mast as our functional centre is radially on successive levels. For reasons into which there is not space here to go, the hexagonal plan results – equilateral triangles in three-point suspension. This plan is capable of expansion radially and in a number of levels, providing flexibility of extent without deviation from the unit principle.

The mast is anchored to the ground by its base, in which are fuel and septic tanks, the heat and power generating units, air pump, air filter, and water supply from an artesian well.

The ground level is utilized only for the storage of automobiles, airplanes, etc., and access to the house is by means of a triangular elevator in the mast. This is propelled by a motor, but if the latter should fail, the elevator may be operated by hand more quickly than one can walk, travelling as it does upon a worm gear, and without counterbalance weights.

The exterior walls, having only their own weight to support between floor levels, as in the modern steel-frame building, are here made of two thicknesses of translucent, transparent or opaque material such as we make from casein, with a vacuum between, providing ideal insulation against both sound and heat. The floors likewise, in tension between their triangular supports, are softened by pneumatic pressure between two flexible shells, the upper one of which might be something like our synthetic approximations of leather.

Illustration of 10-deck *Dymaxion House* in Fortune magazine, July 1932. Right, the *Streamlined Dymaxion Shelter*

A series of light sources in the mast, one at each story level, gives diffused lighting through the whole ceiling areas by means of simple prisms, mirrors, and color screens. In the scale model shown the smallest procurable bulb serves to light a whole floor by this method.

Incidentally, the heat generated by the lighting system is all utilized, the air supplied to the various rooms being passed about the lights on the way. Mr. Fuller believes that, with the almost perfect insulation secured by this construction, the house can be heated merely by the heat generated in the lighting and power requirements. The air is mechanically circulated, cleaned of all dust and obnoxious gases, and replenished.

Through the arterial system in the central mast, all connections naturally would be at standardized points, so that an integral bathroom unit is attached without piping. In the same way the other various utility units are made as integral sections, merely requiring attachment in order to begin functioning. Laundry and grill are thus made, with a degree of mechanical perfection hitherto impossible of achievement. The grill is equipped with electric-vacuum range, electric refrigerator, dishwashing machine, indirectly lighted glass cases for food storage. Every one of mankind's furthest developed appliances for doing away with household drudgery can and will be made a standard unit of the system, an achievement possible only through industrial reproduction on standardized patterns.

In a living-room combination, for example, an integral assembly consists of desk, filing-cabinet, typewriter, calculating machine, telephone, radio-television receiver, dictaphone, phonograph, and safe. Just as has long been the practice in unit bookcase and unit office-cabinet construction, making available a personal choice without affecting the principle of mass production and co-ordination of sub-units, so here the standardized system rather than the old style-complicated, tailor-made system will make every convenience available for all.

Such a house would of course be immediately available, erectable in a period measured by hours rather than by months. (...) A house of the size represented by the model illustrated could be had for about $ 3,000, and the application of the time-payment scheme would bring such a house within the reach of practically all mankind.

(...) Another question that must arise in the minds of this magazine's readers is: What is to become of the architects? Possibly the answer lies in a consideration of the fact that industrial mass production is to-day requiring more designers that were ever found necessary in the periods when arts and crafts were concerned with individul piecework. The architect's efforts to-day are spent in the gratification of the individual client; his efforts of to-morrow, like those of the composer, the designer of fabrics, silver, glass and whatnot, may be expended for the enjoyment of vast numbers of unseen clients. Industrial production of housing, as contrasted with the present industrial production of raw materials and miscellaneous accessories, calls for more skill and a higher development of the design element, not its cessation.

"Architecture" (New York), June 1929, editorial commentary following a text submission by RBF

4

BUCKM
INSTER
FULLER'S
DYMAXIO
N HOUSE

THE HARV
ARD SOC
IETY FOR
CONTEM
PORARY
ART INC

D

139 · *Dymaxion House.* Third model, with illuminated supporting mast and ceiling, floor covering of braced triangular plates.

· Exhibition brochure, Harvard Society, 1929, title page

“He Plans Houses Like Trees”

Chicago Daily News

“Pneumatic House, Built Like Tree, Proposed For Mechanical-Age Folk”

Christian Science Monitor

“A Press-The-Button Home For Our Age”

The World (NYC), Women’s Section

“Houses Made Of Milk”

Daily Mail (London)

“Housekeeping in the Air”

Brooklyn Daily Eagle, 8. 12. 1929

“Houses You Will Carry When You Move”

New York Evening

“House of 1982 Built Like a Ship”

Brooklyn Eagle Magazine, 17. 4. 1932

“When We Live in Circles and Eat in Merry-Go-Rounds”

New Orleans Tribune, 12. 6. 1932

“Machine For Living - Revolutionary House”

The New York Sun, 25. 5. 1932

“Houses that Hang From a Pole”

Modern Mechanix and Inventions, Sept. 1932

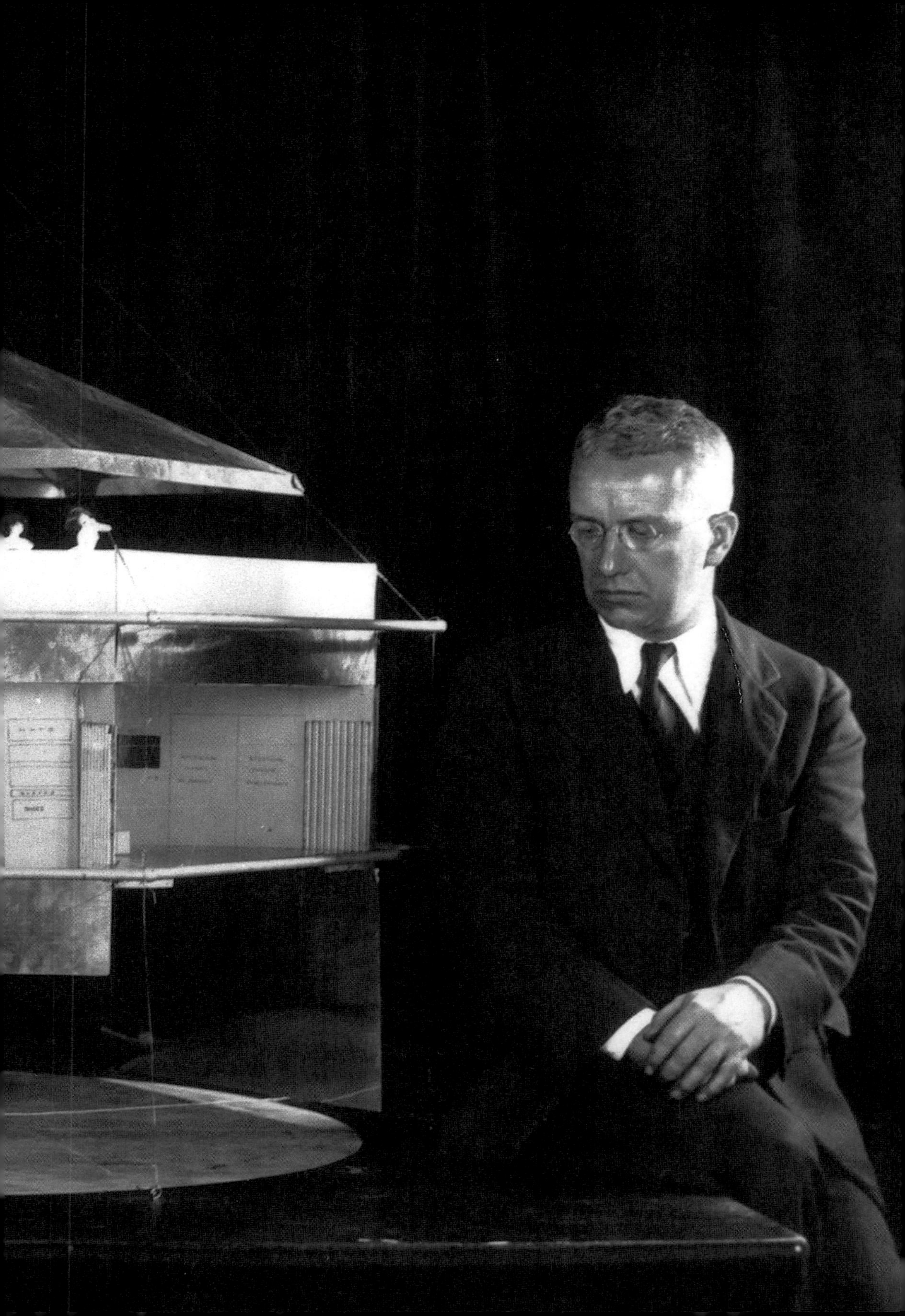

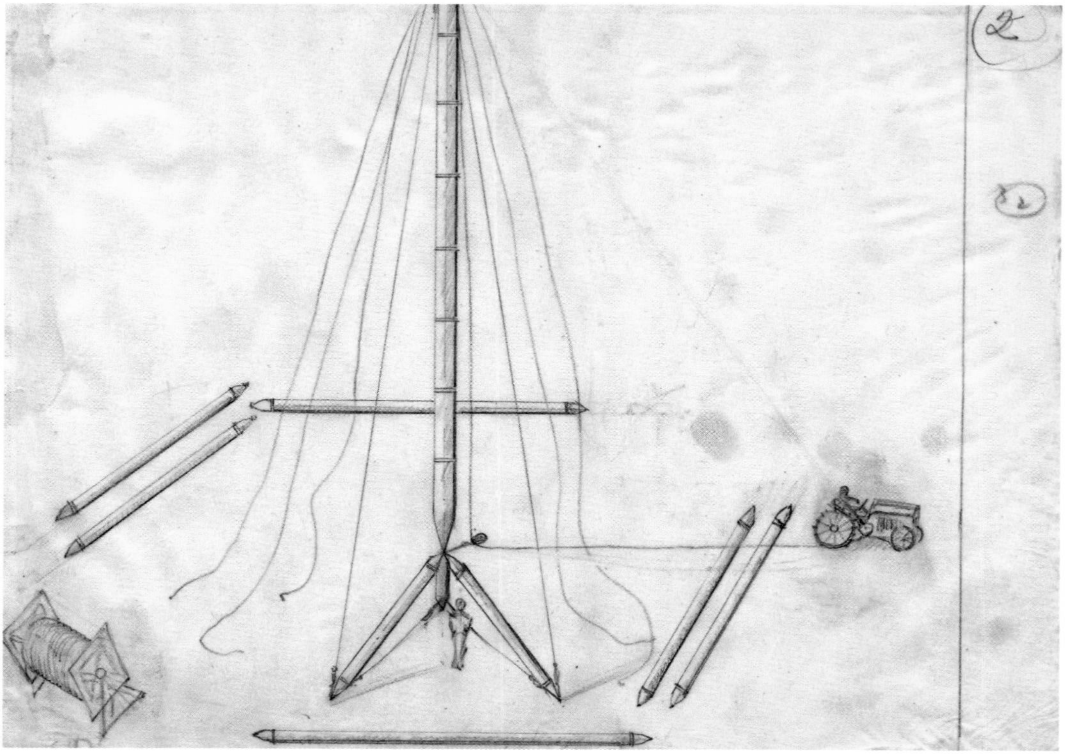

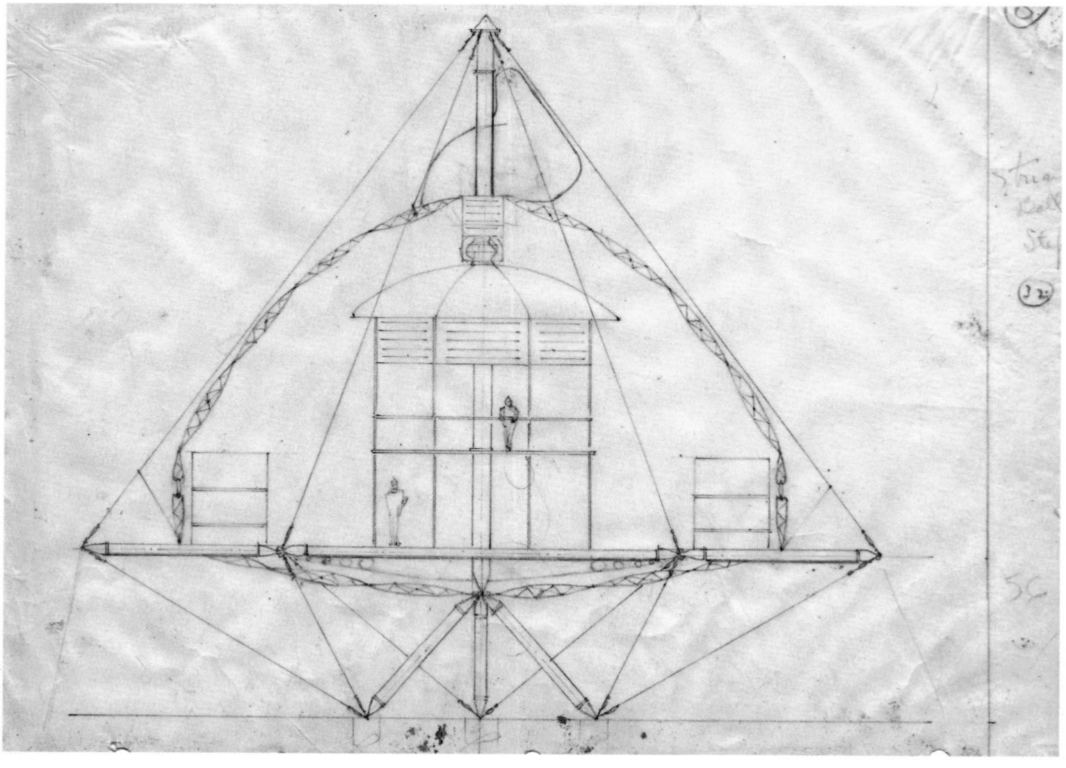

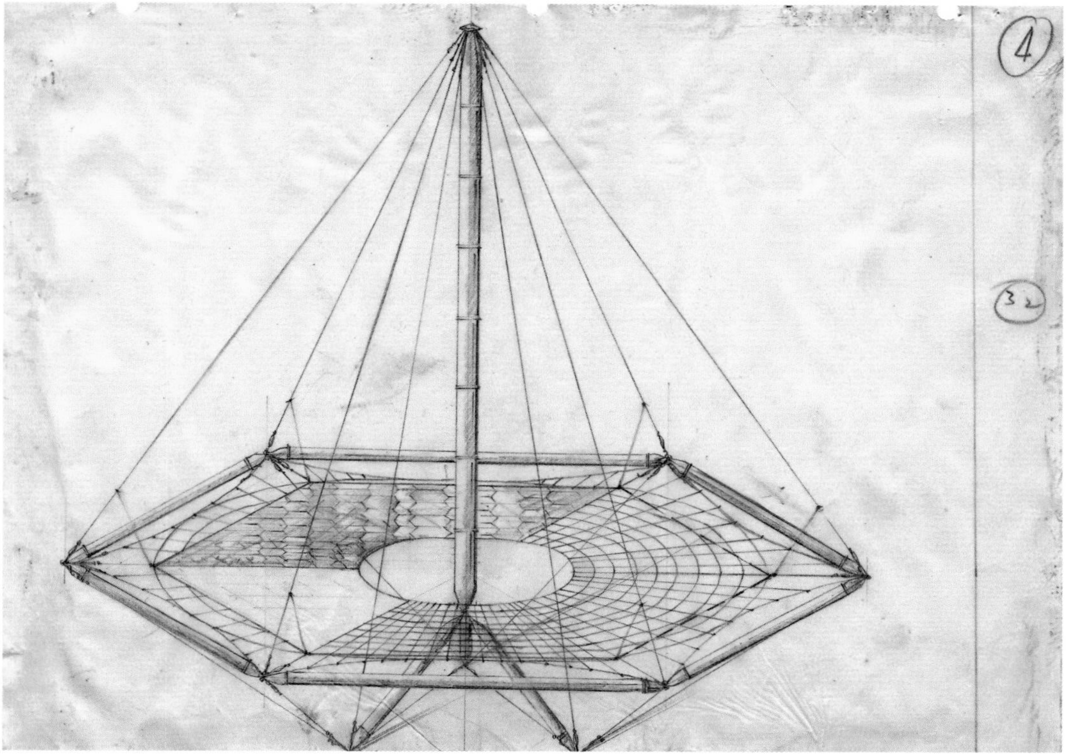

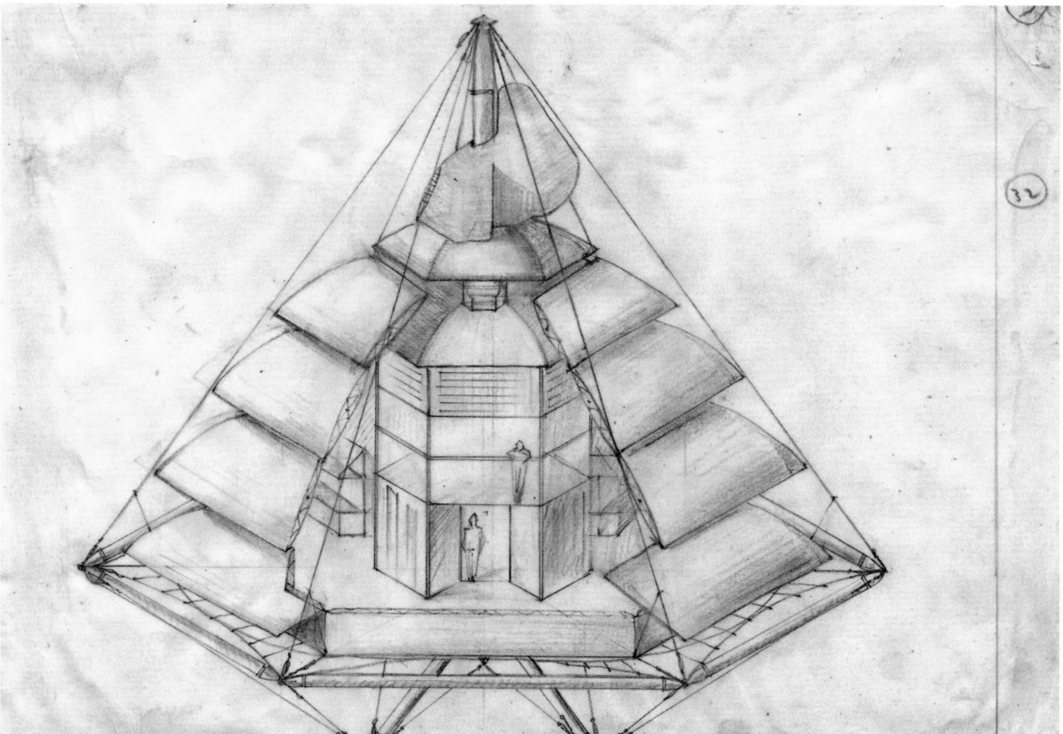

"Now when we came to the studying of winds, and of wind relative to house, it becomes evident that there was the possibility of shaping houses so that we might diminish the amount of drag. It seemed impractical to consider a streamlined house that would revolve around. Therefore I made a number of wind tunnel tests with streamlined shields around buildings, and I was able to reduce the drag on those square skyscrapers by wind stress by at least 87 percent by placing streamlined shields around their models, which also reduced their heat loss by 87 percent."

Designing a New Industry (1946), 33

"Winds are drafted and not blown"

Designing a New Industry (1946), 35

145 ·Structural Studies Associates Dymaxion 20-Worker Shelter for Russian Cooperative Mobile Farming. Series of drawings 1931/32

·*A Streamlined Dymaxion Shelter*. Prototype study. Depicted in Shelter No. 5 (November 1932)

In New York's Greenwich Village, Fuller came in contact with the avant-garde of art and architecture of Europe and North America: Frederick Kiesler, Frank Lloyd Wright, Raymond Hood, Knud Lönberg-Holm, Isamu Noguchi, Paul Nelson, Martha Graham. In the city of extreme density, the Village was a place for decompression. Changes in aggregate states seem easy there: the fantastic becomes real and possible. Bohemia in America: if anywhere, then here, where dreams spoken in thick accents are translated back into American English.

GREENWICH VILLAGE

Many people thought I was some kind of a nut because I was talking about air-conditioning, packaged kitchens, and built-in furniture. But there were some architects who didn't go along with the AIA and my image as some kind of wild man, and I began to get invitations to come and lecture about my ideas.

I decided to make a complete experiment of peeling off from society in general, and started wearing T-shirts which nobody was doing then, went about without a hat and in sneakers – absolutely comfortable clothes. Then when people started getting interested in my Dymaxion House, very nice people with influence, and they'd say, "I'd like to give a dinner party for you" and so forth, I would show up in khaki pants and they'd be very shocked. And when Mrs. John Alden Carpenter, head of the Arts Council in Chicago, gave a beautiful dinner party, I showed up and rudely announced, "I don't eat that kind of food," and was in every way obnoxious.

I was putting self and comfort ahead of my Dymaxion House, and I said, "You're not allowed to do that. You must get over that. You must stop that looking eccentric, with everybody pointing at this guy."

So I decided the way to do that was to become the invisible man, and that means a bank clerk – so I put on a black suit, bank clerk's clothing; then they would focus on what I was saying instead of my eccentricities. I said, "I must get rid of continually making too much of myself."

Anyway, by 1928 I was back in New York and I began to hang around in the Village. I used to go to Romany Marie's, a restaurant I was introduced to by a friend even before I went to Chicago. It was probably the last of the really great Bohemian cafes I know of in the world – very much like the Paris of the '20s. The Village was loaded then with great artists and great intellectuals, and Marie had by far the best place in town. That's where I carried on and developed my ideas. Certainly, in Greenwich Village they took me and my ideas seriously.

What I found so exciting there was that for the first time in my life I had a grand strategy and could try out my thinking against the best of the intellectuals.

Many intellectuals became communists, and I'd say about one-third of the people who came there regularly were communists. I was very apolitical and they were very political.

And they would give me a great battle, and a chance to defend my position. They would really battle me hard. It was probably one of the best conditionings I've ever had.

And they liked the Dymaxion House.

They didn't quite see how we could ever get it going, but they liked the feel of it.

After the great stockmarket crash in 1929 and the Depression that followed, anyone could see that the economic system here in New York as well as everyplace else in America had broken down. People slept in subways; unemployment was everywhere. Some restaurants were down to one-cent meals, and you could buy a whole suit of clothes for a dollar.

Even at that, no one could sell anything. No one had any money!

They were fantastic times.

I made some furniture to decorate Romany Marie's in the Village – she was then at Minnetta Lane – and she would give me a meal every day instead of money. She would keep a big pot of vegetable soup cooking all the time, constantly adding water and pieces of meat. It was delicious, but I didn't want to overdo it so I only came every other night. There I would stay, a table-sitter, all evening, until very late into the night. It was the Greenwich Village of the late '20s and early '30s that generated great new thinking – and I gained many friends for my concepts, and lost none.

Marie was very interested in me and my work. By this time she had moved her place to South Washington Square, down in a basement, and she ran it together with Puck Durant, Will's wife.

 ·*3 Hours for Lunch Club* at Chris Seller's, from left Meridith

·Romany Marie. Patrons of *Romany Marie's Tavern*,

They asked me to have an exhibition of my Dymaxion House. I had a model, d'ye see, a beautiful model that I made for the Chicago show, and in the summer of 1929 I gave my talks down there.

I met a you man there who had just come to New York from Europe where he'd been studying with Brancusi – and this was Noguchi.

"I first met Mr. Fuller, as I used to call him," said Noguchi, much later, "at Romany Marie's in 1929. Some time later I got an old laundry room on top of a building on Madison Avenue and 29th Street with windows all around. Under Bucky's sway I painted the whole place silver – so that one was almost blinded by the lack of shadows. There I made his portrait head in chrome-plated bronze – also form without shadow.

"Bucky was in a continuous state of dialectic creativity, giving talks in any situation before any kind of audience… He would talk to me as though to a throng; walking and talking everywhere – over the Brooklyn Bridge, over innumerable cups of coffee. Bucky drank everything – tea, coffee, liquor – with equal gusto and would often be in a state of wide-awake euphoria for three days straight. Drink did not seem to affect him otherwise.

"He used to drink like a fish. He had become a God-possessed man, like a Messiah of ideas. He was a prophet of things to come. Bucky didn't take care of himself, but he had amazing strength. He often went without sleep for several days, and he didn't always eat either.

"Bucky's zest for life is part and parcel of his creativity. However, he has the capacity and resolution to come to grips in unknown hours and retreats of the mind to fathom new secrets from the universe."

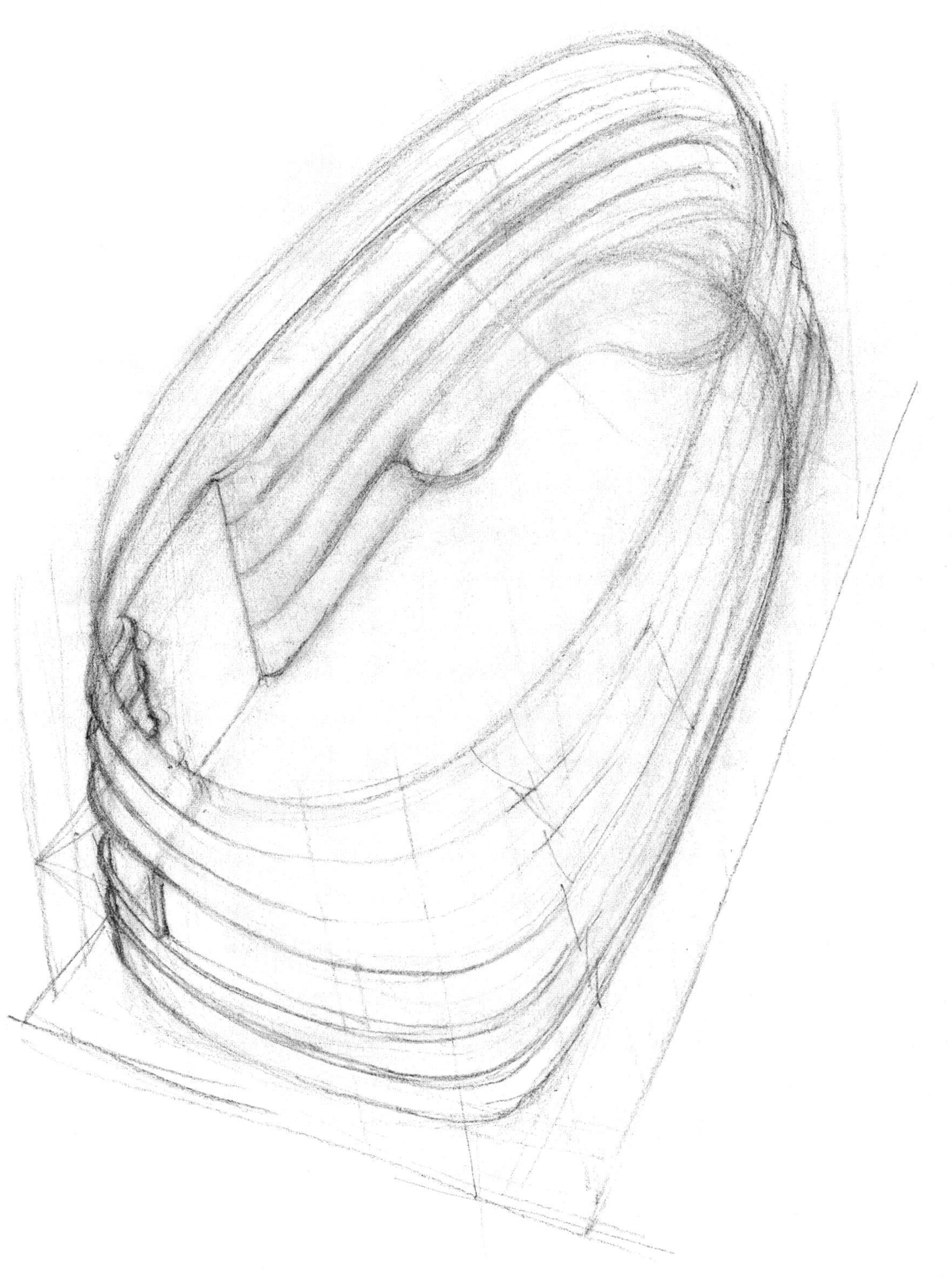

He absolutely fell in love with that house and everything I said;
and he said, could he make a head of me?
and I said I'd be glad to have him do it.

So posing for him day after day gave us a chance to build up our friendship that went on and on from there.

Noguchi used to do real well. He was a bachelor, and he always got big money for his commissions. He made these heads of the most beautiful people. So Isamu was OK; I was continually bust, and he used to let me sleep in his studio – usually on the floor.

When the Depression set in seriously and most of the New York hotels were empty and looking for something to attract people, they'd ask me to come with my Dymaxion house to be on exhibit. They would give me a beautiful room or apartment to use as an exhibition room and Isamu and I would sleep on the floor – no bedclothes or anything. They gave us a bathroom to go to, and we'd literally live on coffee and doughnuts every other day or so.

"We would move in with our air mattresses and a drawing board and that was it. The less the better was his credo. His Shelter magazine was produced under such circumstances (1930–1932)," recalled Noguchi.

It was really tough going. I had Anne and Allegra down in the country. My mother helped me a little with an allowance to look out for them; but I lived on the minimum you can possibly get on with.

In 1930, I sold all my life insurance and took over a magazine called T-Square. I changed its name to Shelter, and published it for the next two years – up to the '32 election of Franklin Roosevelt and his New Deal which followed the absolute disaster of the previous laissez-faire economy. With the New Deal's dedication to "the forgotten man," I decided to cease kibbitzing.

·RBF in Manhattan, around 1929

·Isamu Noguchi: set design for the piece *Frontier,*

Romany Marie

"The friends of Romany Marie wish to announce that she is no longer at Washington Square, but has returned to 15 Minetta Street, and herewith invite you to come and celebrate the opening of her tavern on Saturday, October the fifth. She will continue to serve dinners, Turkish coffee, tea and supper.

There is no tavern in all of New York like the one in which Romany Marie acts as hostess. Always, at her tables, people from all parts of the world, and from all walks of life gather to talk or dream. Old friends meet again and new ones are made. In fact as one watches Romany Marie, colorful, vivid, dynamically strong, move from guest to guest, it is easy to understand why groups of sensitive people constantly circle about her.

And so it is very fitting that Romany Marie – who has ever been eager to sponsor the new life and art – should bravely return to her old tavern and turn it into a new one. No longer heavy oriental draperies and dim lights. A gipsy recognizes no bounds, and when she must renounce ancient customs she does so with a free gesture. Buckminster Fuller, an inventor with vision who has designed the Dymaxion House of the future, a house hung from a mast, has been invited to decorate the tavern. He has covered the walls with aluminum paint and arranged the lighting effects so that there is a sense of space and sunlight and air. He says, 'a room should not be fixed, should not create a static mood but should lend itself to change so that its occupants may play upon it as they would upon a piano.' The result is that the color, form, expression of each individual is accentuated, and there is an interplay of spiritual vibration which is both stimulating and creative. Here paintings are to be hung on the walls, and their intrinsic value allowed to free birth. What is more, exhibitions are to be changed frequently so that the tavern will be largely in the nature of a picture gallery salon.

Come and see for yourselves and tell Romany Marie what you think of it."

Typoskript, *Chronofile* 36/1929

The recognition of modern architecture as an "International Style" was not what Fuller conceived as the new architecture. His slogan "Doing the Most with the Least" was not a formal criterion for him. He turned the small, progressive architectural magazine "T-Square" into the organ of his Structural Studies Associates (SSA), called it "Shelter", and used it to argue a line that in 1938 he would characterize as "In architecture, form is a noun; in industry, form is a verb." "Shelter" was supposed to be a "correlative medium" that would help the world of architecture to reach a breakthrough. The content was original and provocative. Fuller did not feel that the magazine had to pay for itself, but the important issue no. 5 of November 1932 proved to be the final volume.

T-SQUARE/ SHELTER

SHELTER
now now
SHELTER

“T-Square and Shelter”

Despite the great public response to his lectures and demonstration of the model of the Dymaxion House, Fuller’s 4D philosophy was only accessible to a few people and in general re-mained quite obscure. He failed in all his efforts to turn the 4D papers into a book. The publishing house Charles Scribner’s Sons in New York was even willing to try the experiment of producing the book on the basis of stenographic notes of Fuller’s lecture before selected members of the Architectural League in New York (see volume 2). But the book never came about, even though Scribner’s tried to secure Lewis Mumford as a collaborator or ghost writer. As brilliant as Fuller could be in extemporaneous lectures and in his correspondence, his theoretical writings were considered unintelligible and unreadable.

Fuller needed a medium in which to disseminate his theoretical discourse and a discussion platform that would attract a like-minded audience and be suited to explore the controversies with exponents of the new architecture and design. “Shelter” was to play that role.

The slow transformation from “T-Square” to “Shelter” that Fuller began in 1932 was reflected in the removal of all advertising, modern graphic design, and editors that changed from issue to issue. This was at the same time a sign of the reformation of the American architectural scene that resulted from the importation of modern architecture and design from Europe. Thus, Fuller gave editorial responsibility for the April issue to Philip Johnson, who, together with Henry-Russell Hitchcock and Alfred Barr, had organized the exhibition “Modern Architecture” at the Museum of Modern Art, and who, in the exhibition catalogue he wrote with Hitchcock, had attached the label “International Style” to the new movement. “Shelter” documents not only the debate between Americans and Europeans but also the controversy between the “stylists” and the “structuralists” that cut across that division. The latter had formed a loose organization in 1932 in the Structural Study Association, whose guiding spirit was Buckminster Fuller.
The many voices of that position were expressed in the May and December issues of “Shelter”. In his rejection of a superficial, i.e., merely aesthetic or stylistic, modernism, Fuller found a sympathizer in Frank Lloyd Wright and a decided adversary in Knud Lönberg-Holm of Denmark, who was personally acquainted with the protagonists of Neues Bauen (the New Building movement in Germany) and was very familiar with the European context. Lönberg-Holm’s influence was also felt in the modern design of “Shelter”: generous covers, clear divisions, and controversial photo essays defined the new face of the magazine.

In “Shelter” Fuller advocated not only “industrial building but also scientific research of home and household, of the technology and economics of housing.” We find here the beginnings of a theory of climatically controlled shells, and we find here, unmistakably, a new programmatic slogan: ecology. “Shelter means the instrumental safety and the service of ecology and economics combined.”

The overall philosophy of “Shelter”, as a correlative medium, is “to do the most with the least.” Interpretive conditions of this philosophy involve the intelligent use of forces already manifest which to-day might be hampered by a lack of awareness of their potentiality in their relation to the evolution of a world-encompassing shelter industry.

· 157: *Shelter* No. 4, May 1932, back cover

VOL. 2 NUMBER 2

FEBRUARY, 1932

T-SQUARE

Sent by Request of
BUCKMINSTER FULLER
See Page 22

T-SQUARE February 1932
Maxwell Levinson, Editor
L.J. Levinson, Managing Editor

·*Frontispiece* by Diego Rivera
·*Editorial,* George Howe

Articles
·*For All May Raise the Flowers Now, For All Have Got the Seed,* Frank Lloyd Wright
·*Moses Turns Pharaoh,*
George Howe
·*The Effect of the Regional Plan on Urban Architecture,*
Arthur Kallet
·*We Are Entering Upon A New Era,*
Le Corbusier
·*The Myth of Building Laws,*
Irving H. Bowman
·*A Cooperative Dwelling,*
R. M. Schindler
·*Universal Architecture,*
Buckminster Fuller
·*Biographical Sketch: Richard J. Neutra,*
B. E. McLoney
·*Let Us Know What Is Being Done,*
Ely Jacques Kahn
·*I Do Not Believe,*
Henry S. Churchill
·*Exhibition of Modern Architecture*
·*Letters to the Editor,*
Arthur T. North, Frank Lloyd Wright

SHELTER, April 1932, Vol. 2 No. 3
Maxwell Levinson, Editor; George Howe, Henry-Russell Hitchcock, Jr., Alfred Barr, Jr., Philip Johnson, Associate Editors; Leon Levinson, Managing Editor

·*Cover, Photo.*
Typical Unit for Chrystie-Forsyth Area, N.Y.C.,
Howe and Lescaze, Architects
·*Editorial, in Architectural Criticism,*
Henry-Rusell Hitchcock, Jr.
·*Symposium "International Style", Exhibition of Modern Architecture, Museum of Modern Art,*
Lewis Mumford, Henry Wright, Raymond M. Hood, George Howe, Harvey W. Corbett
·*Of Thee I Sing,*
Frank Lloyd Wright
·*Old New Stuff,*
Arthur T. North
·*Two Shows: A Comment on the Aesthetic Racket,*
K. Lönberg-Holm
·*Imagination In Community Planning,*
Henry Wright
·*Housing Development, Chrystie-Forsyth Sts., N. Y. C.,*
Howe and Lescaze, Architects
·*Modernism and Publicity,*
Chester Aldrich
·*Notes on the Manufactured Ring Plan School,*
Richard J. Neutra
·*Creation and Criticism, The Frank Lloyd Wright Autobiography; International Style: Architecture Since 1922, Two Book Reviews,*
George Howe
·*The International Congresses For New Building,*
Richard J. Neutra
·*In Memoriam, Theo. Van Doesburg,*
Frederick J. Kiesler
·*Appeal to Reason,*
Kenneth M. Day
·*Universal Architecture, Essay No. 2,*
Buckminster Fuller

·159: *T-Square,* February 1932, cover. The last issue under the name T-Square.

SHELTER
Keep Article by R.BF Page 30
APRIL 1932 VOL. 2 NO. 3

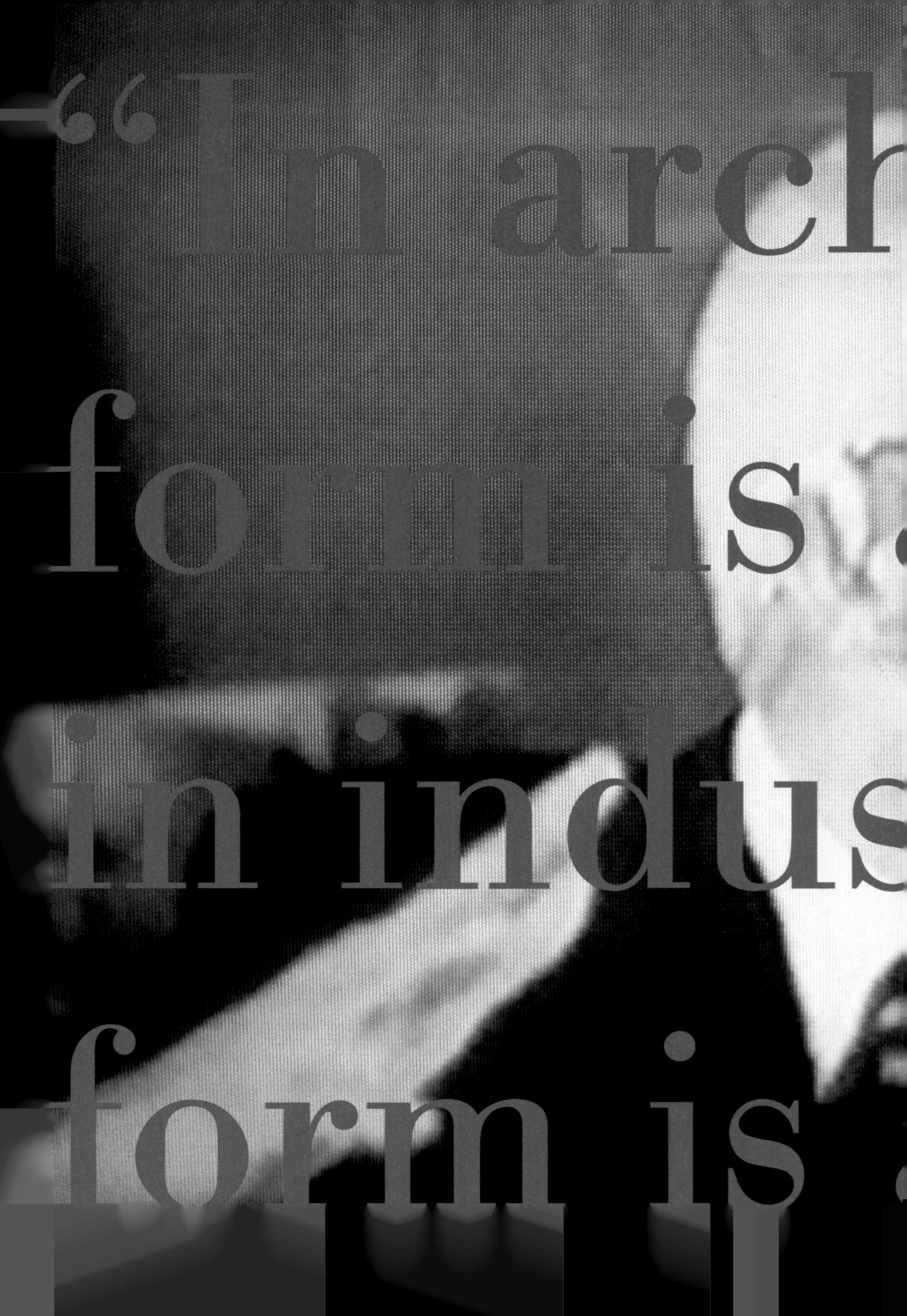

“In arch

form is

in indus

form is

itecture

a noun;

try

a verb."

Shelter, May, 1932, Vol. 2, No. 4
A correlating Medium for the Forces of Architecture
Maxwell Levinson, Editor; George Howe, Associate Editor;
Leon Levinson, Managing Editor

·*Covers* (Photos, Curtiss-Wright, Acme, Vanity Fair)
Knud Lönberg-Holm
·*Shelter's Editorial and Economic Policy*
·*Editorial, Correlation*
Buckminster Fuller
·*Building With Money*
Douglass Haskell
·*Structural Study Associates*
Henry Churchill
·*Monuments and Instruments*
Knud Lönberg-Holm
·*Emergence*
Theodore Larson
·*The Emergencies*
Simon Breines
·*Empire State Apartments*
Simon Breines
·*Transition*
Roger Sherman
·*Tooling Up*
Peter Stone
·*Universal Architecture, Essay 3*
Buckminster Fuller
·*A Festival Shelter*
Frederic Kiesler
·*Critical Judgment*
Eugene Schoen
·*Face The Facts*
Howard Robertson, Director, The Architectural School, London
·*Re-Search vs. Research*
Henry Wright
·*After Functionalism: Function*
Dr. Alvin Johnson
·*At A General Trend*
William A. Delano
·*This Freedom*
Electus D. Litchfield
·*Aluminaire: A House for Contemporary Life*
A. Lawrence Kocher, and Albert Frey
·*4D Essays*
Buckminster Fuller

 ·*Shelter*, No. 4, May 1932, cover; inaugural use of the designation "Structural Study Associates"

·166/167: *Shelter*, No. 4, p. 36, p. 39

SHELTER

STRUCTURAL STUDY ASSOCIATES SYMPOSIUM & 4 D ESSAYS

Doing the Most with the Least—Segregated compression and tension members, flexible joints, stabilized force Triangles — net scientific structure in time annihalating transportation, communication, and power harnessing. (1) The Detroit Bridge—Central Span 1,800 feet, two cables 19-inches diameter each, two masts total 12½ square feet net compression cross section—Able support total traffic city of 500,000. (2) Acrobats' Mast, County Fair. (3) High Tension Power lines. (4) Mast and shrouds of the S. S. "Bremen". (5) Marconi mast of "Whirlwind", cup defense contender. (6) Lakewood Steel Towers and concrete chuting equipment—Pardee Dam—Masts 500 feet high—One of largest ever erected. (7) London, England, to Capetown, Africa, passenger service. Note large compression struts flexible joints and tension diagonals of ephemeral weight.

1 *Aero Digest*

2 *Acme*

3 *Aero Digest* 4 *Acme*

Mobile Shelters—Designed for what they will do, not for what they look like. All products scientific-totality date of issue, no arbitrary withdrawal of efficiency. Giants of Strength in which fractions ounces considered. No Dogmatic Orders. Successive designs evolutionary. (1) New Ford Airliner 14A. Bearing no resemblance to predecessor. Pullman like accommodations. General saloon, 4 private compartments, smoking, baggage, dining and navigational compartments, two toilets, galley, pneumatic furniture, noise-proofed. 15 tons net, 3 tons (40) passengers, 2 tons fuel. Total 20 tons, 2,530 H.P. Note simplified window and streaming design in 3. (2) Aircraft Carrier "Saratoga", mobile shelter and Marine Field favorably comparable non-buoyant Beaux Art Concrete Architecturized Sea Dromes. (4) America's Cup victor "Enterprise" with $50,000 dural mast saving equivalent ton weight hoisted halfmast. Triangular boom. Total masterpiece aero and hydro dynamics all multimillion dollar laboratories for Shelter knowledge not pattern.

Shelter, November, 1932, Vol. 2 No. 5
A correlating Medium for the Forces of Architecture
Maxwell Levinson, Editor, Leon Levinson, Managing Editor

«It is a fact» – there is none.
«It came to pass» – in the continuity language of evolution – 4D.

Covers, Photographs by. F.S. Lincoln, Sculptires by Isamu Noguchi. Front: «Miss Expanding Universe», Back: «Glad Day». *Inside Covers,* Photographs by F. S. Lincoln, A Streamlined Dymaxion Shelter.

Pass-age 1
·*Putting the House in Order*
Pass-age 2
·*Behind the Scenes*
·*Our intimate Journal of Summer Events*
Pass-age 3
·*Fiscal Figs*
Pass-age 4
·*Ecovolution*
·*The Dawn of a New Era,*
R. E. Johnson
·*Journal of the Space Hotel*
·*When Winter comes*
Ralph Leff
·*Conning Tower*
·*Potentials to Initiative*
Pass-age 5
·*Streamlining*
Pass-age 6
·*Pan-Continental Shelter Service*
·*USSR – on the Problems of Architecture*
S. T. Woznicki
·*SSA to USSR*
Simon Brienes
·*An Architect in Russia*
O. K. Fulmer
·*An "international" Ghost*
M. M. R.
·*Wreath to Wraith*
Theodore Larson
·*Our Russian Correspondent*
Mathew Ehrlich
·*Scientific Shelter for U. S. S. R.*
·*Shelters of the Orient*
Isamu Noguchi
·*New Building in Japan*
Richard J. Neutra
·*Basic English*
C.K. Ogden
·*Olympic Village*
C.K. Ogden
·*California Ingenuity – National Steel Homes*
·*A Library on Wheels*
the N.Y. Public Library
Pass-age 7
·*Teleology*
Pass-age 8
·*Colloidals in Time*
·*Isamu Noguchi, F. S. Lincoln*
Pass-age 9
Foil, Epee, Broadsword and Sabre
Concealed Plumbing

 ·*Shelter* No. 5, November 1932, the last issue of the magazine, cover with *Miss Expanding Universe* by Isamu Noguchi

SHELTER

VOLUME 2 NUMBER 5

Expanding Universe

Based on observations at Mount Wilson Observatory, the American astronomer Edwin Hubble discovered that extra-galactic systems can be identified in cosmic nebulae, i.e., other galaxies exist beyond our Milky Way.

In 1929, it was revealed that Hubble could estimate the distance to eighteen additional galaxies based on the apparent brightness of their brightest stars. In the process, he noted shifts in their spectral lines at the red end of the spectrum, which he interpreted as a Doppler effect on the waves that are put out by stars moving away with increasing speed. This simplest expla-nation for the observed phenomenon was the basis for his thesis, still valid today, of an "expanding universe." After Hubble had presented diverse proofs that the speed and distance of galaxies were directly proportional, that the speed with which galaxies moved away divided by their distance was a constant, this "Hubble constant" became a central element of modern cosmology and a topic of public discussion from 1931 on.

Fuller had an extraordinary sense for the significance of such discoveries. His speculative space-time model of the "expanding sphere," which was already described in his manuscript "Lightful Houses" and is reminiscent of Emerson and Edgar Allan Poe's "Heureka," now seemed to be confirmed by science and was thus a topic of discussion in the avant-garde circles for which Fuller was already providing slogans.
On returning from a trip to China and Japan in October 1931, the sculptor Isamu Noguchi met his friend Buckminster Fuller in New York. The exchange he had with his friend was reflected in the title of his latest sculpture, a streamlined flying female figure with outspread arms and legs: Miss Expanding Universe. Noguchi made it for the dancer Ruth Page, who in turn was inspired to create a dance with the title Expanding Universe. A photograph of the sculpture, which has not survived, was used for the cover of the November issue of the magazine Shelter, which contained Fuller's programmatic essay "Streamlining," with preliminary studies for the Dymaxion Car, for which Noguchi had made plastic models. (Page 188)

Noguchi took up the theme again in 1943, this time for an abstract portrait of Buckminster Fuller. The sculpture titled Bucky stretches a body reduced to a torso into a cable structure of rod triangles that do not touch each other. It is an articulation of the principle Fuller would later call tension integrity or tensegrity. The discovery of this principle of the complementary effects of continuous tension and local pressure was something he had always understood, even in his contribution to cosmology. "The universe is a tension integrity," he wrote in 1949.

170 ·*Streamlined Up,* Piccard's Stratosphere Balloon as an example of the "teleological method of design" RBF had in mind, *Shelter* No. 5, p. 105

·172–175: *Shelter,* No. 5, p. 106–109

STREAMLINED "UP"

Dr. Piccard's Stratosphere Transport, an unintentional advantage secured incidental to slack of balloon at starting level allowing for great expansion of gas in relatively low pressures of high altitudes. This streamline observation a teleological correlation.

Similar correlation observation found in similarity this picture to a carrot, also streamlined "up". Carrot moves slowly up through earth, held down as balloon by guy lines (roots) during growth period. Roots giving way at season's end so that with frost motion of ground carrot moves in direction least resistance, up to earth surface to rot and fertilize seed.

Since the last issue of Shelter a multitude of scientific attainments have been recorded of which this incident is but one.

Here are a few more that should eventually be correlated with Shelter service.

X-rays discovered capable of reversing as well as speeding up evolutionary processes. Cosmic radiation has energy of 40,000 million volts, 40 times previous estimates.—Cosmic rays found stronger with increasing distance north and south of earth's equator.—Certain types of diamond found photosensitive and productive of electric current. Effect differs from "electric eye" electron discharge.—Visible light rays of longest wavelength will pass through nearly a foot of water with a loss of half their energy, whereas infrared radiations, with wavelengths only a little greater, have only one ten-thousandth as great a penetrating power. Most effective flesh-penetrating radiation found to be in very narrow band of wave lengths between dullest visible red and invisible infrared.—Marconi "bends" short waves inbroadcasting.—Heisenberg expounds new theory of neutron as "union" of electron and proton.—Einstein estimates age of expanding universe, including earth, as 10 billion years.

Times Wide World

TELEOLOGY

PASS-AGE **7**

"WHAT DO YOU SEE?"—TELEOLOGICAL-DEMONSTRATION

Tension, gravity, timing, streaming, continuity, evolution, rhythm of relative identity, etc.—Typical specific correlation. = The dynamics of the hexagon (force diagram of circle) found in (A) "Centralization" as demonstrated in unification by circular-agglomeration (cable strands); in (B) "Decentralization" as demonstrated in expansion—by crystalization (snowflakes); in (C) "Abstraction" as demonstrated in elemental-synthesis (hydrocarbon-coronene). (Coronene recently developed, unknown in nature. Atomic structure: circle of 6 hexagons ("crown"), each consisting of 6 carbon atoms with 12 hydrogen atoms on circumference.) Pictures above:—(1 and 2) Olympic games events—girls diving, and sailing races. (4) Snowflakes, invariably six-sectored, never identical. (5 and 6) Scientific P. H. diffusing lights.

1

2

3

4

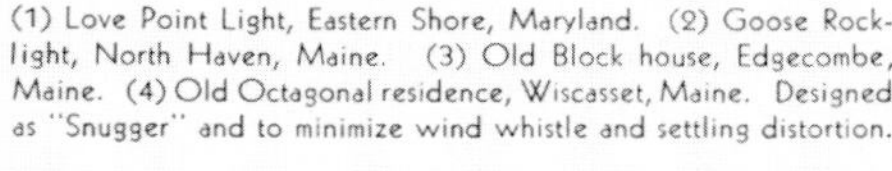

(1) Love Point Light, Eastern Shore, Maryland. (2) Goose Rock-light, North Haven, Maine. (3) Old Block house, Edgecombe, Maine. (4) Old Octagonal residence, Wiscasset, Maine. Designed as "Snugger" and to minimize wind whistle and settling distortion.

5

6

(5) Dymaxion House designed to withstand all forces which seek to penetrate or destroy human life process; and further to abet growth. (6) "Pig Incubator"—Economy Housing Corp., West Chicago, Ill., capacity 6 sows—sixty pigs—"saves your pigs in coldest weather" "farrows two litters per pig per year." Little pigs can huddle 'round heater in central ventilator—windows of cello-glass—allowing majority ultra-violet ray penetration—Price, $175.—Hogs have "money" value therefore have] had a scientific "break". "This little pig went to market,—."

1 Keystone-Underwood

2 Acme

3

4 International

HABIT BOUND INADEQUACY

(1) Hurricane—Texas.

(2) Earthquake—Argentine.

(3) Fire—Philadelphia.

(4) Flood—Georgia. Thousands dead and homeless—2,000,000 annual human victims these elements, 1929 estimate. Yet it is scientifically feasible design and fabric relatively "proof" Shelters against these forces as well as against marauders, gas attacks, tornado, pestilence, selfishness.

FORCE POTENTIAL BALANCE

(5) Note able design Horton Spheroid. Its shape found most ideal for combination of gas and liquid pressures—and for this reason it requires the least amount of stiffening and can be constructed of thinner material than other shapes with the same safety factor—"Most with the Least"— shape arrived at by filling large rubber ball with liquid and laying on ground allowing it to develop dynamic-bulge.

5 Chicago Bridge & Iron Works

WORLD EMBRACING FORCE UTILIZATION

(6) WABC's 600-foot radio—CBS Station, Wayne, N. J. (7) New self propelling telescopic dirigible mooring mast, U. S. N., Lakehurst, N. J. (8) A 220,000 volt rotated tower to stand broken line distortion. (9) Hetch Hetchy transmission line. (10) Snow type dead end suspension tower. (11) Snow type. (12) Eagle-bell trans-line Pacific Coast Steel—Mast erection by halves.

"DON'T FIGHT FORCES, USE THEM"—4D

6 *Western Electric*

7 *Science Service*

8 *American Bridge*

9 *Hetch Hetchy*

10

11 *Great Western Power*

12 *Pacific Coast Steel*

Conning Tower

A mechanical layout for the new "industrial conning towers", showing how, with currently available mechanics, information routing and correlating activity in general may be expedited with a hitherto incredible acceleration, accuracy, and lack of human effort. For instance, it is suggested that all industrial headquarters be equipped with a conference room, to be designed as the 4D "Hoop Skirt" for broadcasting stations as compounded with the latest mechanics of movie headquarters, that all conferences may be recorded by film, visually and orally, The duplication to be eliminated by such mechanics is extraordinary. Conferees who tend to recite acknowledged history, convergent in the specific conference involved, may rapidly be enlightened by a talkie, quite interestingly runnable in a few moments, as the net result of hours of conferences in which there is multitudinous lag for cigarette lighting, banalities, etc. There could be mechanical hookups of industrial until production headquarters by teletype, telephoto and television with central publishing headquarters of industrial units, who in turn would be tactically hooked up in like manner with information sources such as Bureaus of Standards, Navigation, Department of Commerce, etcetera or corporations such as Standard Statistics, Consumers' Research, Science News Service, etcetera, as well as university hook-ups.

The conning tower is free of private records, superfluous accounting, checkup methods etcetera, as the "bridge" of a giant liner or dreadnought, with simple "log" or "continuity notes" of O.D. Economic and elemental mechanical indicating and reporting systems of universal interest are incorporated, such as intramurally integrated ticker services, weather forecasting instruments, economic traffic indicators, elemental availability indicators, storage indicators, storage pile-up indicators, dominant news indicators, harmonic trend indicators.

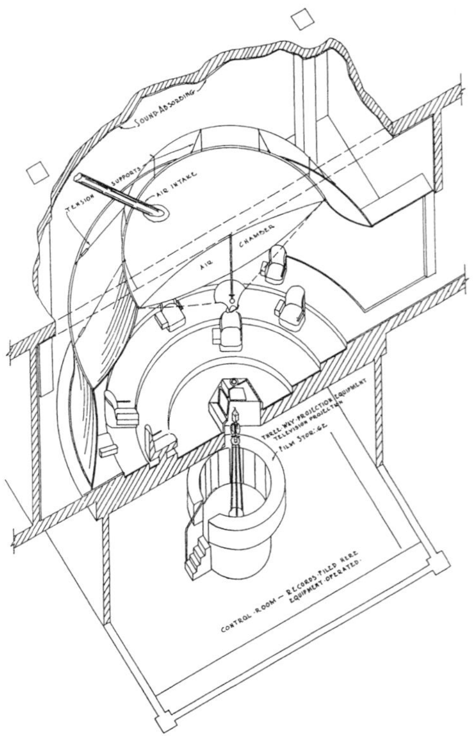

 ·Conning Tower - "Hoop-skirt" Room, design of a communications control-center for contemporary firms, text and design RBF, *Shelter* No. 5/1932, beginning on p. 64

·*Isometry:* air-conditioning by means of fabric cover – no reverberation – tent – diffusion of light from a central apparatus using convex tenting made of metallic silver fabric

Section and Plan of Conning Tower showing two deck relationship of tented conference room above and mechanical control room below. "Conference" divided into 3 team-sectors on principle "3's a crowd", and that "correlation of 3 forces essential to stabilization." – 4D.

"Hoop Skirt" designed to compensate cubical structure "interference" factors of sound and light control. Light diffusion enables operation of 3-way movie mechanism concealed in triangular shield of central recording device containing also microphone. In same device are housed speaker for "incoming" sounds and 3-way screens for trans-lux visible displays as required by conning group. Air intake through top envelope – exhaust through step risers.

Circular stair leading from mechanical control room to tent exterior enabling operators to circulate around conference without disturbance to conferees. Two-deck segregation obviates visiting conferees conduct alteration as result exhibition-consciousness if aware of recording devices. Mechanical operating room allows of television "in" or "out" operation and provides for storage pertinent film – for immediate conferee reference.

1. Conning Tower designed to preserve the integrity of intellectual progress by freeing the enthusiasm momentum of hitherto stultifying effects of constantly repeated disclosures of "purposes" and "results" which through medium of "conning tower" require but one disclosure for their potential social diffusion.

2. To obviate possibilities of misunderstanding not possible in conference records by written word only – sight, sound, continuity, and delicate gesticulation punctuation made available for reference.

3. To allow of new perspective "to see ourselves as others see us". Not only would this be potential to extraordinary individual development, but also to whole human welfare as conferees would incline to elimination of esoteric idiosyncrasies, and obvious selfishness.

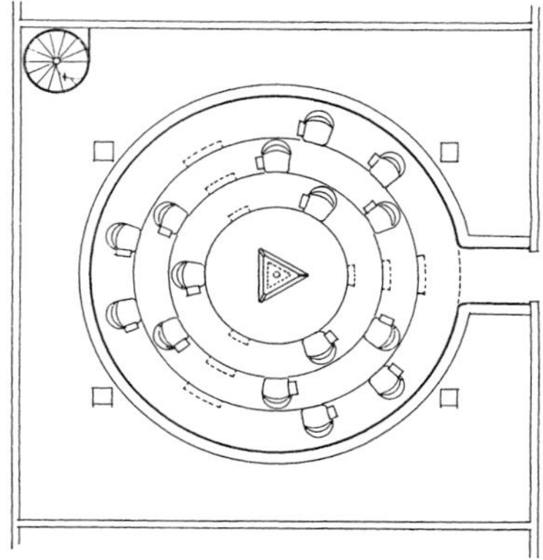

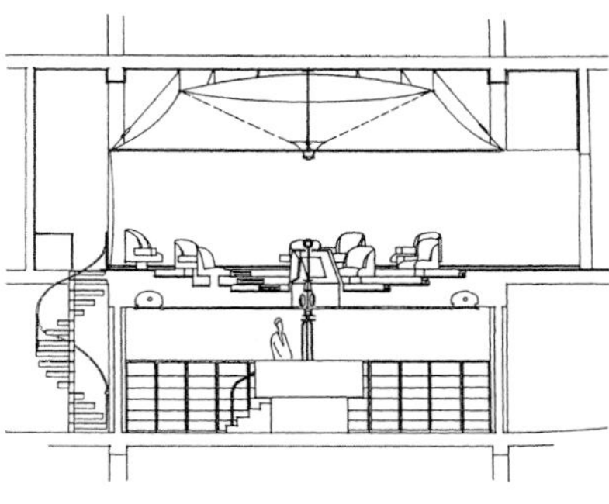

The 4D world lies at our feet. Gone are the leisurely times when an average individual lived in an area with a diameter of twenty miles. Our forms of transportation carry us wherever we are going. Flying is best, but if not flying, then something as much like it as possible. The three-wheel Dymaxion Car developed from the idea of a means of transport guided by side rudders that would roll in front and fly in back. Fuller enlists Starling Burgess, a recognized expert in the construction of high-performance yachts. The form is sensational. The car causes a stir; it is unusually fascinating; it is "amazing." The second coup for Dymaxion, even more resonant with a broad audience than the house was. Not a finished product, but a built sketch with a long life.

DYMAXION CAR

NO
DYMAXION TRANSPORT
STARLING BURGESS
BUCKMINSTER FULLER
F V 453
CONN. 1933

4D
TOP
HIGH SPEED
SHIMMY AND HEAVY
ATED

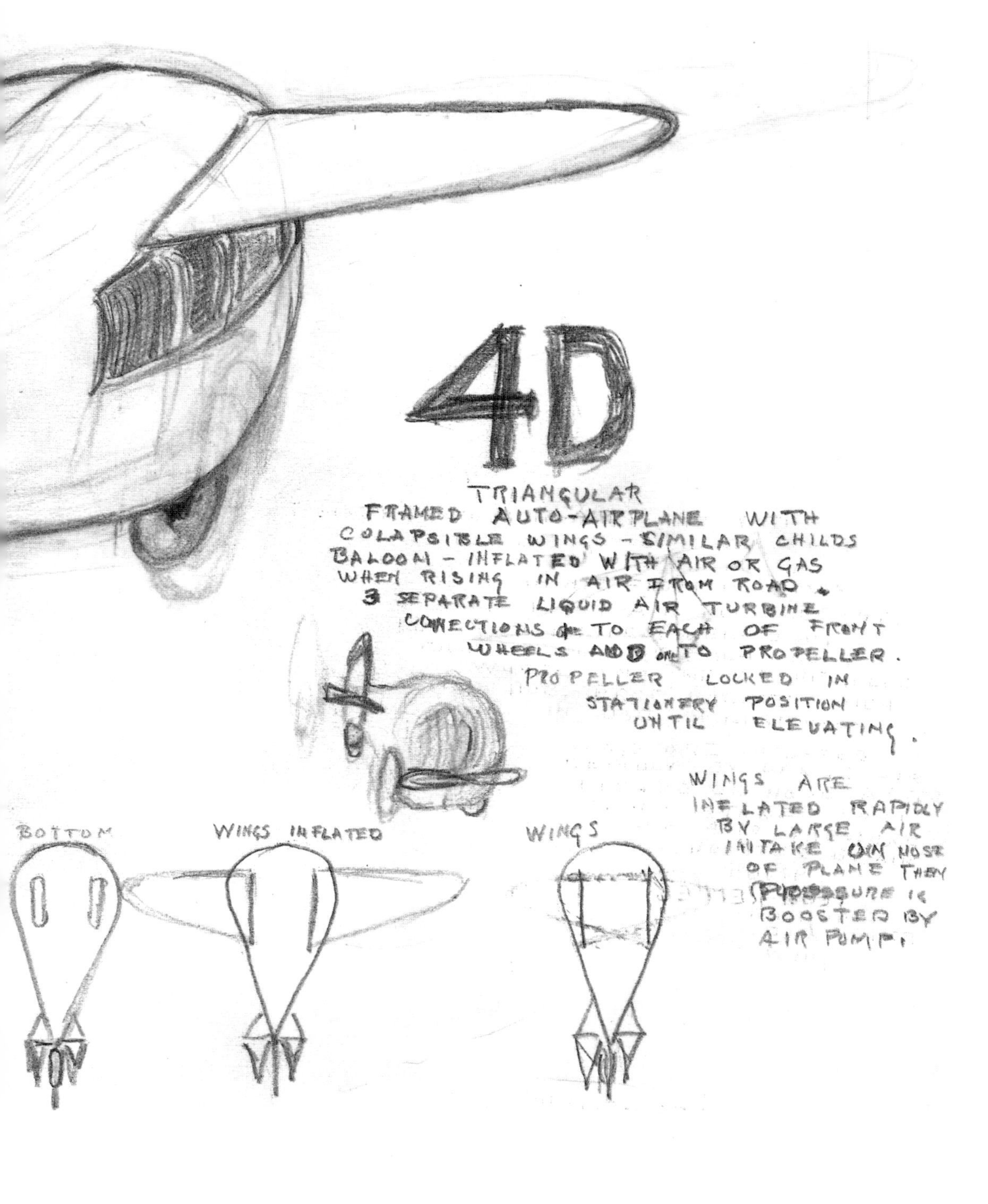
4D
TRIANGULAR
FRAMED AUTO-AIRPLANE WITH
COLAPSIBLE WINGS - SIMILAR CHILDS
BALOON - INFLATED WITH AIR OR GAS
WHEN RISING IN AIR FROM ROAD.
3 SEPARATE LIQUID AIR TURBINE
CONNECTIONS TO EACH OF FRONT
WHEELS AND TO PROPELLER.
PROPELLER LOCKED IN
STATIONERY POSITION
UNTIL ELEVATING.
WINGS ARE
INFLATED RAPIDLY
BY LARGE AIR
INTAKE ON NOSE
OF PLANE THEN
PRESSURE IS
BOOSTED BY
AIR PUMP.
BOTTOM
WINGS INFLATED
WINGS

1850 – 1950

1950 – …

Man on earth

Shown above are the three most distinct pictures of the history of man's years on earth. Picture one may be called, "The first half million years were the hardest." Ignorant and isolated, man was unaware of other men and of the potentials of friendship, integrated resources, and mutual survival. The first picture is of a microbespeckled, enormous sphere, an arithmetical isolation, a physical impasse, escapable only through intellect, instrumented – through science.
Picture two shows man linking up resource and survival by lines of transport and communication. Wealth is generated astronomically. Standards go up. Health and life expectancy tend to double. But in these 100 years the velocity of integration and increased energy flow leads to an arterial cloggage and explosive high pressure. The two-dimensional picture is a neat limited equation, fulfilled – and again occurs an impasse escapable only by intellect.
Picture three shows the intellectual answer – a new volumetric and dynamic dimension – wireless, trackless, omnidirectional. It is a high-frequency interaction of time synchronized, relaying from resource to logically dispersed processing centers for physical separation, reintegration, and unimpeded direct flow to next function. It bypasses all constrictions, yet in every way facilitates man's range and frequency of voluntary assembly and separation in a continuity of ever higher standards of environment and process control.
Picture three is a moving picture. Everywhere its physical facilities move with ever-increasing velocity and synchronized knowledge, allowing man to choose when and how and where he wishes to move. He specifically controls his own accelerations and decelerations. Picture three's scientific key is: to serve an ever-increasing number of functions of more people more of the time, with an ever-decreasing investment of energy, matter, and numbers of parts per unit of function, by ever greater intellectual re-investment of man's unique capital asset, "hours of time of his life."
Picture one is very long, as it was "against forces." Picture two is very short, as it was "transitional." Picture three will again be very long, as it is truly "natural." It synchronizes with the dynamic universe. The evolution of its "transport species" is multiplying. Here we go … but you need not hold your hat… It won't blow off.

Man on Earth (1950), in L. Sutnar: Transport (Preface RBF)

 ·179: *Dymaxion Car* No. 1, 1933
·180/181: *4D Transportation,* sketch, undated (1928)

·Ladislav Sutnar: *Transportation,* brochure, 1950.
Three illustrations for the preface by RBF
·*4D Houses/4D Transport,* sketch, undated (1928)

 · *4D Transport,* sketch, undated (1928)

· *4D Transport,* sketch, undated (1928); when driven

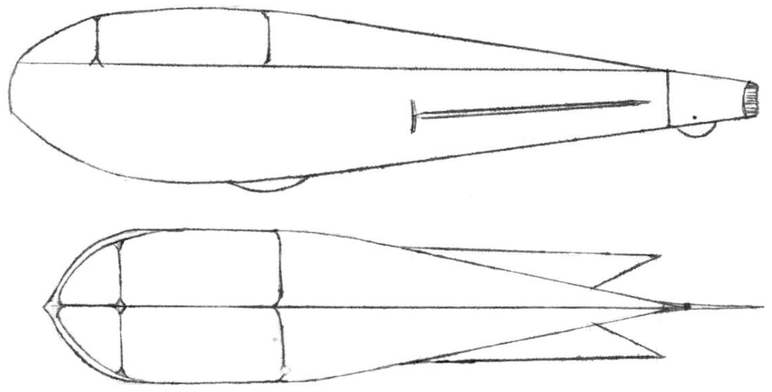

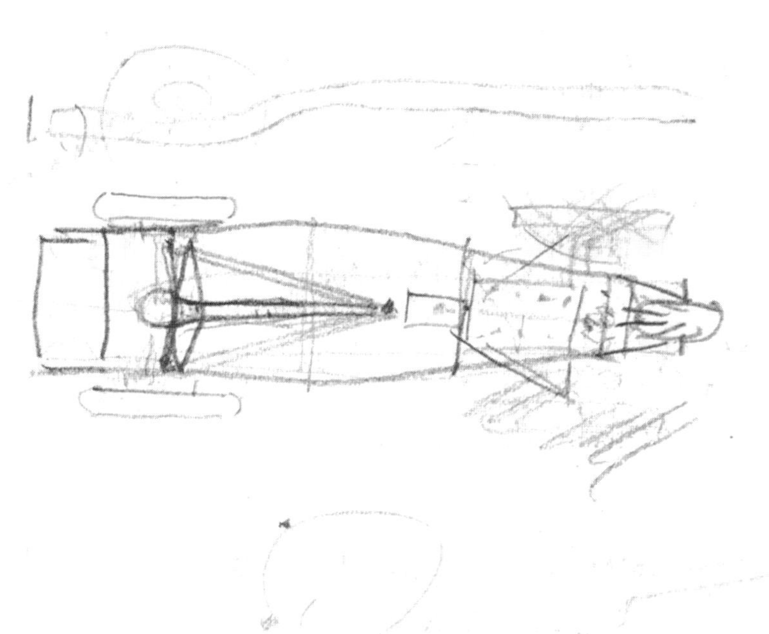

 ·4D Transport, sketch, undated (1928)

·Chassis of the *4D Transportation Unit,* sketch, *Chronofile* 44/1933
·186/187: *4D Transport,* gypsum models following ideas of RBF, by Isamu Noguchi, undated (around 1932)

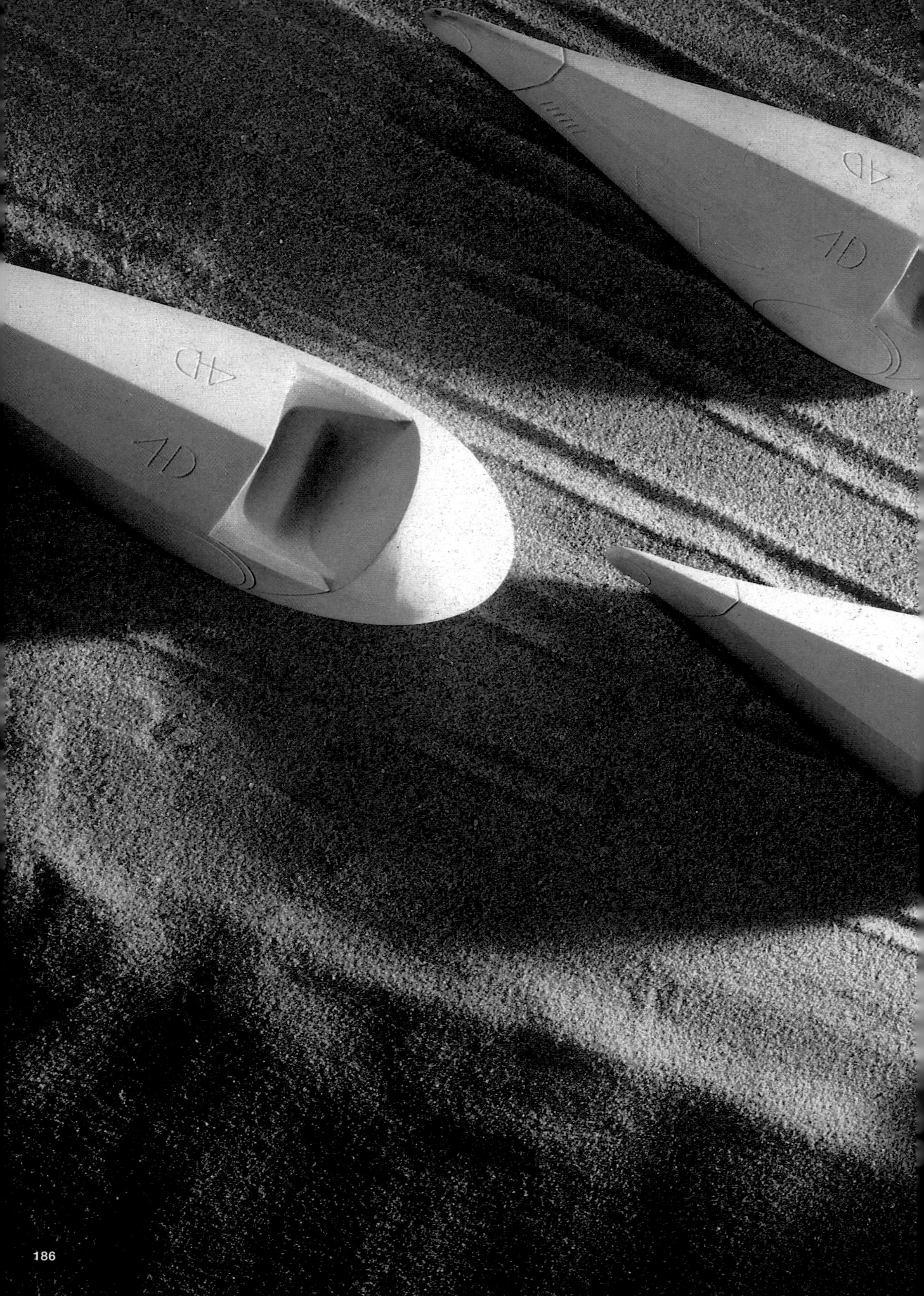
4D

4D

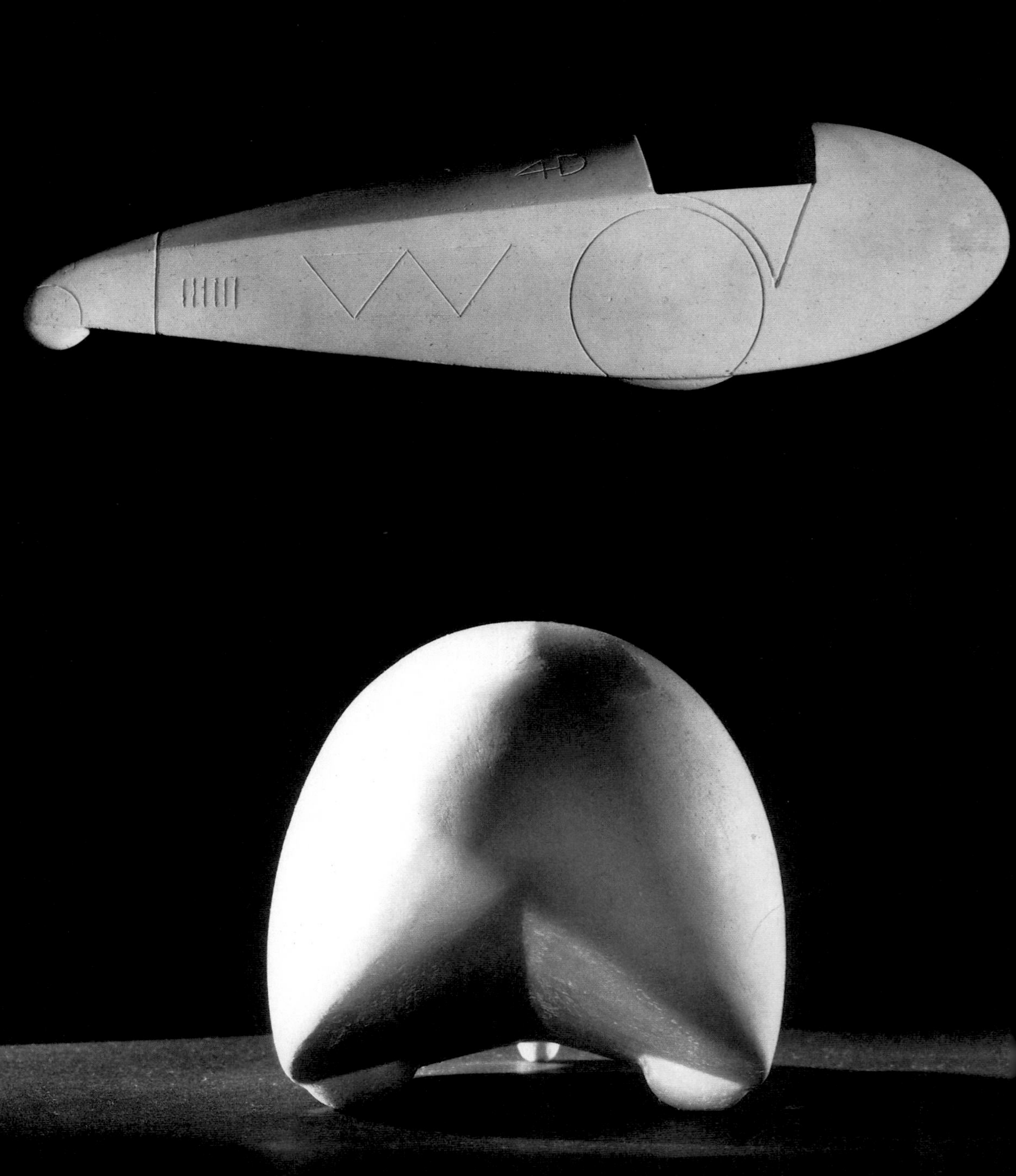

 ·*4D Transportation Unit,* gypsum model, side and front views; negative keel should generate air cushion and lift

·RBF: “Streamlining,” page from article in *Shelter* No. 5, 1932, 75

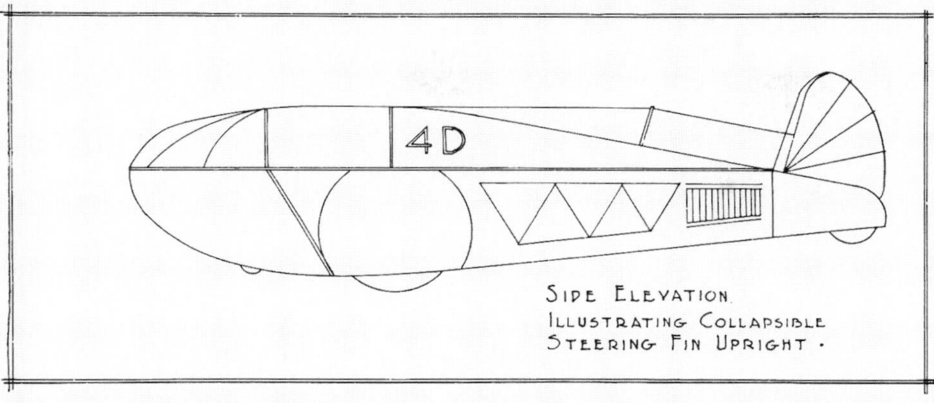

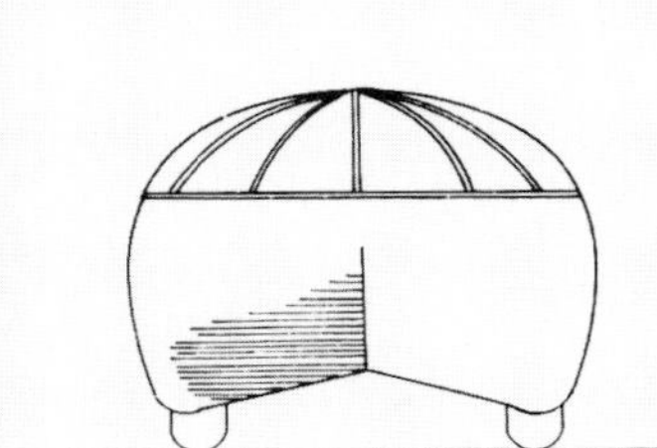

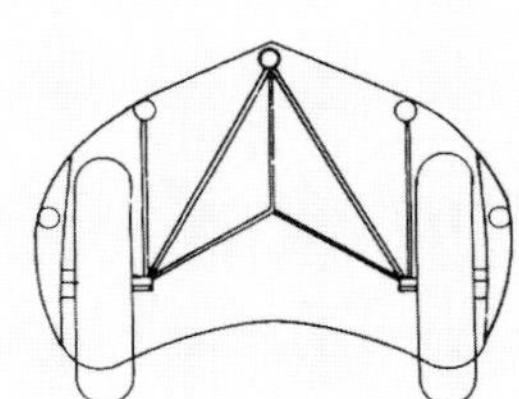

FRONT ELEVATION

TRANSVERSE SECTION

(c) Wm. Beebe, N. Y. Times.

Courtesy Hamburg-American Line.

4D TRANSPORT - DIAGRAM ABOVE SHOWING FAN-TAIL RAISED X SECTION AND FRONT VIEW

Note similarity to luminous fish (below) large fish streamlined contour even with mouth open by inclusion victim; also (lower right) streamlining of palm tree as well as of Graf-Zeppelin in Germany to Brazil 3 Day Service.

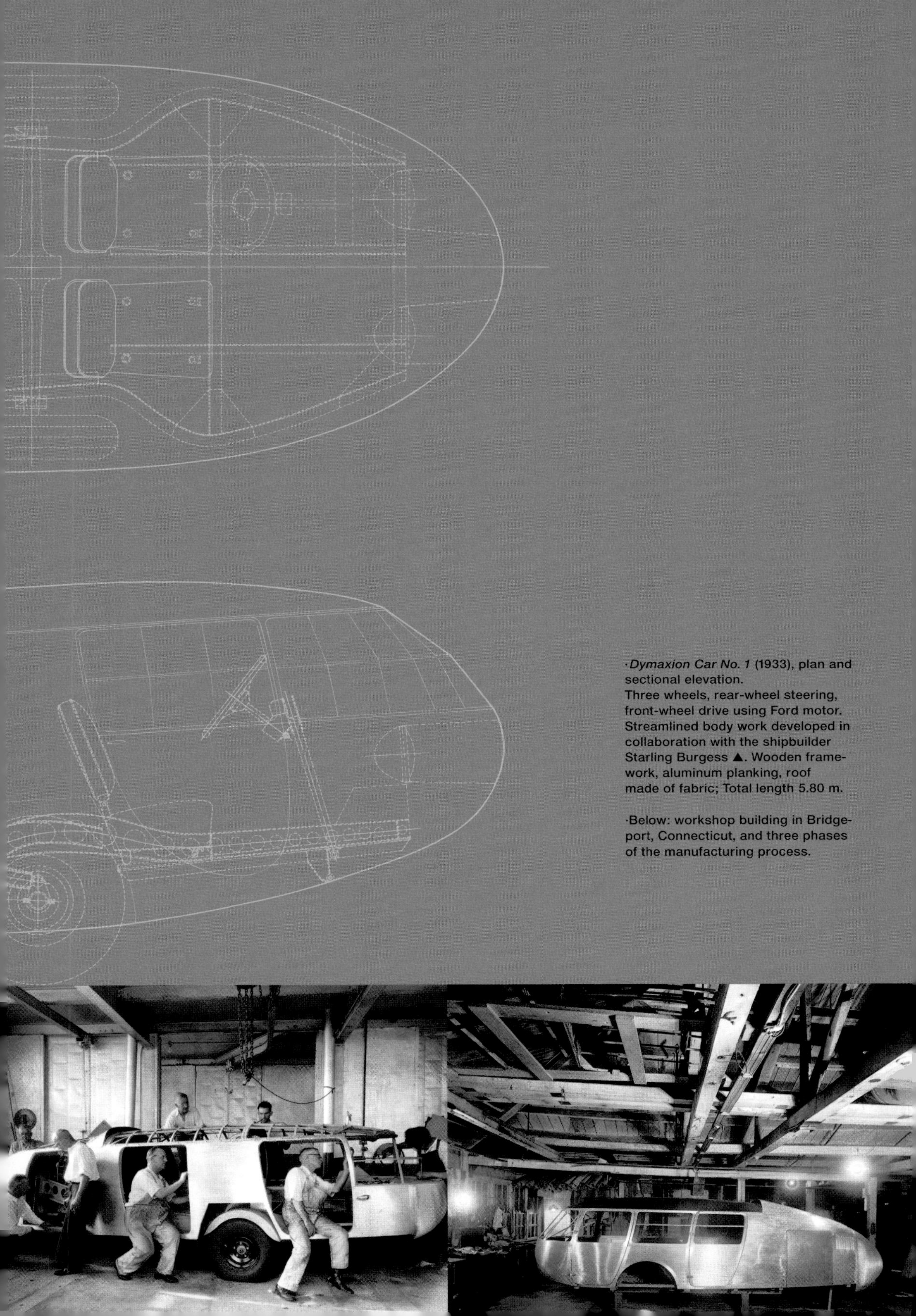

·*Dymaxion Car No. 1* (1933), plan and sectional elevation.
Three wheels, rear-wheel steering, front-wheel drive using Ford motor. Streamlined body work developed in collaboration with the shipbuilder Starling Burgess ▲. Wooden framework, aluminum planking, roof made of fabric; Total length 5.80 m.

·Below: workshop building in Bridgeport, Connecticut, and three phases of the manufacturing process.

If this car had been the only thing Fuller designed, then he would have been just one among many who wanted to give the automobile a fundamentally new architecture – and this included other exceptional engineers. As a technical construction, the Dymaxion Car was not a success, but it is still significant as evidence of his wide-ranging search for innovation. Fuller was seeking a scientifically determined streamlined body. In his essay "Streamlining" in "Shelter 5" (November 1932), he established a direct connection between the ships, planes, and streetcars of the future: "Planes and ships are steered from behind; if they had a rudder in front, it would break off." The main feature of the Dymaxion Car is that it has three wheels – the rear wheel is for steering, and the rigid front axle is driven by a motor in the rear. This scheme is problematic, first, because of the adverse distribution of weight, and second, because of the direct transfer of physical laws of fluid dynamics – of liquids and gases – to a vehicle that is limited to streets, a type of motion in which no drift can be tolerated and reliable traction is the first requirement of safety. Three of these cars were built, all of them different. None of them is a "prototype" in the usual industrial way of thinking that was so important to Fuller. Nevertheless, the external appearance of the car was electrifying; its aesthetic is still strong today (the contribution of the ship builder Starling Burgess), and the work as a whole had an enormous impact on the subsequent questioning of the norms of the "horseless carriage." As late as 1943, Fuller designed a small car for the industrialist Henry J. Kaiser using the same principles (it was not built). The dance of the car around the traffic policeman in an intersection on Fifth Avenue, recorded in 1934 for the weekly newsreel, was an early sign of the showman in Fuller.

 ·*Dymaxion Car No. 1,* front and side views

193 Two cross-sectional views (with aerodynamic profile)

194/195: Dymaxion Car No. 1 after its completion

NO PARKING IN
FRONT OF BUILDING
EMPLOYEES CARS →
← VISITORS CARS

DYMAXION

DYMAXION

The First Stream Line Car in the World

THE FIRST STREAM LINE CAR IN THE WORLD IS HERE ON ITS TRIAL TRIP FROM BRIDGEPORT, CONNECTICUT PRIOR TO ITS EXHIBITION AT THE CHICAGO WORLD FAIR.

ITS DESIGNERS WILL TAKE PASSENGERS ON A SHORT RIDE AT 50c FOR ADULTS AND 25c FOR CHILDREN FROM 2.30 TO 5.30 P. M.

Friday, July 28

Under the Auspices of the

Damariscotta Mills Library

BISCAY CHARITY FAIR

Bremen, Maine

· First test and demonstration drives, July 1933; RBF at the wheel

· Sign announcing the public demonstration

· 198/199: View from the interior; Schematic drawing detailing specific steering characteristics, 1933

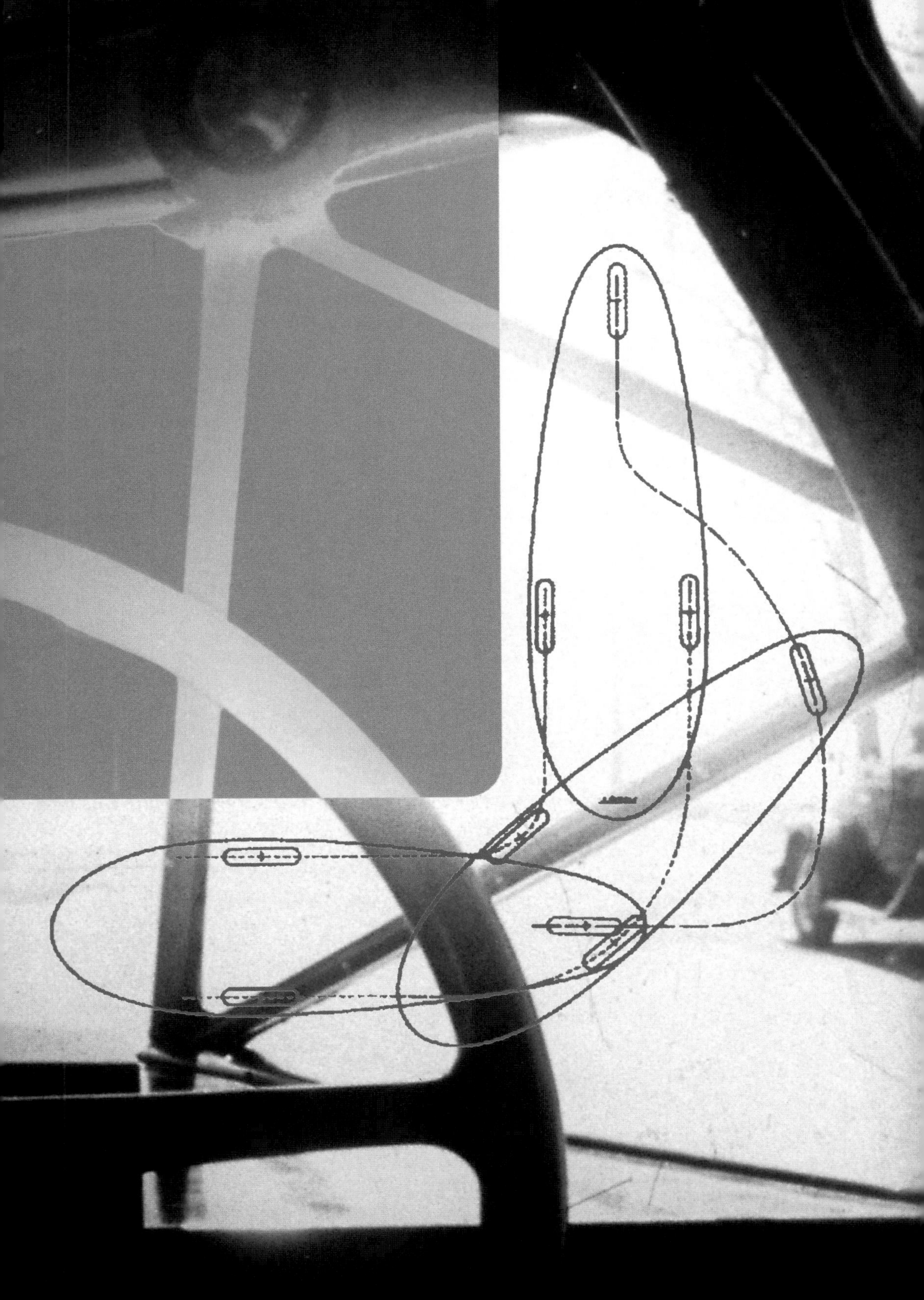

House = Porch Dwelling Machine

On the old farmstead there were a great many buildings to be seen – the great barn, containing hay and cows, the stables, corncribs, silos full of wet fermenting ensilage, the woodshed, pigsty, the carriage house, the cold cellar and the warm cellar. All these buildings and many others on the farms are disappearing or have disappeared because machinery in the house has displaced the functions carried on by the so-called "buildings." The small electric refrigerating device took the place of ice, the icehouse and icebox system. The electric current took the place of the wood, the woodshed and stove system, etc. In two decades the windmills, formerly on every farm, have gone.

In this way, we discover that the buildings, which controlled energy conditions of heat, cold, dry, and wet, were in effect machines because machines process and control energy. All those machines known erroneously as "buildings" have now been replaced by machines more readily recognized by us as machinery. Now, however, the recognizable components are decreasing as technology employs more and more of the invisible capabilities of electronics. What we are witnessing is the disappearance of the ever less economic "housing" or slow-motion phase of machinery as its functions are taken over by the high-speed machinery that brings about and maintains the preferred environment conditions at ever less cost and personal effort. This evolution is well underway, but we hide it from our awareness through semantic error, typical of which is society's noncomprehension of what Le Corbusier meant when he said, "A house is a machine for living."

(...) When tools, and more tools, came to shorten the time taken to do a given job, the farmer gained more time of his own. Finally, he had enough time before twilight to sit and look at the scenery, and he built porches around his house. As he began to have more and more time, he began to put screens on the porches. With ever more time, he began to put glass windows on the porches. Sitting on his porches, he watched other people go by. Then came the automobile, which in effect put wheels under his glassed-in front porch, so instead of waiting to see people go by he drove down the street to see the people. In a very real sense, the automobile was part of the house, broken off, like hydra cells going off on a life of their own. The young people who used to court in the parlor, then on the glassed-in front porch, now began to do their courting in the automobile, or the porch with wheels. Today, the young people do their courting in their parlor on wheels, driving it to the drive-in theater. Because we are conditioned to think of the house as static, we fail to realize that the automobile is as much a part of the house as is the addition of a woodshed.

Utopia or Oblirion (1969), 357

CRYSTAL HOUSE
CRYSTAL HOUSE
A SPACE STRUCTURE
SEE THE
DYMAXION CAR
ADMISSION 10¢
CHILDREN UNDER 12 FREE
DYMAXION
H349

The project for an industrially manufactured wet cell came closest of all the Dymaxion projects to implementing industrial production in series, but it foundered because the plumbers resisted it, fearing for their jobs. The wet cell was designed to work both as a part of the Dymaxion House and in renovating older buildings. Its elements (bathtub, toilet, sink, integrated lighting, ventilation, and plumbing) were designed with an eye to minimizing the number of parts used in construction. The design shows what industrialization meant for Fuller: high investment cost and low part cost thanks to mass production. A built-in version without the upper section was also completed and patented.

DYMAXION BATHROOM

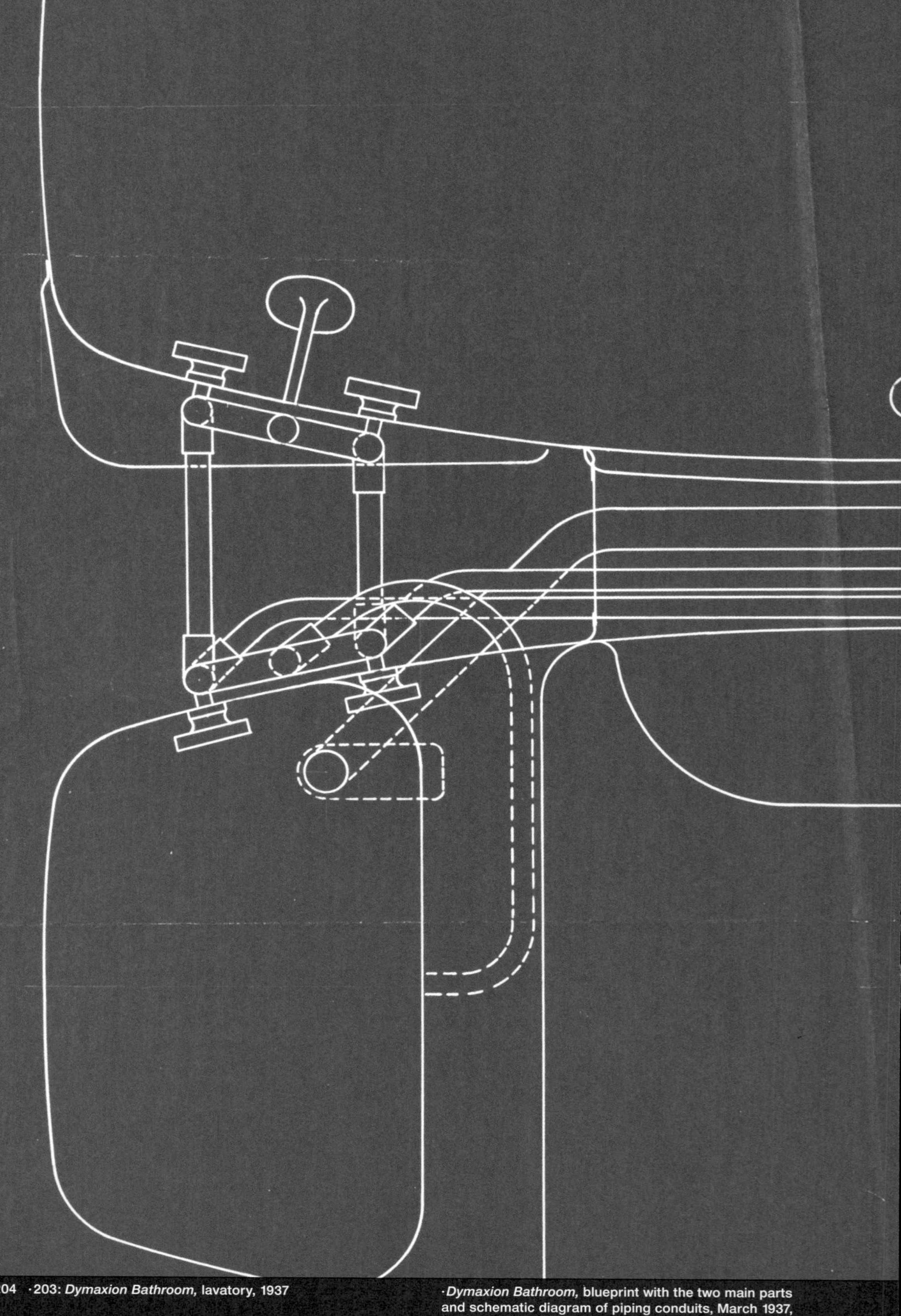

 ·203: *Dymaxion Bathroom,* lavatory, 1937

·*Dymaxion Bathroom,* blueprint with the two main parts and schematic diagram of piping conduits, March 1937, ground surface area 2.50×1.50 m

In 1936 I developed a much improved unit for Phelps Dodge Corporation which weighed less than a single bathtub of that time ... However, I have never felt that this project was really worth pursuing because there are other better ways of cleaning ourselves. Today we have learned from industry that when we clean a watch or a gyroscope we do not have to dunk it in a bath tub ... We found that it was possible to get along without a piped-in water supply by using atomized water under air pressure. In this manner we can do a very successful personal cleaning job with a pint of water brought to our residence.

RBF: Architecture out of the Laboratory (1955)

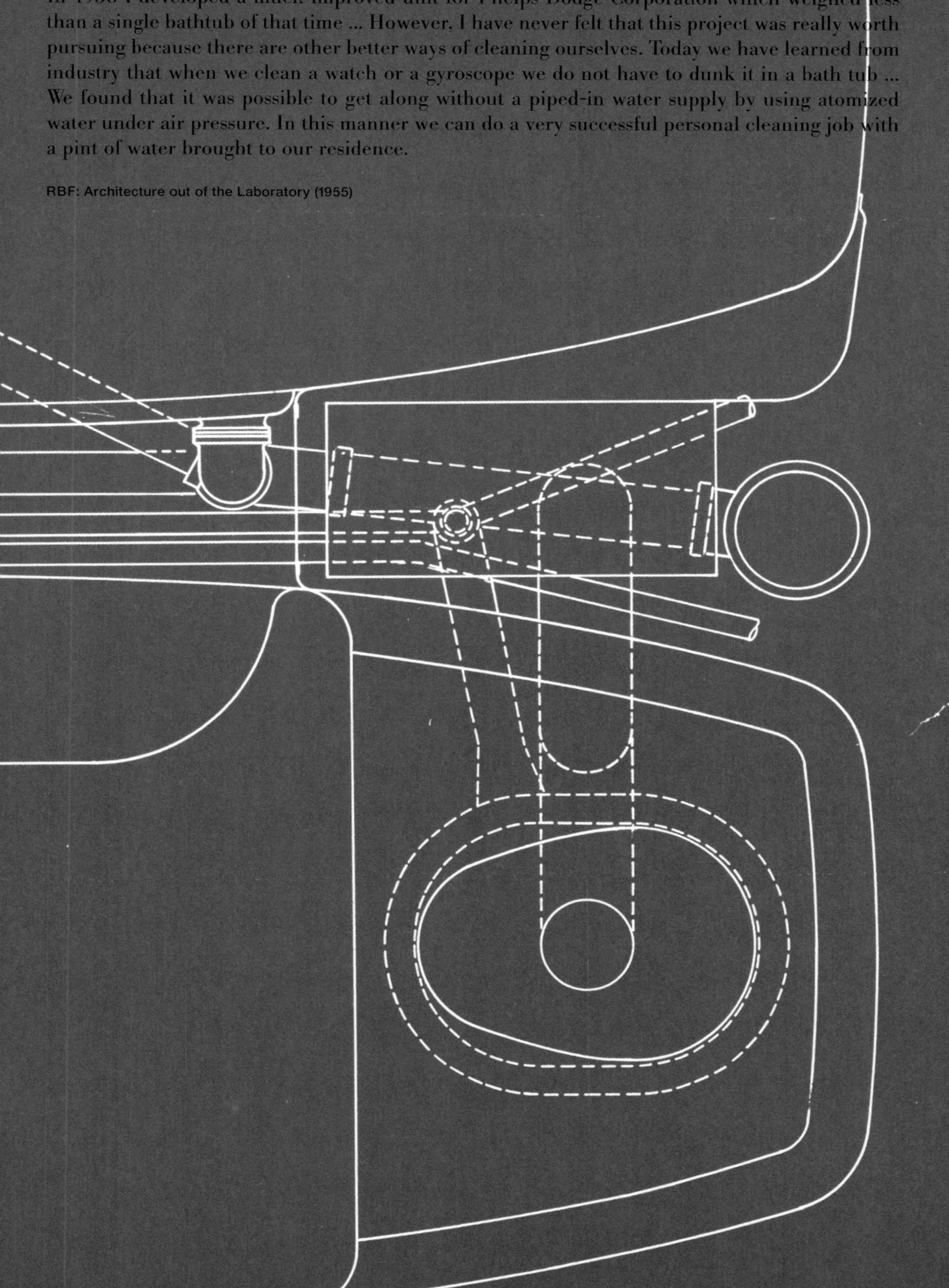

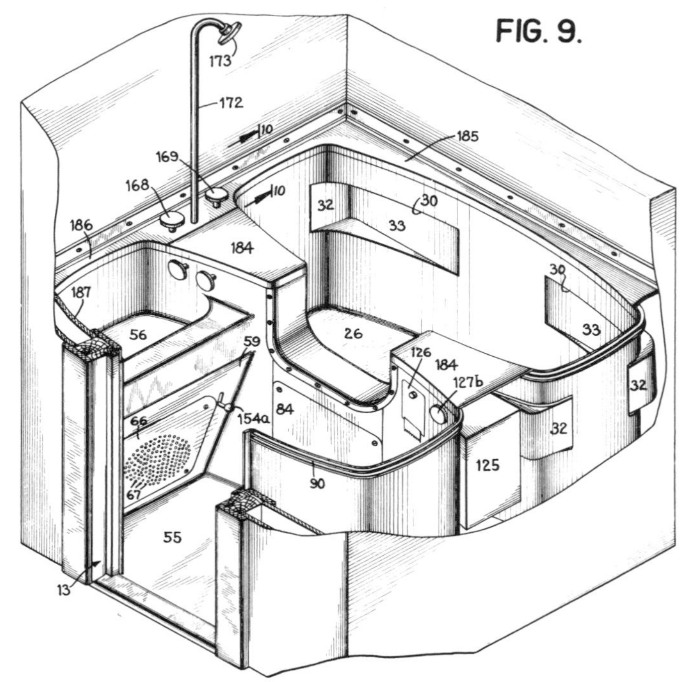

 ·(above) *Dymaxion Bathroom,* prefabricated version (without hood) to be inserted into existing structures, ·(below left) *Dymaxion Bathroom,* Detail: toilet seat, sketch from Chronofile 1936 ·(below right) *Dymaxion Bathroom,*

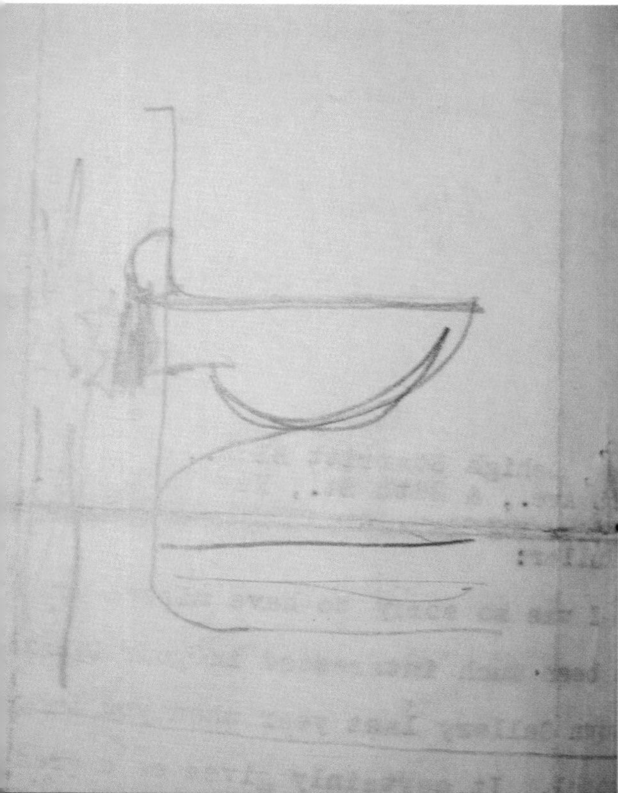

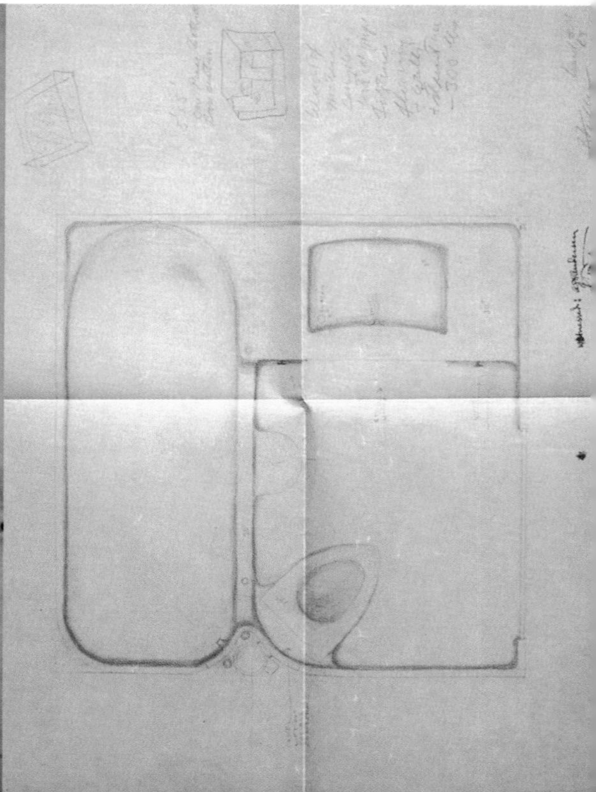

207 · *Dymaxion Bathroom,* 1937, photograph of interior. The two sections of the bath, each die-stamped in one piece, are joined along seam by U-shaped connector component

Filed Oct. 8, 1934 8 Sheets-Sheet 2

Fig. 6.

Inventor:

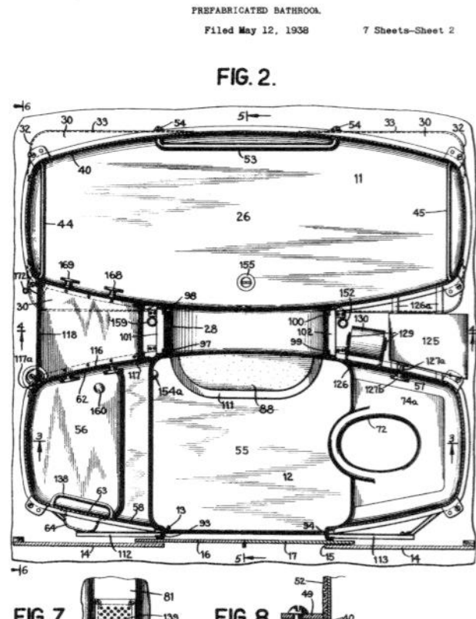

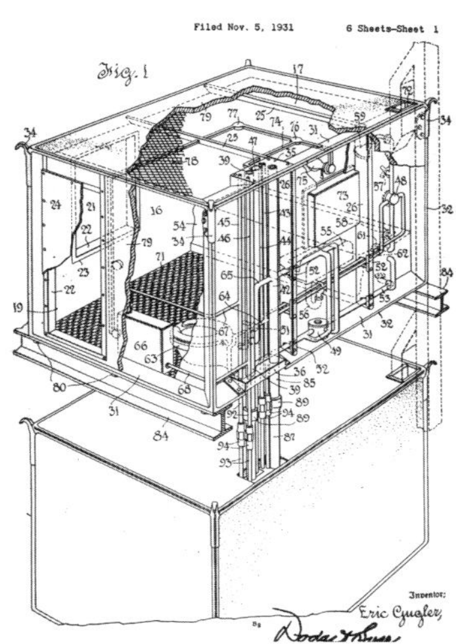

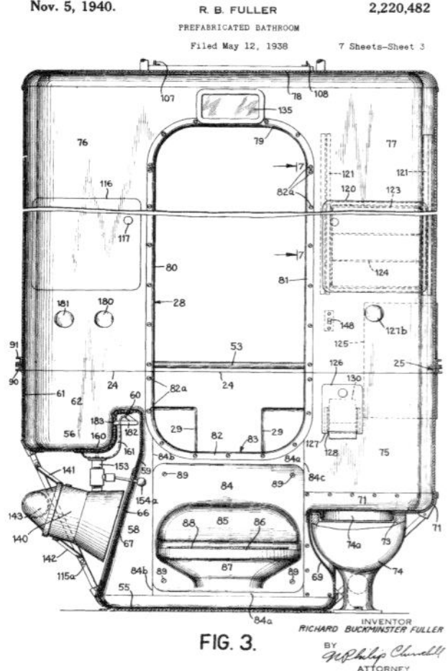

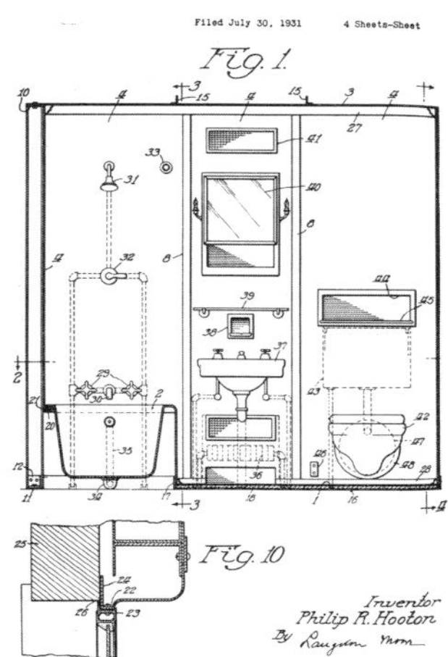

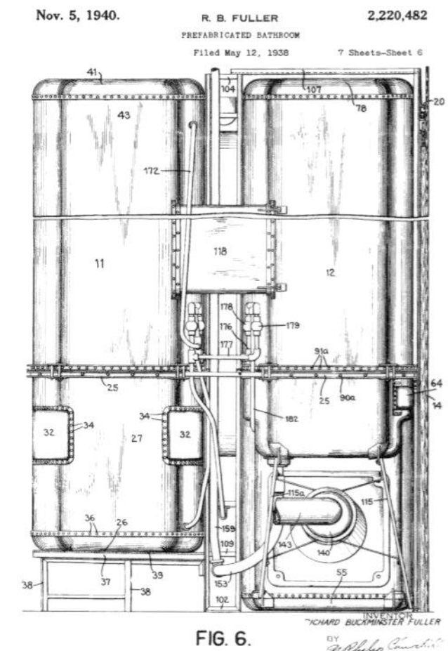

 ·Three contemporary competing patent designs, above Samelow (1934), middle Gugler (1931), below Hooten (1931)

·*Dymaxion Bathroom,* view of the prefabricated unit (with hood, as a built-in element for new buildings),

The Dymaxion wet cell is an unsurpassably clear expression of Fuller's understanding of design. It is not a matter of giving something form but the recognition of the possibilities for mass production inherent in construction. In contrast to other contemporaneous efforts to design a serially manufactured bathroom, the Dymaxion Bathroom was not conceived as a serially produced addition of individual parts but as the integration of partial aspects into an archetypal, seamless unit. The wet cell consisted of four main parts, two die-stamped tubs for the lower part, two laminated hoods as a cover, but the basic shaping process of die-stamping is more highly developed and more complex than simply trimming and folding sheet metal. It requires high investment in the production of tools and the necessary machinery; on the other hand, it provides the means for genuine industrial fabrication, so it has a low part cost. Moreover, the shell is self-stabilizing and light, and is thus itself an optimizing factor. These features were central to Fuller's conception of industrial design. The resulting look anticipates some of the features of the toilets used decades later in today's airplanes and trains.

It seems to me then that the factor of enormous improvement in performance inherent in transfer from the craft to the industrial system was a very important factor and that the fantastic increase in performance required to dislodge the old way of doing things, was as important in principle as the other principles already cited. You must remember that man didn't think he wanted other men to go five miles an hour, or possibly ten miles an hour, but no faster. It took a lot of courage to buy a car. [...] In short, the gain had to be enormous to offset the inertia of man against himself. Therefore, in principle, if we are to get houses into industrial mass production, they will have to be so much better than anything man ever thought of having before as to be able to offset that inherent mental inertia.

"Better than" brings us abruptly to the matter of performance of house. We find that people haven't classified to any important degree the standards of performance of house, and for a very good reason. As engineers you certainly understand that man is born "inside the frame of scientific measurement reference," therefore, it is impossible unless he gets "outside" of the whole phenomenon house for him to be very critical of his performance standards. [...]

Man starts life crawling around on the floors and associating the smell of the floor and the color of the walls with the most important events in his life which are warmth of understanding and love. I found it very hard to discuss house scientifically with people. Architects were very tactful with people and knew how to discuss these delicate things, but compromised science at every turn to please the ignorance of a well-dollared patron. You could, however, discuss big buildings and hotels scientifically, which man had evolved from houses and which had perforce to discard the idiosyncracies of the one for service to the common needs of the many. Because the hotel "guest" found public buildings inpersonal affairs, he also found himself inadvertently "outside the frame of reference" which quickly aroused his critical faculties and advanced his desired standards. He might for instance, have a privy at home, but he insisted on his private bath with modern plumbing at the hotel, or he would not think of having the sponge and lace curtains in his bathroom at the hotel that he found at home. When he went traveling he manifested to his own surprise a high set of standards, but at home, no, you could not get him to see that critically. So we had to take advantage of his inadvertent out-of-house behavior in learning about converting the house into the advantages of industrial technology.

Designing a New Industry (1946), 16

1.
THUD

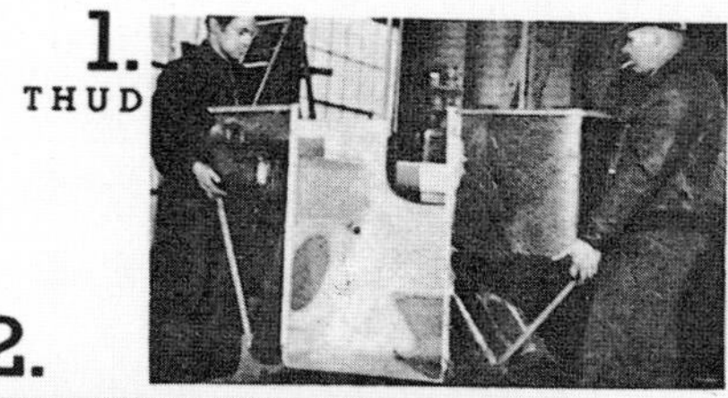

2.

THE NEWEST
prefabricated bathroom
IS ALSO NEAREST

CLICK

On the 29th floor of 40 Wall St. there sits the finished model of a new plumbing fixture that might well bug the eyes of any bystanding master plumber, a fixture that to all intents and purposes constitutes a one-piece bathroom. Designed by Architect Buckmister Fuller (Dymaxion House, Dymaxion Car) it accomplishes, by the simple connection of four basic parts, a complete bathroom weighing 404 pounds, with integral lavatory, toilet and bath. First known as the "Five by Five" (because that's the space it takes up), the official designation is now "The Integrated Bath".

In the research department of the Phelps Dodge Corporation it sits, ready for moderate production (100 units) in 1937. Architect Fuller has assigned his patents to the PD organization, and rumor discerns a new manufacturing and marketing subsidiary in the immediate offing. The range of uses for the unit is broad: pullmans, planes, trailers, trains, but mainly small homes. In fact, Mr. Fuller hopes this light, compact, complete bathroom will even inspire renters to install copies in their apartments, and remove them when they move. All of which lies in the realm of speculation. For the immediate future the device will probably induce fewer orders than conversations. None the less the fact remains that this prefabricated bathroom comes closer to commercial reality than any of its predecessors.

The Integrated Bathroom consists roughly of two oblong sections that form a partition where they join, which conceals the piping and other mechanical appurtenances. The sections (each a monometal stamping) are each split in the middle, the top being aluminum and the bottom 272 pounds of sheet copper unmetallized and tinted by a coating of silver, tin and antimony alloy. The bottom of one section is the lavatory and toilet, of the other a flat-bottomed tub.

4.
PRESTO!

SNAP
3.

The toilet, though reminiscent of the old backyard one-holer, is fully sanitary. The seat lifts and remains upright by compression against the walls. Underneath is a standard form of bowl (though chrome nickel bowls are also available).

Two men can handle an installation in three hours, for all piping except a minimum amount of connection material is integral with the unit. So are electric connections, ventilation equipment, etc. Fresh air is drawn by a motor under the lavatory from the nearest room, and exhausted—wherever circumstances permit.

Miscellaneous features: A composition Venetian blind gives privacy to the bather, and, while permitting the escape of steam, prevents the escape of water. The door-frame between the two sections is six inches thick, permitting use as seat. Complete cleansing of tub is easily attained. The plumbing layout was devised in collaboration with a local master plumber, copper tubing being used for water lines. Particular care was used to avoid back-siphonage possibilities. Sliding doors conserve space. The metallic finish has a "hammered" appearance while at the same time being thoroughly sanitary, the inventor claims. Under surfaces of the base metal are covered with Dum-dum, a sound deadening material. An electric heating system between the two units warms the metal itself, radiating heat to occupant of bathroom. Removable panels permit access to plumbing traps and connections under toilet and lavatory.

Reprinted from the April, 1937, issue of THE LADLE
official publication of the New York State Association of Master Plumbers

In anticipation of the bombing of British cities in 1940, Fuller was asked by the British War Relief Organisation to design an emergency shelter. He created the design for the Butler Company of Kansas City, which manufactured grain silos of galvanized steel. The refinement in the move from the epitome of the rural economy to housing for people is characteristic. The conic roof is converted into a highly sophisticated shell; the bent tin of the cylinder cover becomes curved galvanized steel. It is designed to be easily set up and taken down. The British government did not, however, make metal available for its construction, as it was needed for the production of armaments.

DYMAXION DEPLOYMENT UNIT

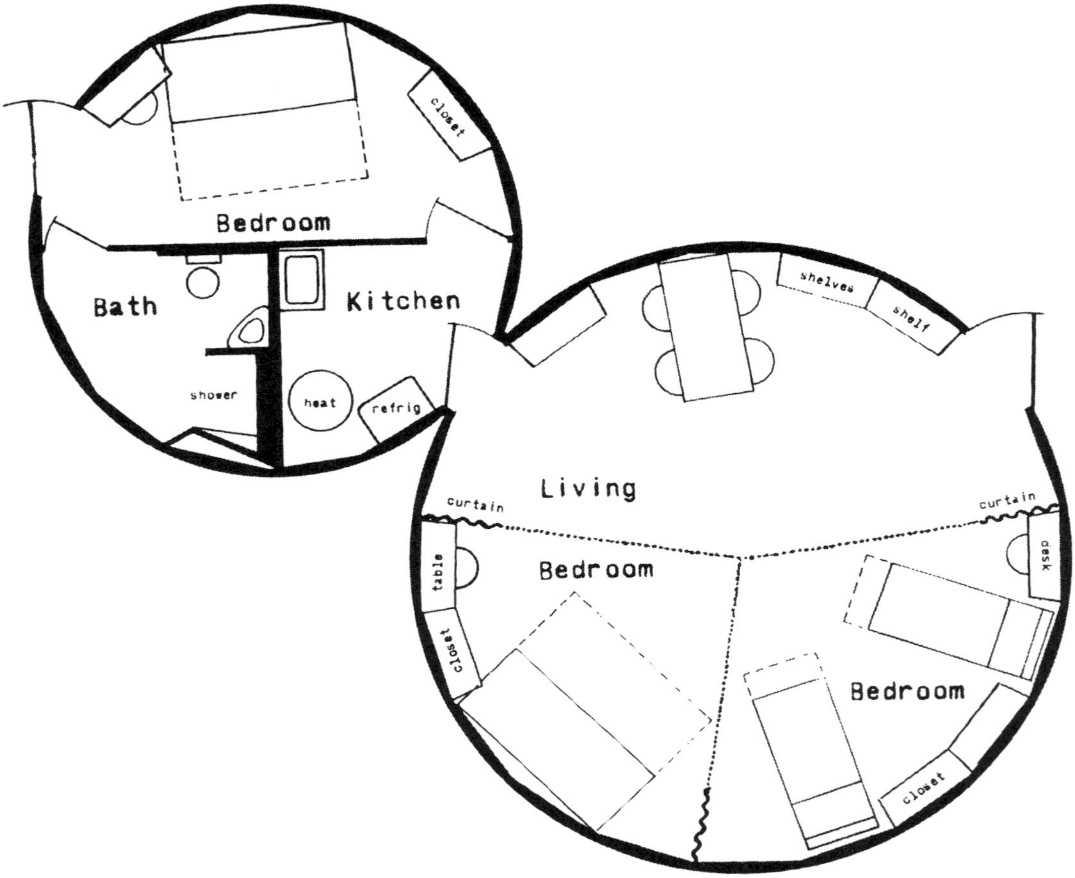

Bedroom
closet
Bath
Kitchen
shower
heat
refrig
shelves
shelf
Living
curtain
curtain
Bedroom
table
closet
desk
Bedroom
closet

 ·213: *Dymaxion Deployment Unit (DDU),* 1940, interior view, looking from the connector opening into the secondary unit.
·Double unit, ground plan: room partitioning by means of curtains

·Butler Manufacturing Corp. grain bin, 1930s
·Assembly of DDU's

Fuller originally developed the Dymaxion Deployment Unit beginning in 1940 as a small house for families. In the course of developing it, the possibility of using it as emergency housing for members of the army became more important; with the entry of the USA into World War II, the development took a new direction. A number of units were produced and employed – in the Persian Gulf, for example – but they were not produced on a large scale. In its conception, Fuller started with an existing industry that he hoped to upgrade technologically. The Butler Company produced grain silos of curved corrugated metal. The round plans of these containers was the right starting point for both static and economic reasons: bent sheet metal is self-supporting, and the circle has the most advantageous relationship of circumference to interior space. The sanitary unit is enclosed in a separate cylindrical element. Two units can be attached directly to each other. The pylon is important, though originally Fuller conceived of it merely as a simple way to set it up: it holds up the dome that is built beneath it. From a constructive perspective, and even more from a structural perspective, Fuller was breaking a path that years later would lead away from the idea of a central supporting pylon to the supporting shell of the geodesic domes.

215 ·The *DDU* was envisioned not just for military purposes but also for civilian use, from an advertising prospectus

·216/217: *DDU*, overall view, curved sheets of galvanized corrugated iron provide high stability; central rooftop air-vent

LOUISIANA
STREET
STREET

“Reshape environment;

RBF: New Forms vs. Reforms (1963)

on't try to reshape man”

From 1938 to 1940, Fuller worked as a technical advisor for the editors of the renowned magazine "Fortune". The articles he produced had elaborate layouts and contained thoughtful treatments of themes from the world of technology such as the electronics industry, plywood, the Sperry gyroscope, television, Chrysler Fluid Drive, and so on. For the magazine's tenth anniversary issue, he put together elaborate statistics to show in essence that the old economic data for coal and oil requirements should be replaced by production figures for electricity. Fuller's analytical interest in all experiments also applied to models for the development of industrialization. His attentiveness to historical developments was guided by an eye to the future; he used the retrospective as the driving force for the prospective.

FORTUNE

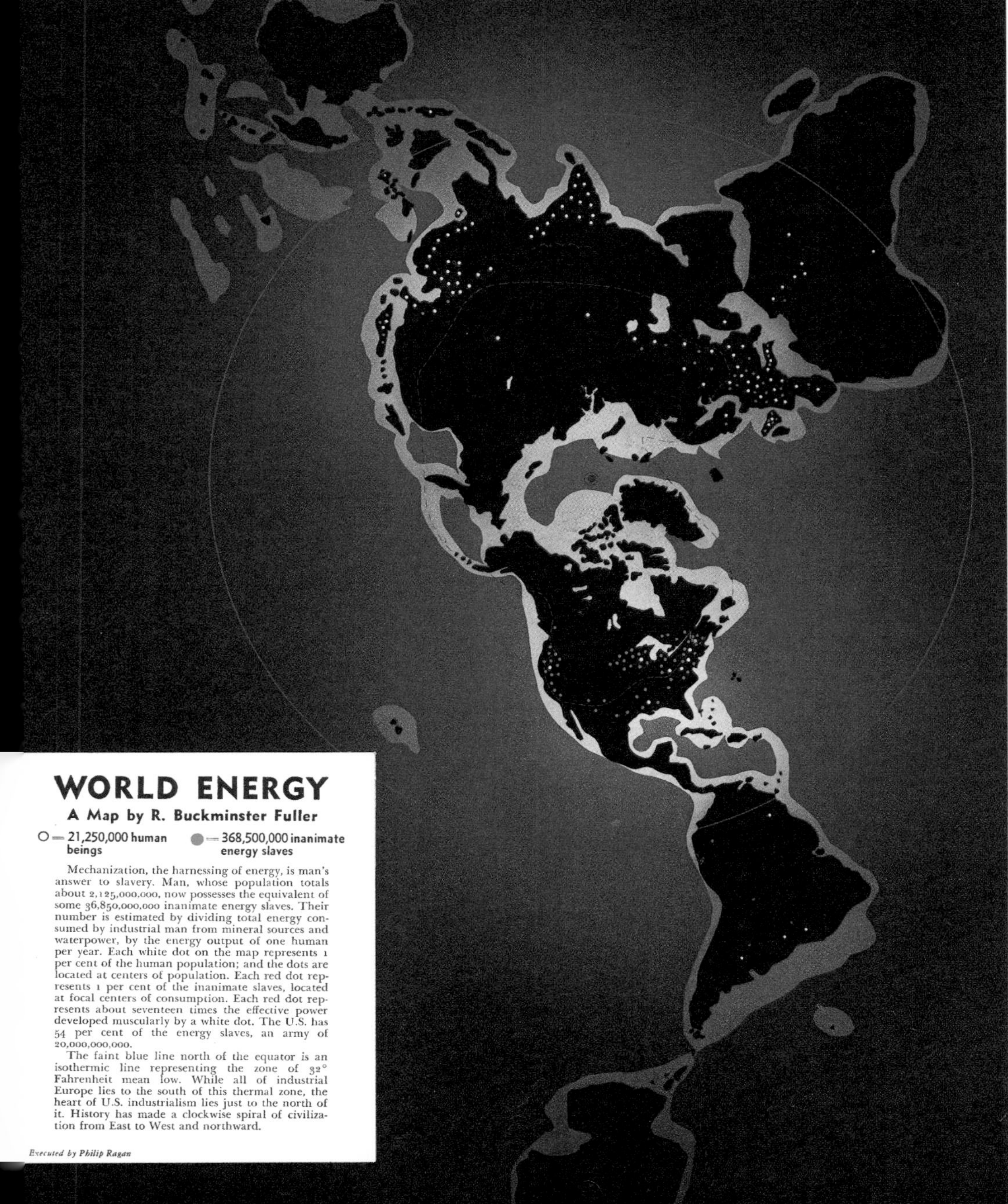

WORLD ENERGY

A Map by R. Buckminster Fuller

○ = 21,250,000 human beings ● = 368,500,000 inanimate energy slaves

Mechanization, the harnessing of energy, is man's answer to slavery. Man, whose population totals about 2,125,000,000, now possesses the equivalent of some 36,850,000,000 inanimate energy slaves. Their number is estimated by dividing total energy consumed by industrial man from mineral sources and waterpower, by the energy output of one human per year. Each white dot on the map represents 1 per cent of the human population; and the dots are located at centers of population. Each red dot represents 1 per cent of the inanimate slaves, located at focal centers of consumption. Each red dot represents about seventeen times the effective power developed muscularly by a white dot. The U.S. has 54 per cent of the energy slaves, an army of 20,000,000,000.

The faint blue line north of the equator is an isothermic line representing the zone of 32° Fahrenheit mean low. While all of industrial Europe lies to the south of this thermal zone, the heart of U.S. industrialism lies just to the north of it. History has made a clockwise spiral of civilization from East to West and northward.

Executed by Philip Ragan

1

Metals

If you are not familiar with the world of metals, you might not be familiar with the fact that our metals do not go down a one-way street. When I was young, I was brought up with the idea that we could and probably would use up all the world's iron. I was brought up to think that after we finished using something made of iron that the iron went into the trash heap to rust away to dust in the city dumps and could never be used again. The fact is that it came as a surprise, even to world industry, in the nineteen-thirties of this twentieth century, that all of the metals are continually being melted up and recirculated. For instance, out of all the copper mined in all the history of man, only 14% is not at present in an averagely recirculating twenty-two year cycle of use; and the 14% which is not in present recycling use is now in munitions ships lying on the bottom of the ocean. We know where that copper is, and in due course, it will be brought back into use, and within decades, as much as 98% of all the copper that has been mined by all man will be in continual recirculation. The recirculation of the entire copper inventory of man, as well as approximately all the other metals, occurs at an average rate of twenty-two years.

The chief engineer of the American Telephone Company was able to state to me in 1935 that while the telephone company had so much copper in wires and dynamos, etc., the telephone design science was increasing the messages per cross sections of wire so rapidly (going for instance from one message to twenty-eight and then to a thousand per given cross section of a copper wire circuit and finally from wire to wireless), that the telephone company is continually increasing its copper scrap inventory out of old and obsolete equipment. He said, therefore, that the telephone company would not have to buy any more copper or take any more copper out of the mines in order to expand the telephone service from serving only the population of the United States, to serving all of humanity, at an equal per capita frequency of high standard communication. The American Telephone Company's chief engineer went on to say that the company's copper resources, either in use or in scrap, were in such constant surplus through improved performance per pound that the company would be continually selling excess copper while expanding its service from a domestic to a world service without, in effect, having to buy any more copper. It would, he said, however, be commercially more feasible to buy new wire and sell scrap due to the non-matching of the geographical operations of scrapping, briquetting, smelting and fabrication operations.

Sixty-five per cent of all our steel is now made out of scrap. That is very roughly the ratio of recirculating metal to new mine production metal in all of the metals categories. It is perfectly practical to think about taking the metals out of obsolete automobiles, taking all the two-ton automobiles off the road, melting them up and making twice as many higher performance one-ton automobiles from the same metal. You may say that you don't want more automobiles, that the parking problems are too great. In speaking of automobiles I have chosen an industrial tool that you are familiar with. I am not advocating more autos. I am simply considering the feasibility of the principles involved through which we can, by design science, take care of twice as many people in a given function with a given obsolete scrap resource. I have learned by experience that it is possible and feasible for the world of architectural students to undertake an amplification of the functional effectiveness of the world's resources through design science.

I realized a year ago, from my own experience, and from the frustrated attempt on the part of enthusiastic architectural students in various schools around the world to get going with the world redesign, that it is not going to be a practical matter for the world's architectural students to take a world inventory of resources as well as an inventory of all the trend patternings and needs of men in order to learn how to reorganize the designed use of the total resources to highest advantage. Such economic intelligence harvesting is not within the present training or even the extra-curricula experience of architectural students.

 ·221: World Energy Map, *Fortune Magazine* (New York), anniversary issue February 1940. Compares areas of

"energy slaves" (replacement of human labor through machines; red dots). Indicator of the level of industrialization in a part of

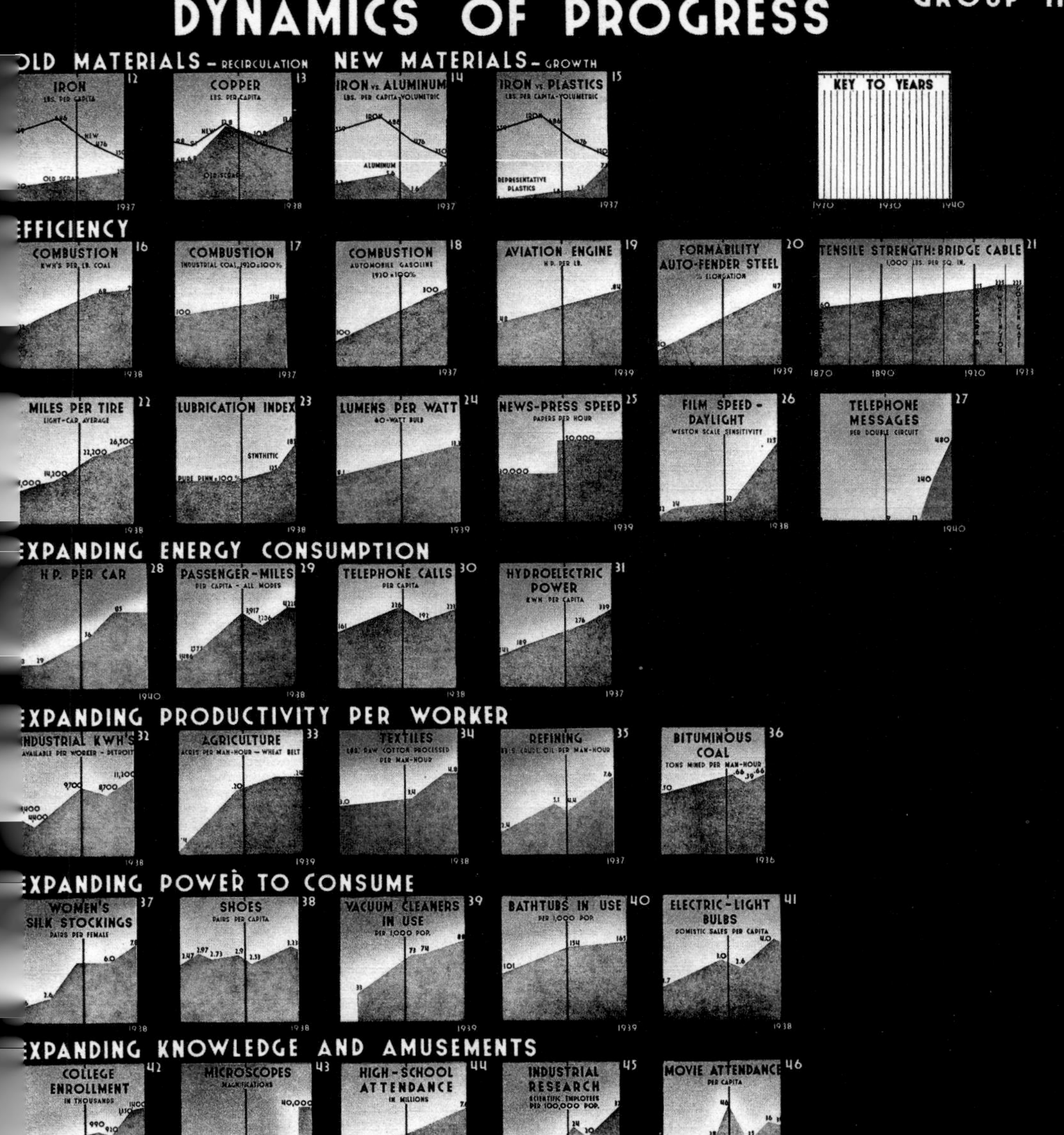

· "Dynamic of Progress," *Fortune*, February 1940. Supplied by RBF, information graphics referring to numerous parameters of US industrialization. From above: Old vs. New Materials; Efficiency; Expanding Energy Consumption; Expanding Productivity per Worker; Expanding Power to Consume; Expanding Knowledge and Amusements.

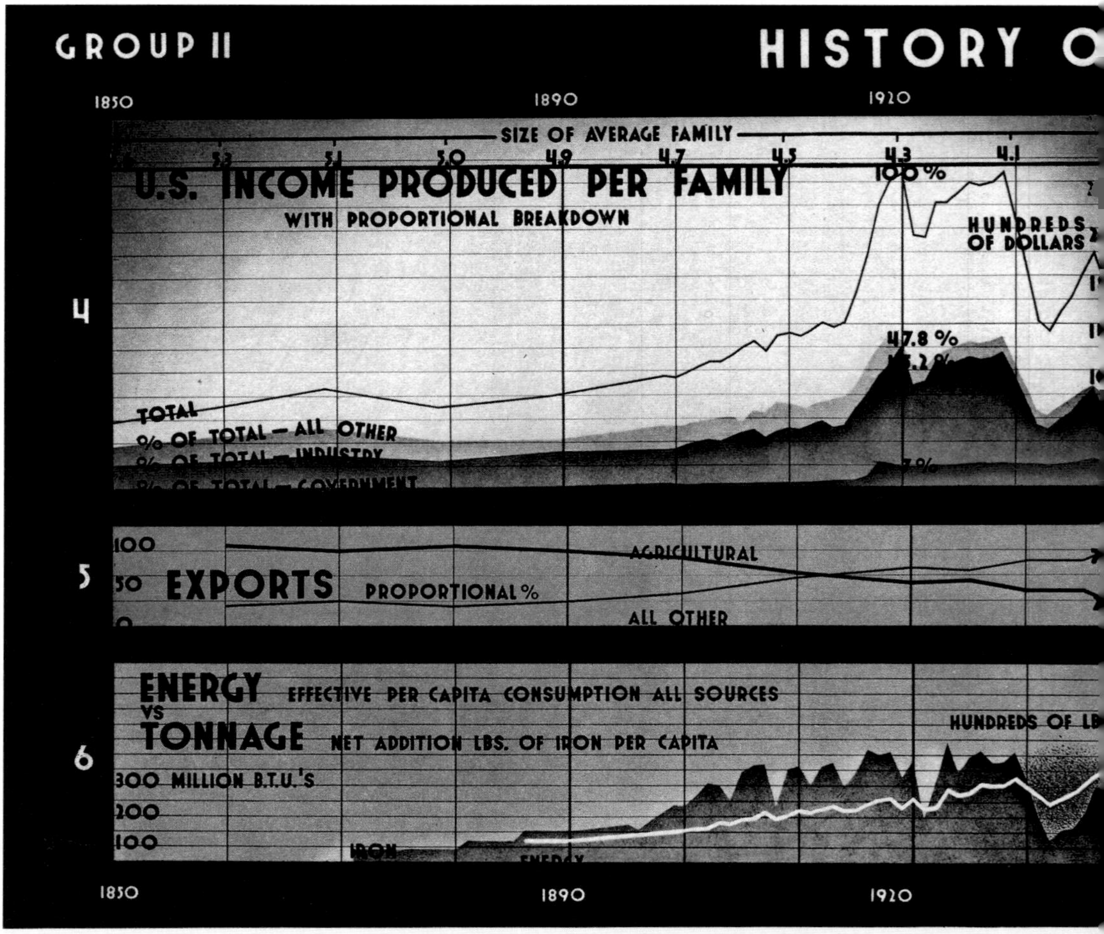

THE EVOLUTION OF AN INDUSTRIAL ECONOMY

The above charts provide a capsule description of the U.S. economy. Chart 4 (top) shows income produced *per family*, 1850-1939. Chart 5 shows the relative gain of "all other" (chiefly industrial) exports over agricultural exports. Chart 6 provides the key to the present article. Abandoning dollar figures, it shows the net addition of new iron to the economic system (black area), in hundreds of pounds per capita; versus the effective consumption of energy (white line) from mineral fuels and waterpower in B.T.U.'s per capita. The decline of the iron curve, coupled with the rise of the energy curve, is symptomatic of the basic change from a tonnage, or producers', economy, to an energy economy dependent for its growth upon the consumer.

solution for the present "stagnation" would be an increase in capital investment. There is no question that this would help the economy; money would be spent, and this would increase the actual power to consume. But an increase in the rate of investment, or any other such measure, would not necessarily increase the potential power to consume. It would not necessarily advance the progress of industrialization in this most basic sense. The progress of industrialization will depend rather upon what kind of investment is made. The potential is not at all increased when a railroad buys new cars that are obsolete in design; it is very little increased when the federal government builds a new post office where the old one had served: but it is very much increased when the chemical industry puts a fraction of the above investment into the development of a new solvent; it is very much increased (or *was*, during the growth of the automobile) when the government builds concrete highways. Some capital investments increase *both* the potential and the actual power to consume. Others increase only the actual. And projects can be imagined that might have relatively little effect upon either.

Thus, in measuring our industrial system and in proposing solutions to its problems, it is necessary to use industrial measurements as well as dollar measurements; it is necessary to measure with such elements of progress as units per capita, tons, thermal units, kilowatt-hours, watts, and engineering indexes of viscosity, flux, torque, and efficiency. To talk about income and expenditure, price, capital promotion, or investment, without taking the engineering factors into account, is both futile and confusing. For while the power to consume—in both senses—is certainly affected by the economic factors, it does not spring from them.

The measurement of industrial progress is the purpose of the accompanying portfolio of charts. Beginning with the simplest static charts on page 51, and ending with the extraordinary illustration of U.S. industrial power on page 57, the portfolio

52

224 · "History of Progress," double-page spread from *Fortune*, February 1940, with charts showing development of: (oil and coal); Cotton vs. Rayon; Transportation (railroad, automobile, airplane); Communication.

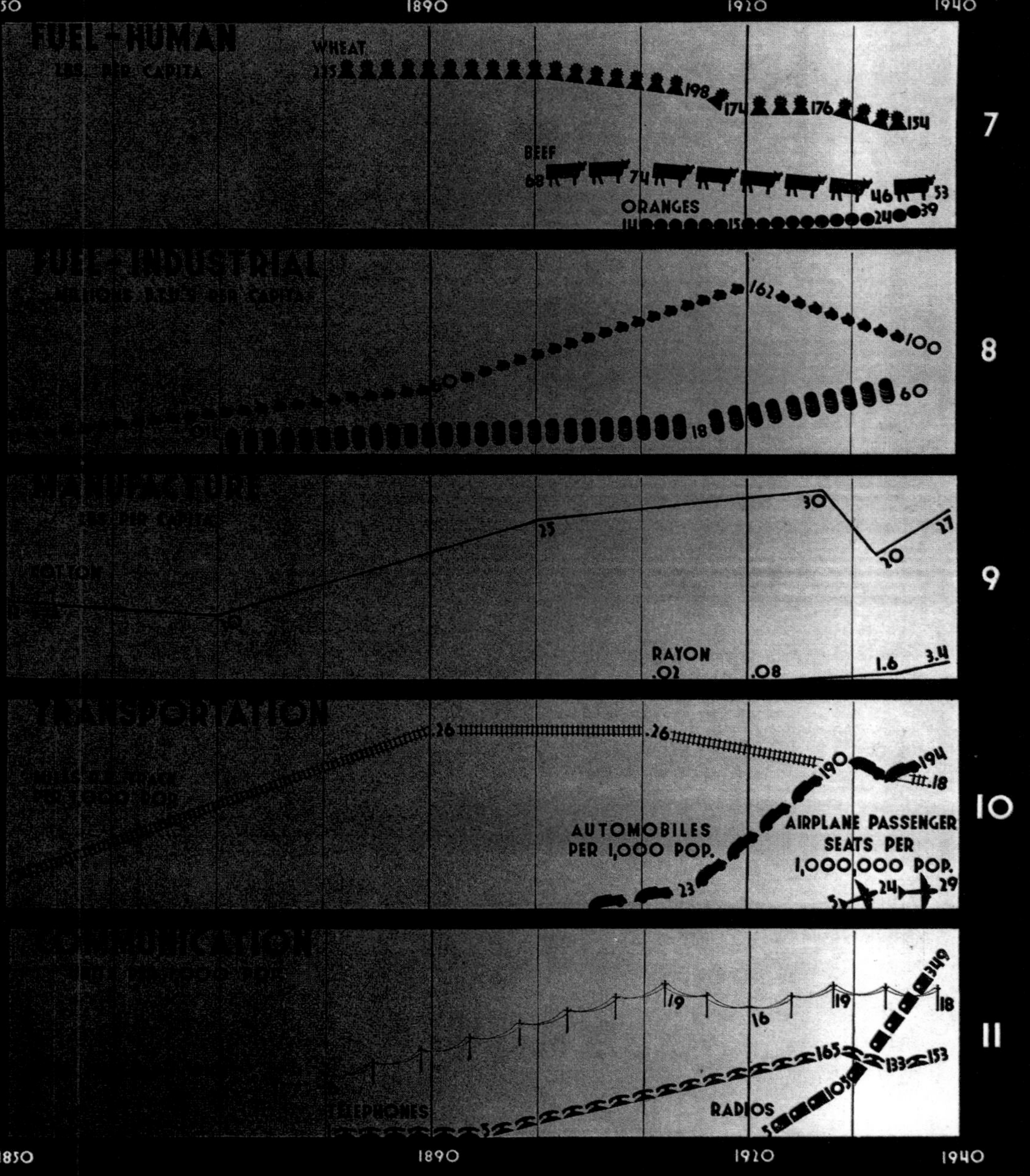
PROGRESS
GROUP II
1850
1890
1920
1940
FUEL–HUMAN
LBS. PER CAPITA
WHEAT
198
174
176
154
BEEF
68
74
46
53
ORANGES
15
24
39
7
FUEL–INDUSTRIAL
162
100
60
18
8
MANUFACTURE
COTTON
25
30
20
27
RAYON
.02
.08
1.6
3.4
9
TRANSPORTATION
.26
.26
190
194
.18
AUTOMOBILES PER 1,000 POP.
23
AIRPLANE PASSENGER SEATS PER 1,000,000 POP.
5
24
29
10
COMMUNICATION
19
16
19
18
349
165
133
153
105
TELEPHONES
RADIOS
5
11
1850
1890
1920
1940

THE U.S. vs. THE WORLD

OF THE FOLLOWING, THESE ARE THE RESPECTIVE SHARES

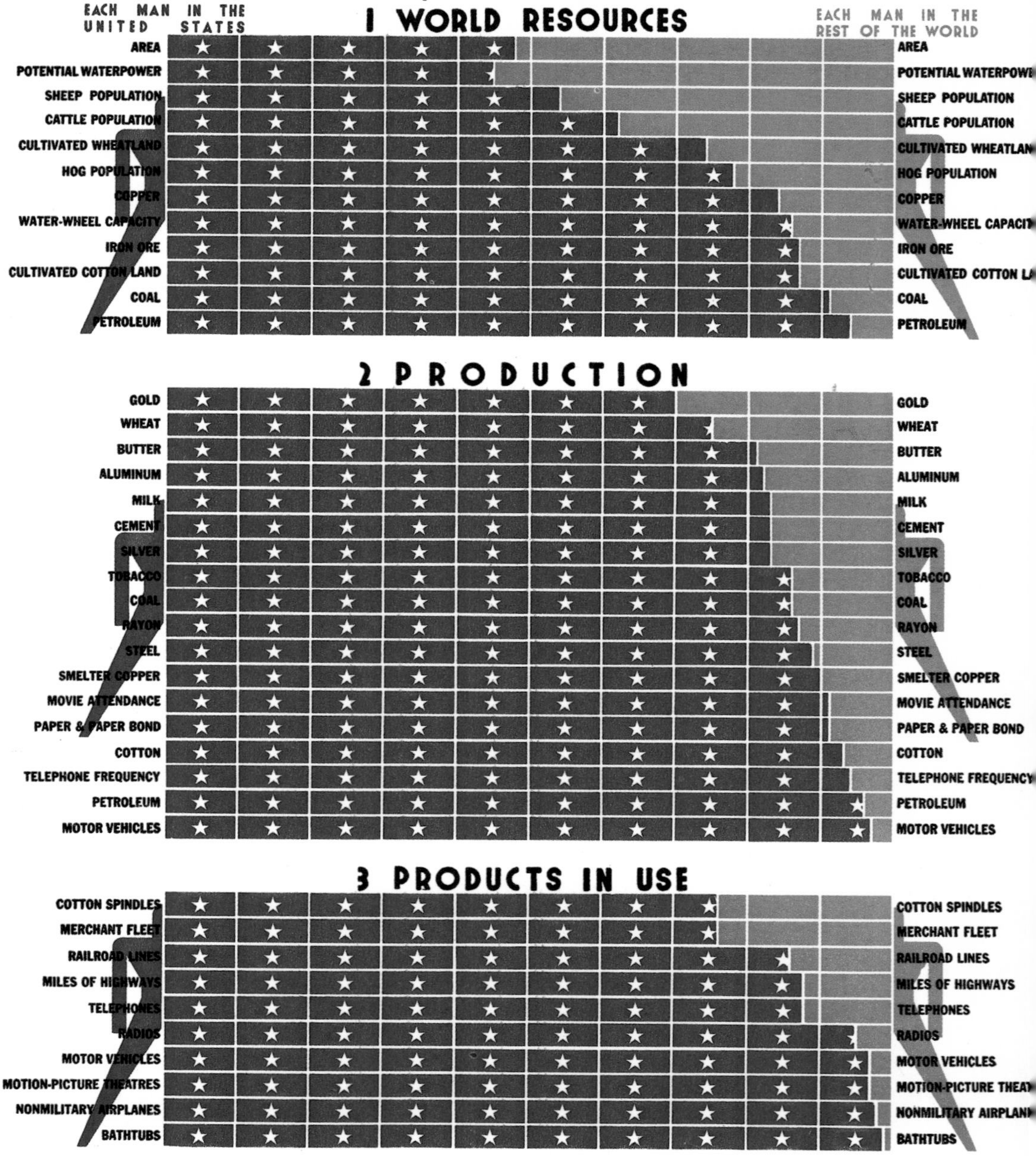

EACH MAN IN THE UNITED STATES

EACH MAN IN THE REST OF THE WORLD

1 WORLD RESOURCES

AREA
POTENTIAL WATERPOWER
SHEEP POPULATION
CATTLE POPULATION
CULTIVATED WHEATLAND
HOG POPULATION
COPPER
WATER-WHEEL CAPACITY
IRON ORE
CULTIVATED COTTON LAND
COAL
PETROLEUM

2 PRODUCTION

GOLD
WHEAT
BUTTER
ALUMINUM
MILK
CEMENT
SILVER
TOBACCO
COAL
RAYON
STEEL
SMELTER COPPER
MOVIE ATTENDANCE
PAPER & PAPER BOND
COTTON
TELEPHONE FREQUENCY
PETROLEUM
MOTOR VEHICLES

3 PRODUCTS IN USE

COTTON SPINDLES
MERCHANT FLEET
RAILROAD LINES
MILES OF HIGHWAYS
TELEPHONES
RADIOS
MOTOR VEHICLES
MOTION-PICTURE THEATRES
NONMILITARY AIRPLANES
BATHTUBS

On the other hand I have had extensive experience in making such inventories – in 1936 – for the world copper industry, in 1940 for "Fortune Magazine", and in 1943 for the United States Board of Economic Warfare. Therefor I have undertaken with the help of Southern Illinois University and the assistance of my colleague, John McHale, to prepare for this congress and for the world architectural students, a very complete inventory, not only of the world's resources, but also of the patternings of men's trendings and needs.

World Design Science Decade: 2 (1965): Mexico Lecture, 19

During World War II, Fuller was the head engineer on the Board of Economic Warfare. With the end of the war in sight, he realized the theme of his lifetime in a light-construction (and thus cost-effective) dwelling for everyone. He found a technologically advanced site for development and production in the Beech Aircraft factory in Wichita, Kansas, and he surrounded himself with a highly motivated team. The project continued to develop the technological background of the concept behind the Dymaxion House (for example, a round floor plan now replaced the hexagonal one). The development lasted from 1941 to 1946. The reactions to the prototype were extraordinarily positive; nevertheless, it was not put into industrial production, because Fuller insisted on the kind of perfection necessary for mass production, and they lacked the financial means for an uncompromising tooling-up and technology. The onset of the Cold War put a quick end to this conversion project.

WICHITA HOUSE

For new air inlet and old air outlet. use radiator similar to stack of auto radiators ~~testing~~, which will transfer the temperature of the out going air to that of the incoming air.

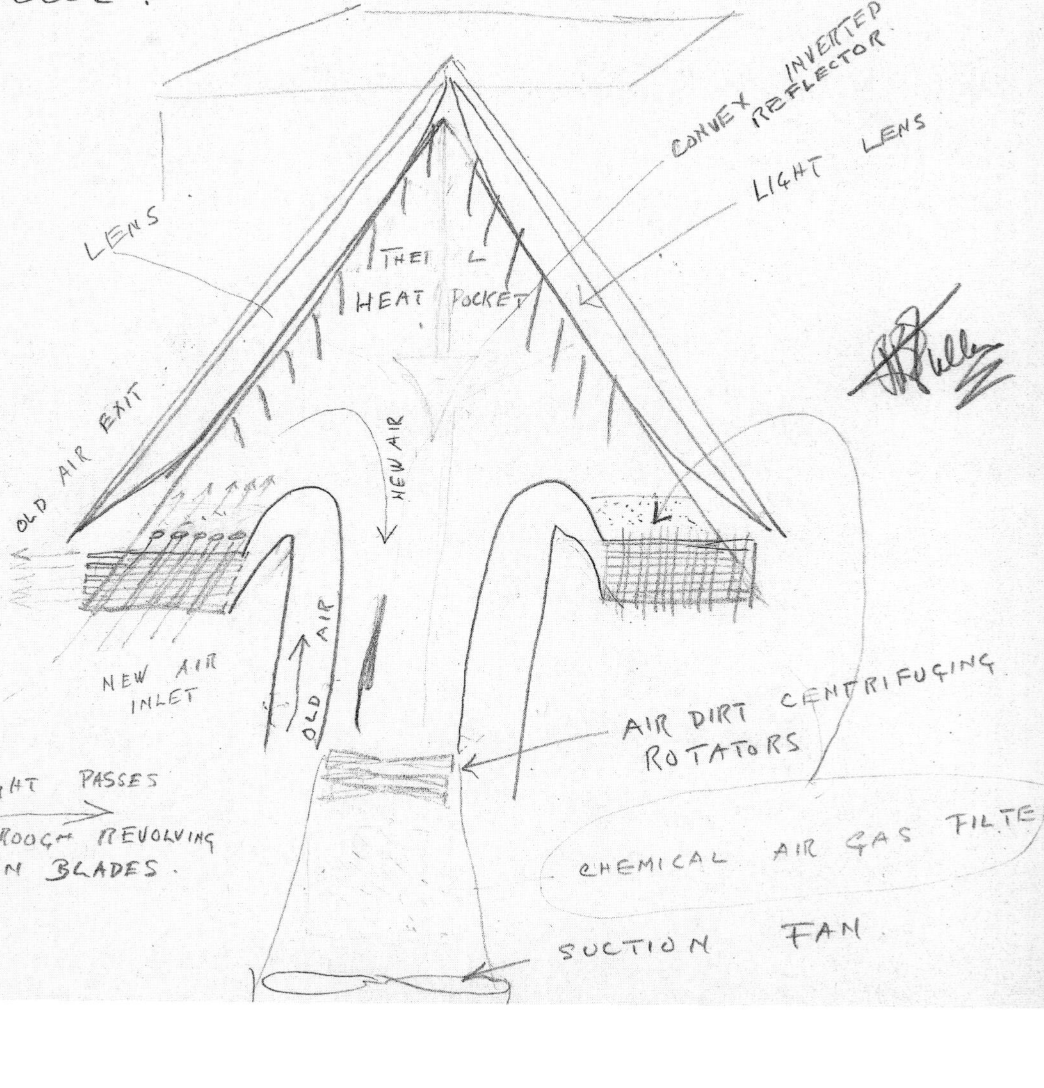

CONVEX INVERTED CONE REFLECTOR

LIGHT LENS

LENS

THEI L

HEAT POCKET

OLD AIR EXIT

NEW AIR

NEW AIR INLET

OLD AIR

AIR DIRT CENTRIFUGING ROTATORS

LIGHT PASSES THROUGH REVOLVING FAN BLADES.

CHEMICAL AIR GAS FILTER

SUCTION FAN

229: “Board Chairman R. Buckminster Fuller with

230/231: *Wichita House*, photograph of model, *Fortune*

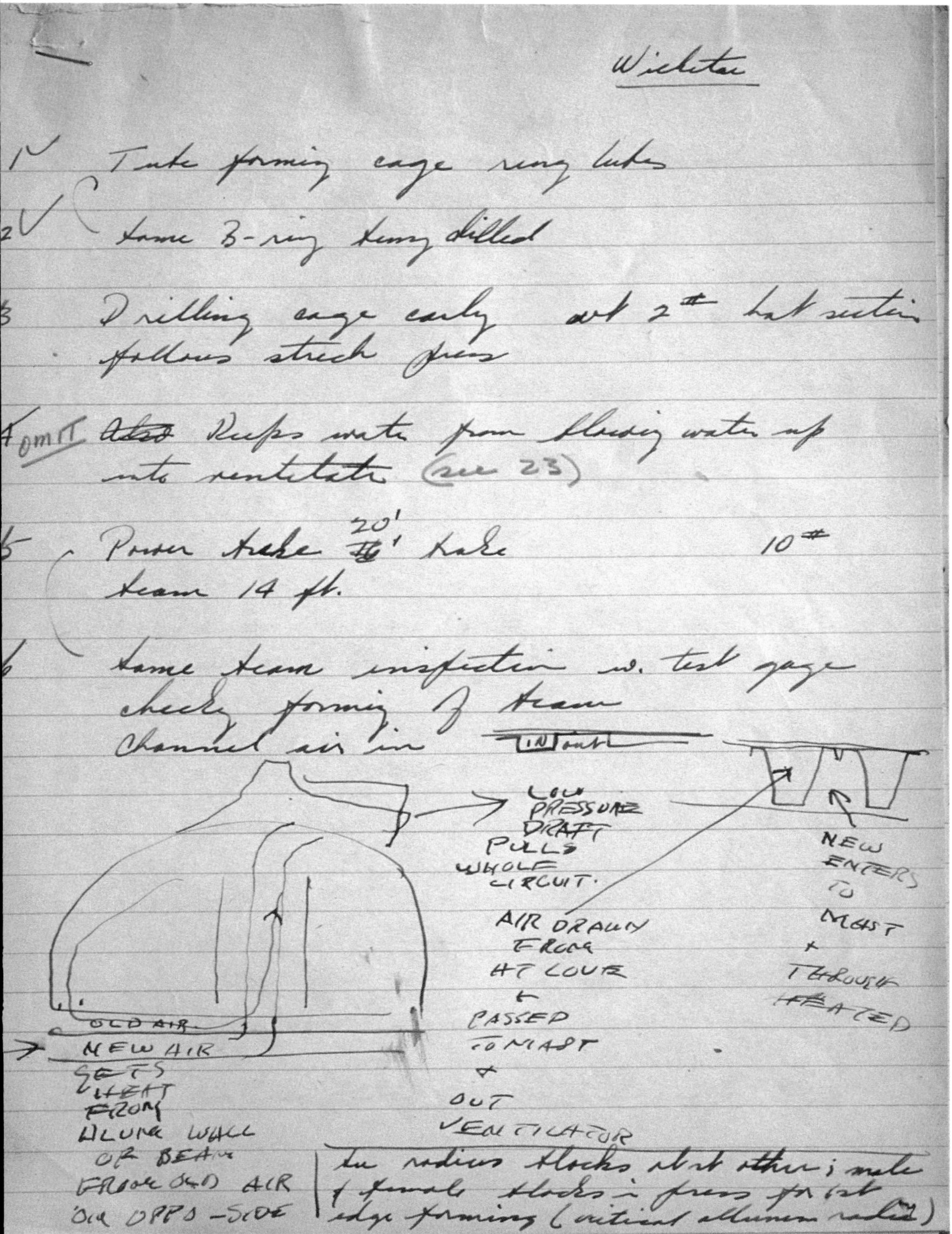

Wichita

1✓ Tube forming cage ring tubes

2✓ Same 3-ring being drilled

3 Drilling cage early at 2nd hat section follows stretch press

4 OMIT ~~Also~~ Keeps water from bleeding water up into ventilator (see 23)

5 Power brake 20' ~~16'~~ brake 10#
beam 14 ft.

6 Same beam inspection w. test gage checking forming of beam

Channel air in IN out

LOW PRESSURE DRAFT PULLS WHOLE CIRCUIT.

AIR DRAWN FROM HT LOUVER & PASSED TO MAST & OUT VENTILATOR

NEW ENTERS TO MAST & THROUGH HEATED

OLD AIR

NEW AIR GETS HEAT FROM WARM WALL OR BEAM FROM OLD AIR ON OPPO-SIDE

Use radius blocks at rib others; make female blocks in press for 1st edge forming (critical aluminum radius)

·Pencil sketch of *4D Houses/Dymaxion House:* the house as 'valving.' Detail of the top of the mast, undated (1929).

·Manuscript sheet with sketch of *Wichita House,* undated (1944)

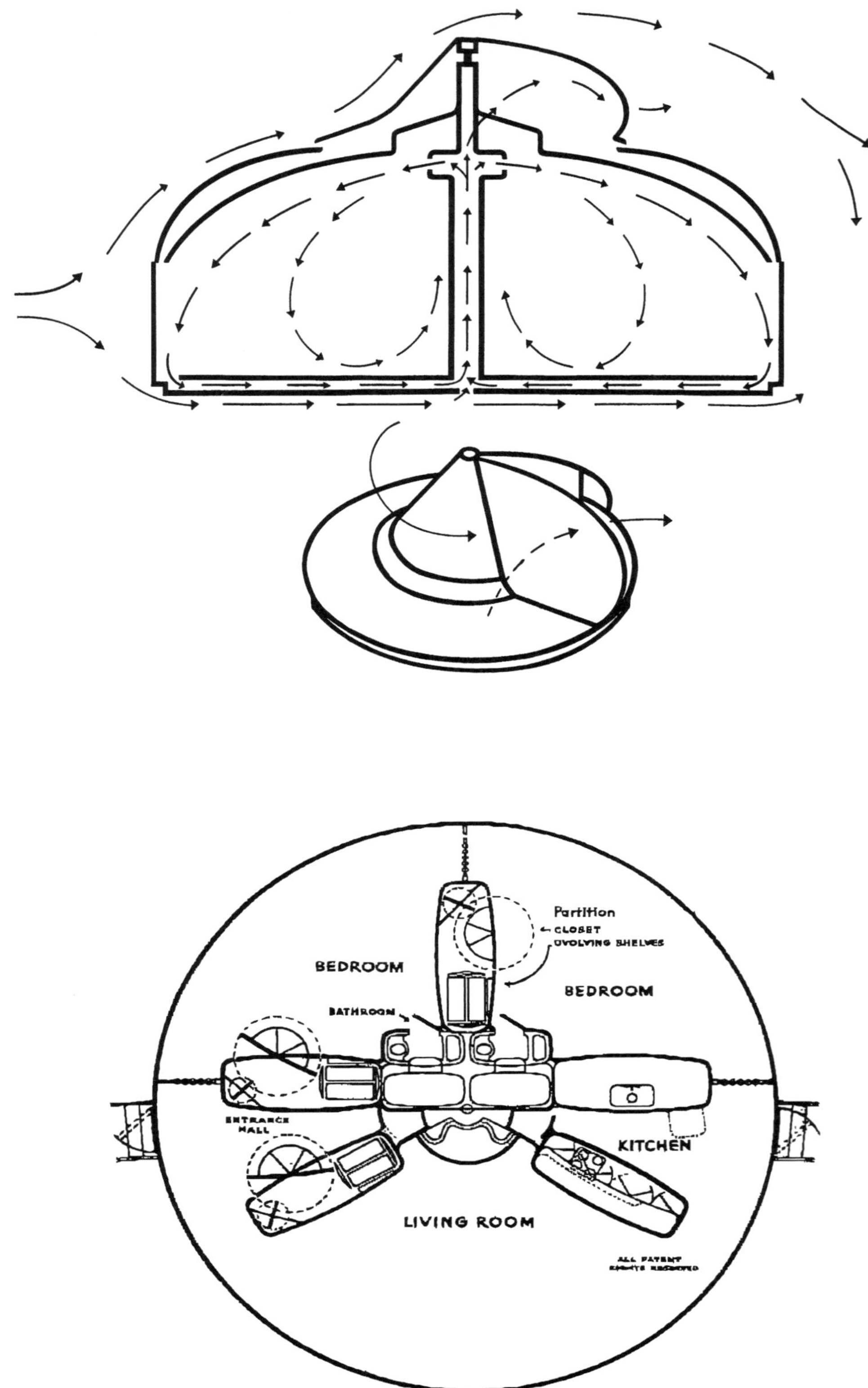

234 · *Wichita House:* schematic illustration of the air circulation. Ventilator mounted on supporting mast. Floor plan with two entrance ways, two bath cubicles, kitchen unit and electrically operated "ovolving shelves" modelled after the paternoster principle.

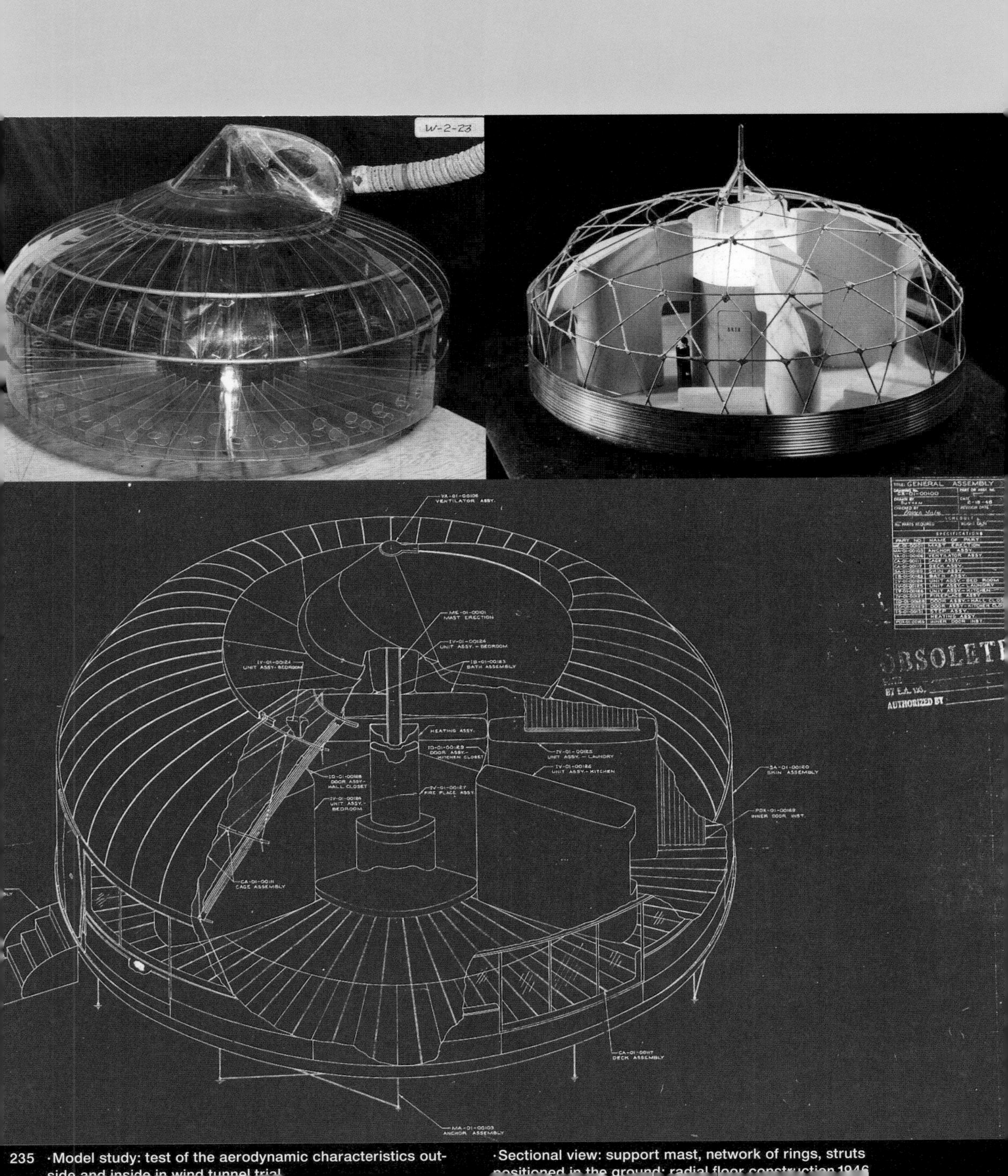

235 ·Model study: test of the aerodynamic characteristics outside and inside in wind tunnel trial

·Sectional view: support mast, network of rings, struts positioned in the ground; radial floor construction 1946

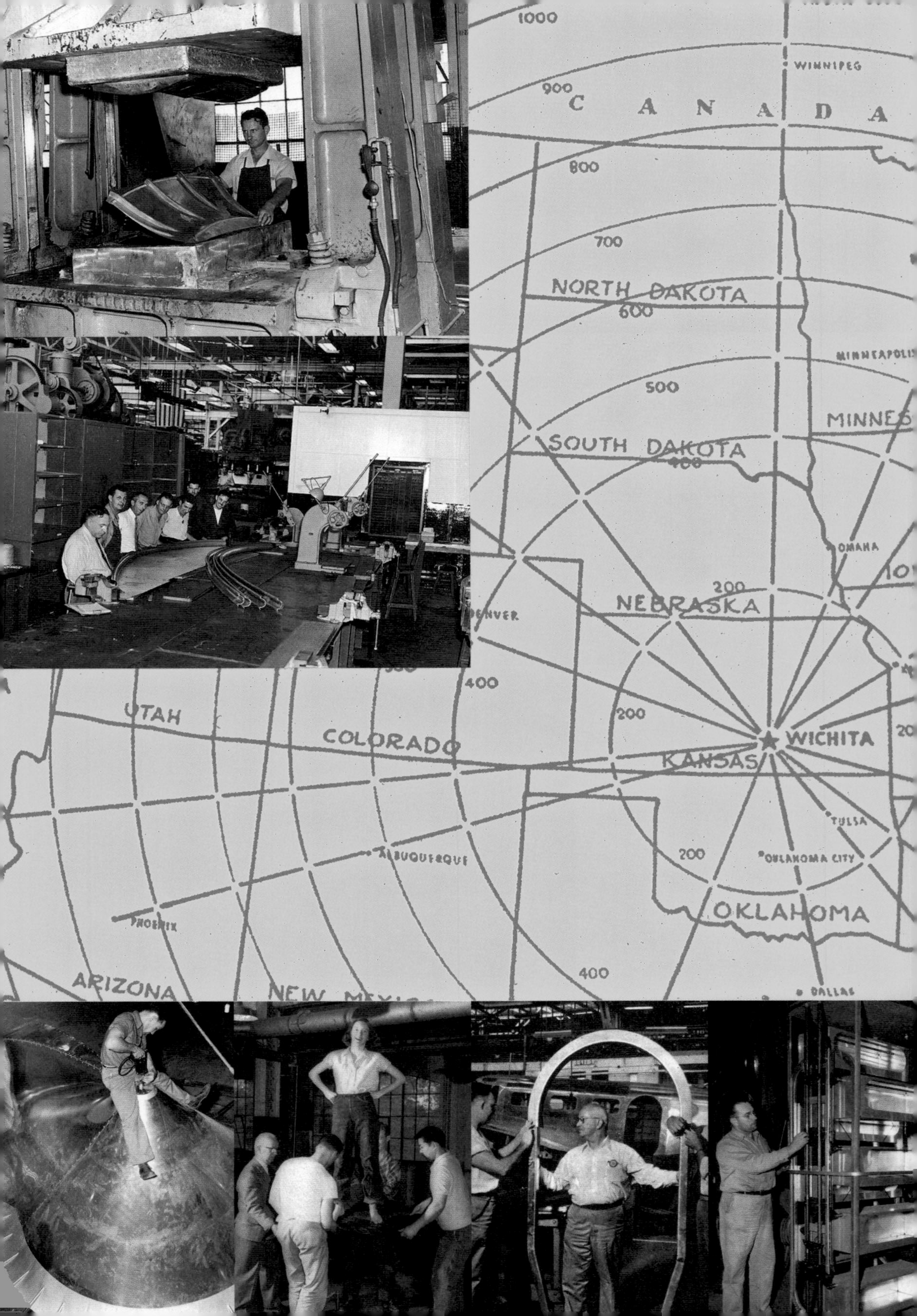
1000
WINNIPEG
900
CANADA
800
700
NORTH DAKOTA
600
MINNEAPOLIS
500
MINNES
SOUTH DAKOTA
400
OMAHA
200
NEBRASKA
DENVER
400
UTAH
200
COLORADO
WICHITA
KANSAS
TULSA
ALBUQUERQUE
200
OKLAHOMA CITY
PHOENIX
OKLAHOMA
400
ARIZONA
DALLAS

237 ·Photos of tooling-up in the *Beech* Aircraft Company, Wichita. The leading role of the aircraft industry for all of society. Stationery for the *Dymaxion Dwelling Machine Inc.* showing the *Dymaxion Industrial Strategy Map:* Wichita (Kansas) in the center of the USA. Assembly of Beech Aircraft, Wichita.

(...) The first house I designed in 1927 was a hanging one, but I had to keep my sling pitch very sharp because the tensile strength of the materials available was then so inferior. My mast and tension members therefore had to protrude out beyond the enclosure as those who have seen pictures of the 1927 Dymaxion House will remember. However, as I started this new house in Wichita in 1944, the increase in tensile strength of the materials available was so great, it became very practical to enclose the tension members by having them at relatively flat sling pitch, so you didn't even see those tension members because they could be enclosed within the roof. I didn't do it for looks, though. I did it because as we increased the angle of sling from vertical toward horizontal the stress was increased very rapidly, and as we increase the stress, we increase the rigidity.

I wanted to have a relatively rigid affair, a composition of tension and compression that would not give like a bed spring. The rigidity advantage was gained by positioning a compression ring, i.e. a horizontal arch, at a considerable distance outward from the mast and at as flat an angle as was reasonable within the limitation of tension metallurgy. So that is what we did here in Wichita in 1944 and 1945. By this method we can enclose a large space very rapidly, that is why circuses and armies employ tents. However we think of tent not as a principle but in association with the relatively flimsy materials with which the armies and circuses have occasion to build them. Golden Gate Bridge and a wire-spoked auto wheel layed horizontally are tent structures in principle. We don't think of them as flimsy nor is our house flimsy. It is designed to withstand fire, tornado, hurricane, and earthquake.

It was next important to develop a structure from the top of the mast downward which would be triangulated all the way for rigidity – thus we criss-crossed our tension members, as are the spokes of the wire wheel, out from the mast head to the first horizontal compression arch and downward again criss-crossing to the successive horizontal compression rings and finally all the way to the ground anchoring to the ground and using the inertia of the earth. (...) Anchor rods must be both vertical and diagonal in order to avoid swing or torque of the whole building.

We come now to a second argument about circle. As the steel world took over progressively from the old wood world by providing stronger structural shapes it went on also to provide rolled steel to replace boards. At first they rolled steel thinly enough to make steel plate which was used for ship siding and for trains, but they were able to roll the steel much thinner than plate which at minimum is 1/8 of an inch thick. Below 1/8 of an inch steel is called sheet, and steel sheet, if thin enough, was as light as a board per square foot. Plate was, however, no substitute for wooden boards in their function as sheathing, siding, boxing or partitioning for if thick enough to equal a board in stiffness it was much too heavy and if not too expensive as bulk, it was much too expensive to handle and to support structurally because of the great weight.

(...) The sheet principle can be used when curved into an arc. In curving the sheet it is made to fail as a column which throws the outer side of the arc skin into tension, which in effect is wrapped about the inner surface which represents a compression arch composed of the individual molecules of the material acting separately as do stones in an arch. Such a curved sheet then presents a strong, stiff, cylindrical form; strong against the exterior pressure of wind and enormously strong against interior pressures which explode conventional structures by hurricane's or bomb's or tornado's low pressure outside.

(...) Then we discover additional mutual advantage to our house inherent in combination of the geometry of circle and the curving of sheet, not by virtue of the respective unique properties but by their interactive virtue when associated for purposes of housing. This new advantage is discovered as we consider the second major stress applied to house by nature. Second to gravity is wind stress.

The average wind speed over the United States as computed by widely reported recordings is considered aerodynamically as little ships whose standard cruising speed is 12 miles an hour, but which suddenly are accelerated to 30 miles an hour, and then suddenly again have to go 50 miles an hour, and sometimes suddenly they have to go 70 miles an hour and then the flat planking begins to fly off as flat boards develop lift in parallel with the wind, which lift is opposed only by the friction of the nails amounting to but a few pounds in tension as nail pulling experience confirms.

Designed to look secure by guess and by prayer to the gods of inertia, conventional houses are not engineered from measured data to cope with the greater wind speeds which they sometimes encounter.

Looking for chances to take advantage of high tensile ability of the new sheet, I became interested in the effects of wind stresses on houses and discovered in wind tunnel tests that a cube and hemisphere of equal volume indicate a drag advantage of ten-to-one in favor of the hemisphere, that is the drag is ten times greater on the cube. That indicates that we might either cut down the size and weight of our original structural members in hemisphere to maintain equivalent wind strength to that of the cube or we might take part of the increment and turn it to greater strength advantage. (...)

Another interesting discovery in the wind tunnel was that the heat losses were in direct proportion to the drag. It was indicated that you might be able to reduce your amount of heat necessary to heat the building, to a very high degree, by employing efficient aerodynamic shape. Shape factors are used very little today in heating and ventilating, they have been toyed with to date only by higher mathematics which concludes that there are great potential efficiencies to be had by measured evolution of shape control. In our own tests we had discovered the relative degree of that shape importance to heat savings in the coincidence of the drag and heat loss curves.

Having been in the building business in the '20s, I was particularly interested in heat losses because my material, as I told you, was a very good insulating material. Insulation was a brand

new selling word, in 1922, something I could talk about in trying to sell my building system despite the traditional handicaps. I frequently had heat loss tests made at different laboratories. We discovered that we had an advantageous heat loss "factor" in our material, almost as favorable as in an 8 inch cork wall. However, in 1926 some of the weather stripping people had tests made to see what savings they could make with their material, and they suddenly shocked the whole insulation world by proving that savings two to three times as great could be obtained by putting weatherstripping in the window and door cracks as could be effected by an 8 inch cork wall. The wall insulation people had to kind of shut up.

(...) The phenomenon seemed well explained when we discovered that the losses of buildings of comparable tolerance of joint fittings was directly proportioned to the drag. Houses re-presented very large obstacles, and the low pressure tails on buildings are very long. They stream out two to three hundred yards in relatively mild winds. Houses are usually surrounded by trees and other houses and those tails get mixed up with each other but if you discover a house out in the open you can observe this 300 yard pattern in the snow shaping as seen from an airplane.

Air being highly compressible as it goes around large objects like houses you get a fairly high degree of compression at the widest beam of the obstacle to the wind. The result is a very long low pressure tail because the pressured air shoulders dissipate their pressure outwardly as well as into the wake. You have a long lag in the rate of reestablishment of pressure equilibrium in the wake and much energy is required from somewhere in the form of high pressure to satisfy the long low pressure streamer. (...)

The heating energy inside buildings is converted in the air of rooms into the work of expanding air, and expanding it within enclosed chambers necessarily develops an increase in pressure. So you have a high pressure on this side of your house wall and an enormous low pressure tail on the outside to be satisfied, and your high pressure inside the house simply is extracted by

successive energy conversions right through a masonry wall to satisfy the low pressure as does gravity pull the water down Niagara Falls. The high pressure is drafted directly through chinks or cracks, ergo the fine showing of weatherstripping. (...) However, people still say that cold comes into houses. (...)

One way you ought to talk about the phenomena house and cold is that cold and vacuum are in physics almost identical – that is you have energy in the presence of cold and in the presence of vacuum and when your energy – either as heat of kinetically accelerated gas molecules or as radiation – is eliminated, cold or vacuum alone remain. That is the best way for you to look at it. You see, when they say, "cold is coming in," it is because energy as heat, is dissipating so fast as to leave cold gases in your presence. Air that is cold because low in energy, content moves to you so you seem to feel cold draft but there is no physical entity "cold." Temperature should be thought of as relative heat concentrations or dissipations.

(...) We discovered that the motion of wind along the surface of the earth is a turbulent affair rising here and hitting the earth's surface there and rising again. We discover that the wind instead of being considered as blown horizontally along the earth from a god's mouth or by a mysteriously hidden blower, should be recognized as an enormous system of many up or down drafts (similar in picture to many water spouts) converging to form bigger single up and down drafts. We discovered then that these columns should be recognized as enormous convection columns in the thin atmospheric layer surrounding the earth and caused by a warm earth in the presence of a cold outer space. (...)

As a net result of these convection column tests it was discovered that the rising winds tend to greater velocity than that of the lowering air (...), and therefore that the resultant of focus of wind stress near the earth's surface is in an upward direction in the lee of an obstruction – such as a house at the earth's surface. The large low pressure area in the lee of the house, of course,

“Efficiency of um

RBF: handwritten notice (1928)

View inside the mast showing the mounting point for the

Detail of the interior room partitions: window with adjustable

rella and bicycle”

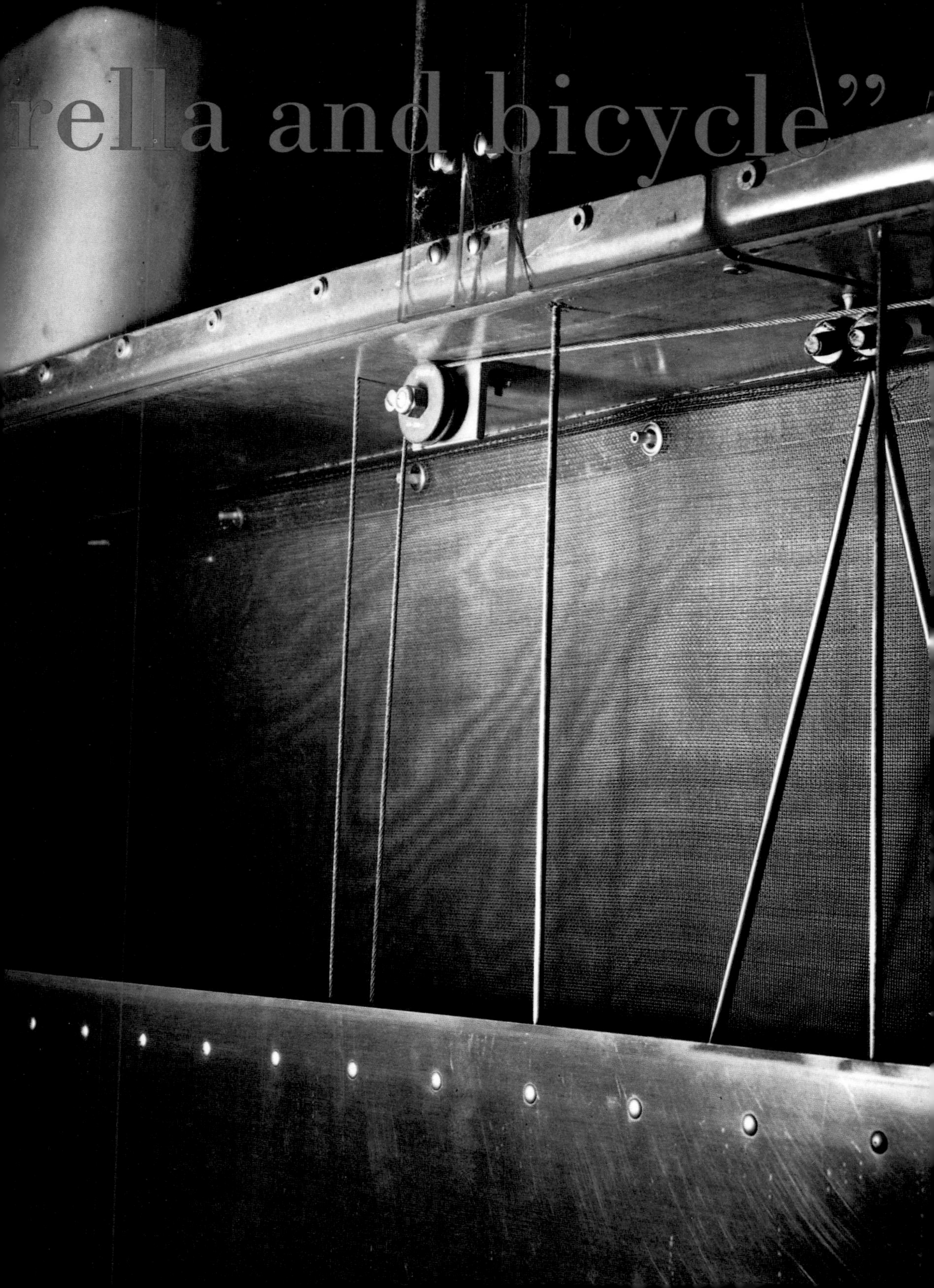

tends to float upward thus adding to the upward angle. Trees are also aerodynamic design structures to permit a large frontal area necessary to the functioning of the trees. Trees tend to avoid destruction by the wind by rounding their lower frontal branches to the approaching air and by coning their upper branches to point in the direction of the leeward and upward draft, thus reducing drag to a vital degree.

In the same way it became evident that a large ventilator could be designed to rotate upon the top of the house in such a manner as to focus the low pressure – caused by the air passage about the building – at a point about 45 degrees leeward and upward from the center of the house. The ventilator was developed through a series of assumptions and tests in the wind tunnel, until successful design was arrived at which reduces drag to a minimum and prevents oscillation of the ventilator, while at the same time putting the focused low pressure to work in pulling a draft out of the building through a duct system that induces the draft to create an air conditioning circuit, as well as to remove dust from sweeping traps in the floor and odors form the kitchen and bathroom, etc.

(...) With the central vantage point for generating air, light, sound and work services, we discover that those services when in operation describe fountain-like flows upward, outward, downward in all directions with concentric flow for recycling below. We discover also that this fountain flow can be reversed but in either case, maximum coverage with least distance is effected.

The fountain flow is appropriate for maintaining relatively warm atmospheric flow in winter, and reverse fountain is most efficient in maintaining relatively cool atmospheric flow in summer. In neither of these fountain flow cases does energy set up a chaotic echo system as we find it doing in the indiscriminate, cubical, squash racquet court shaped chambers in which we now live. (...)

Complimentary handling of dynamic flows teaches that the same principles control the structuring of energy flows as controlled the structuring of our building members: that in effect the principles of push and pull and their unique characteristics of distinctly limited compression behavior and almost unlimited tension behavior hold true also in hydraulics, pneumatics, sonics, and electronics. You can pull or draft air over vast distances but you can push it only a few feet by blowing. Winds are drafted and not blown; we should speak of the southeast draft instead of the northwest wind. (...) In the same way, visible light is a pushed phase of radiation and is limited to relatively short distances through atmosphere and requires enormous power to push it, while what we call electricity is tensed or pulled radiation and the distances over which radiation can be drawn by wire is very great compared to searchlight beaming, but, drawn by wireless tension control the limit of distance is unpredictably great. Even as a tension controlled lasso can be gyrated and thrown and wave impulses can be sent out controllably over it, as a snake whip may receive a wave by the wrist to hit an object and return the wave as a tension circuit again to the sender, so does radio and radar tensively induce circuits to pull radiation phenomena over almost unlimited distances.

By simple attention to and comparison of this phenomena of push-pull, enormous advantages can be gained by man over his environment through the initial advantage provided him to by our dwelling machine.

We associate the idea of air conditioning with very large apparatus and when anybody talked about air conditioning for small houses they found it way out of their price range, because of the apparatus and expensive quantities of energy to accomplish that task. Now with our ventilator employing outside air motion to accelerate interior dynamic fountain motion it becomes an inexpensive feat to provide excellent air conditioning.

Designing a New Industry (1946), 31

Wichita Dwelling Machine

For Fuller, design did not present an aesthetic problem but only the question of a teleologically clearly formulated task and its logical solution. There is probably no other plan for a model home that has moved as far from traditional architecture and as close to industrial design as the Dymaxion Dwelling Machine, as Fuller called this project. With its round form, the surrounding band of windows in the thin-skinned wall, its airspace between the ground and the floor, and the ventilator in the roof, it is like a civilization station inhabited by people outside of any urban context. Architectural journals reported extensively and with great fascination on the house. Still, as radical as it was, the house still did not make the step from a built sketch to a serially produced model: only two prototypes were built at full size, and it never made it to production. There are several reasons for this. First, the responsible parties at Beech Aircraft could not or would not make the immense investment that would have been required to achieve true serial production. Second, and not unrelated, the rapid development of a Cold War between the former allies hindered any hope of retooling the armaments industry for civilian uses. Third, probably Fuller himself, having immersed himself in his energetic-synergetic research on structure, realized that he did not really want to become a business.

· Photograph of model with furnishings.
· The prototype in Wichita after completion. Weight including ter 36 feet, floor space 1,017 sq. ft., interior volume 12,000 cubic ft., maximum interior height 16 feet, overall exterior height 22 feet.

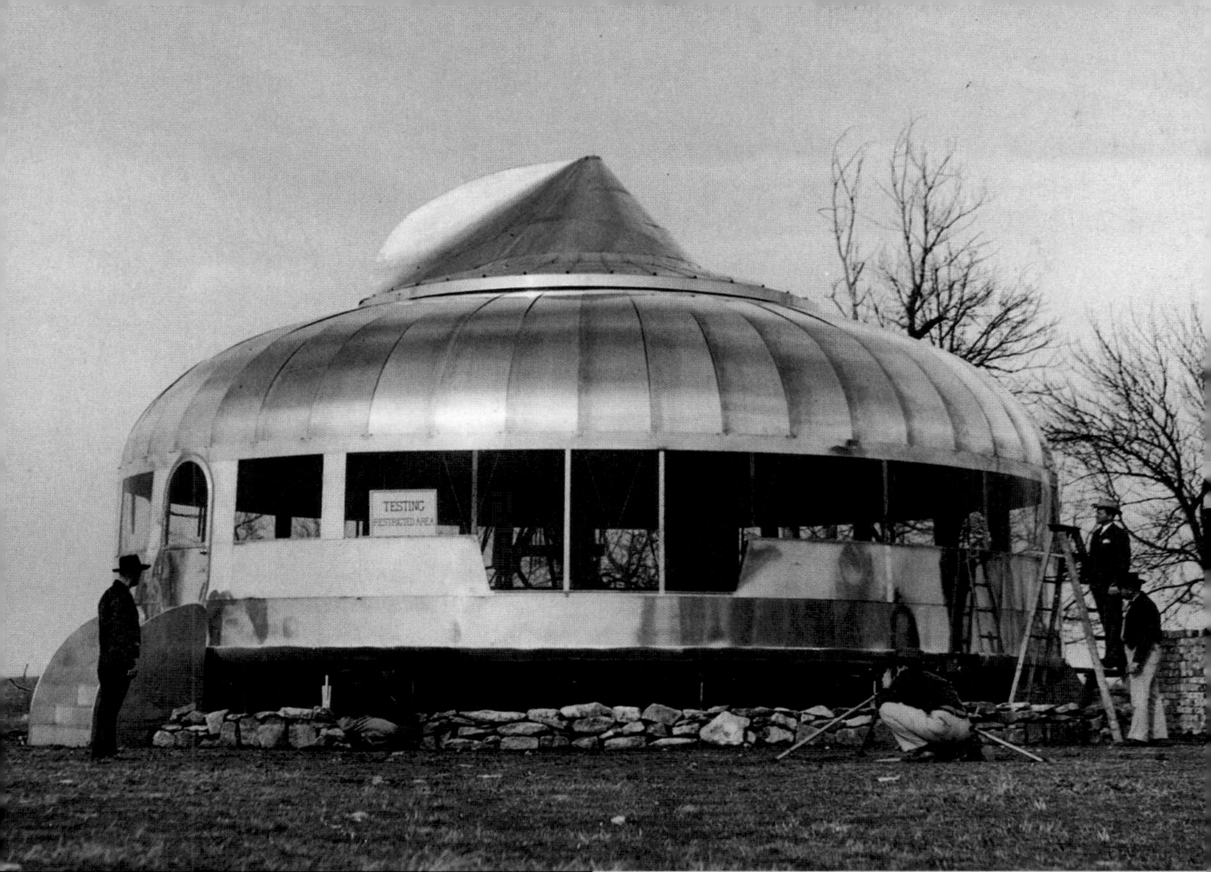

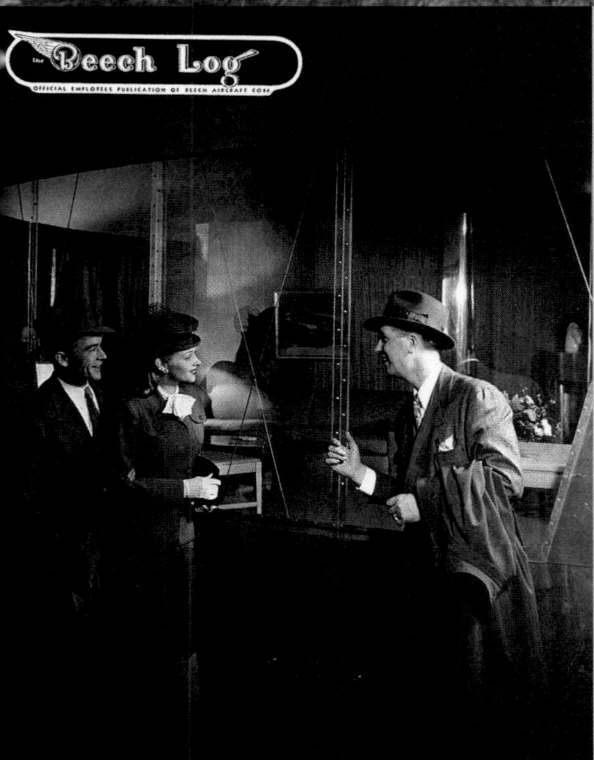

Prefabrication

Fuller considered it important that the Wichita House was not a prefabricated house in the usual sense. The building trade defines as prefabrication the fabrication of panels and parts, that is, semifinished products, that are used at the construction site, that is, subassembled. The Fuller House, by contrast, consists entirely of industrial prefabricated units that are assembled at the site. The difference between subassembly and assembly is fundamental. In his view, only the assembly of fully stand-ardized parts and complete units will result in the costs savings that come from mass production.

247 ·248/249:Interior of the prototype with standard commercially available, tastefully modern furniture. Following

House, *Fuller Houses, Inc.*, received more than 30,000 expressions of interest from all over the country.

Ever since the statistical research into global distribution of resources he conducted during the thirties, Fuller had been working on a design for a world map that would be suited to represent global data and relationships. At the beginning of World War II, he noted a dangerous gap between the conventional picture of the world and a reality defined by air travel and "blitzkrieg." He wanted to correct the mental map that had been formed by four hundred years of using the Mercator projection. The Dymaxion Projection was intended to produce the key to an all-round logistics on the globe. As a topological way of transferring data from a globe to a flat surface by means of great circle grids, it was a magnum opus for Fuller: the fundamentally important weigh station on the way to energetic-synergetic geometry and the geodesic domes.

DYMAXION WORLD MAP

DYMAXION PROJECTION

Dymaxion Projection

Suddenly, within months, people have come to realize that they can girdle the planet in an infinite number of directions. The world has been surprising itself by coming in its own back doors and down its own chimneys from every unlooked-for direction. This has called for a revolution in map making and in cartographical principles such as history has never seen. A need has risen for new methods of peeling data off the globe and for assembling the peelings in such a manner as to gain useful knowledge of the spherical coursings.

Fluid Geography (1946), in: Ideas and Integrities, 143

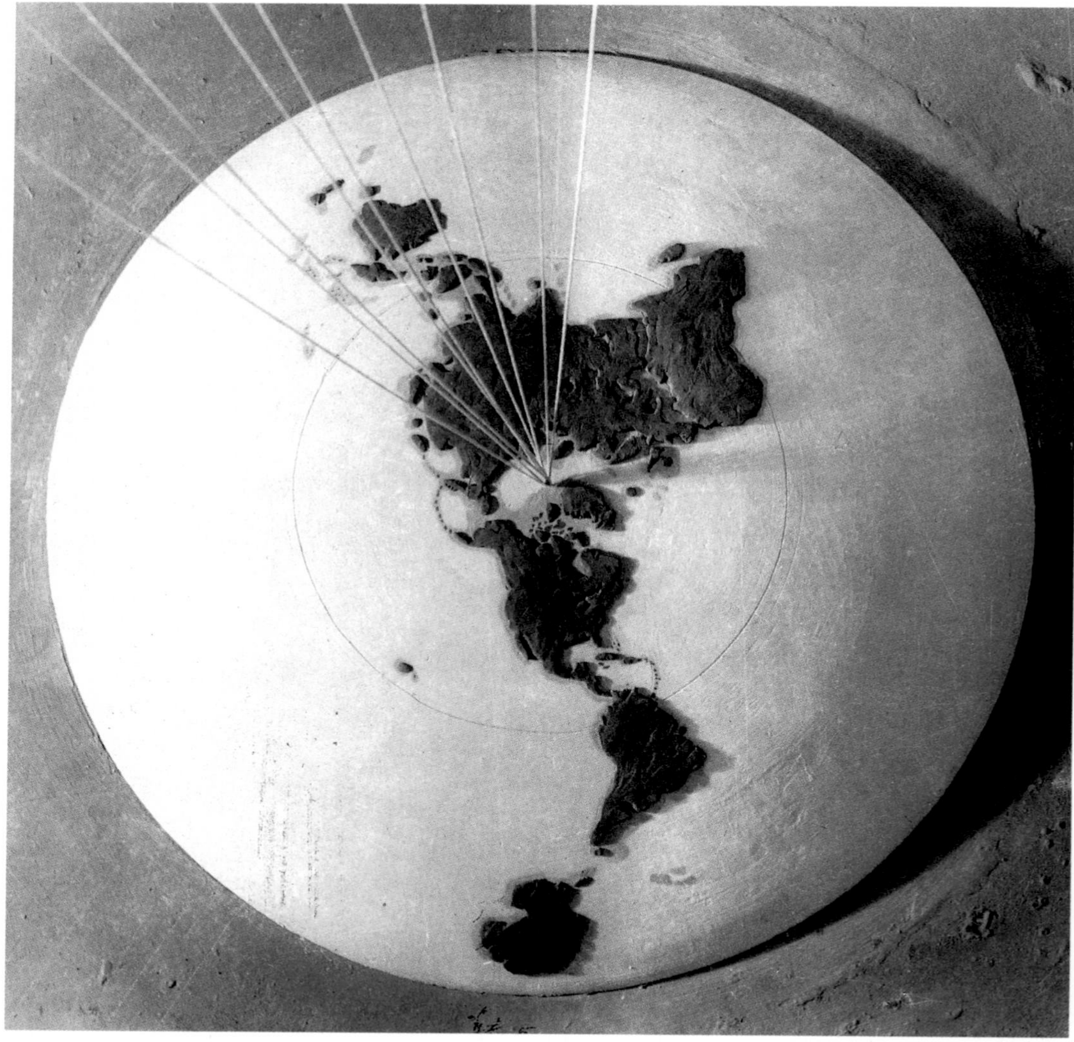

 · 251: RBF presents the Dymaxion Projection of the · RBF'S World: "one-world island in a one-world ocean,"

RESOURCES AND RESERVES OF COPPER
SMALL SCALE CORRELATION
WORLD TRENDS
COPPER DEVELOPMENT
WORLD
LARGE SCALE COMPARISON OF TRENDS USING U.S. COPPER HISTORY OF DEVELOPMENT AS KEY TO WORLD COPPER DEVELOPMENT
REST OF WORLD

Dymaxion World

Fuller's Dymaxion World embodies his effort to resolve the dilemma of cartography: how to depict as a flat surface this spherical world, with true scale, true direction and correct configuration at one and the same time.

Orthodox cartography demands that the world be projected correctly in at least one of these aspects, even to the sacrifice of the other by distortion. This is a reasonable requirement in maps that are to be used for navigation, statistical work and other precise purposes.

On his map, Mr. Fuller adjusts the requirements of scale, direction and shape in a compromise which distributes distortions evenly between them and around the globe. For the layman, engrossed in belated, war-taught lessons in geography, the Dymaxion World map is a means by which he can see the whole world fairly and all at once.

President's Big Globe rests in ball-bearing mount. It can be revolved in any direction and studied in the changing perspectives of war's strategy.

Dymaxion Globe, a cube with its corners cut off, has shape of irregular solid first constructed by Archimedes. Its scale, constant on edges of segments, is similar to that of a 12-in. globe.

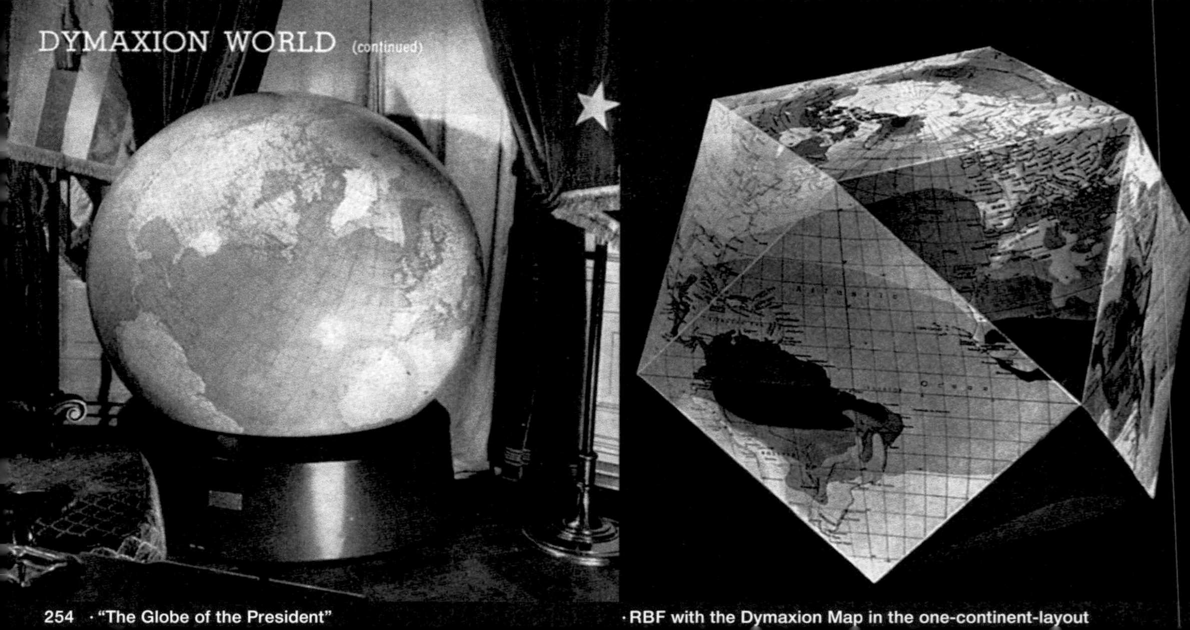

 · "The Globe of the President"

· RBF with the Dymaxion Map in the one-continent-layout

It is Designed for Political Geographers

The President of the U.S. keeps a 50-in.-diameter globe close enough to his desk so that he need only swing his chair to consult it. As a political geographer, the President knows that no standard flat map can give him all the information he requires. The student – and master – of political geography is interested in the true, relative geographical locations of the Great Powers, and in the strategy of communications on the great-circle (shortest distance) routes between them. He must be able to visualize the world's geographical layout not only from his own vantage point, but in the divergent perspectives of other nations and their political geographers.

R. Buckminster Fuller designed his Dymaxion World map to fit exactly these requirements. Like all flat maps, it evidences the distortions that result from the translation of the three-dimensional surface of a sphere to a flat plane. Its distortions, however, are distributed proportionately within each of its 14 segments and are nowhere extreme. Greenland on the Dymaxion map appears close to its relative size, in contrast to its inflation to six times that size on the Mercator projection. The segments, plotted on an entirely novel grid of great circles, are constant in scale along their edges. Distortion increases toward the center of each segment.

Thus, though it does not easily yield precise calculations, the Dymaxion World used as globe or flat map, satisfies the need, short of a perfect sphere, for a visually correct picture of the earth.

Life, March 1943, 41f.

HOW TO ASSEMBLE THE GLOBE

Here demonstrated is the simple procedure by which the segments of the Dymaxion World map are assembled into a visual approximation of a round globe. The opposite page is the reverse side of the second of the two heavy center sheets on which the map is printed.

First step, removal of center sheets from magazine, must be taken with care to avoid tearing map on staples. Segments cut out are best fastened together by paste or mucilage. Because they warp the paper, pins should not be used. For neatest product, sequence of assembly here illustrated should be followed. Marginal letters of triangles match marginal letters of squares.

The map, thus assembled into a 14-faced solid, has many of the advantages of a globe. Like a globe it can be viewed from any perspective to bring geographical relationships into new relief—to show that the Southern is the water hemisphere, that Chicago and Sverdlovsk are fairly close together over the top of the world, that Dutch Harbor lies closer to the shortest San Francisco-Tokyo route than Pearl Harbor.

Before they are hidden inside globe, statistics on reverse of each segment are worth inspection. For example, the North Pole square's 8.9% of world population contrasts dramatically with the South Pole's .0004%.

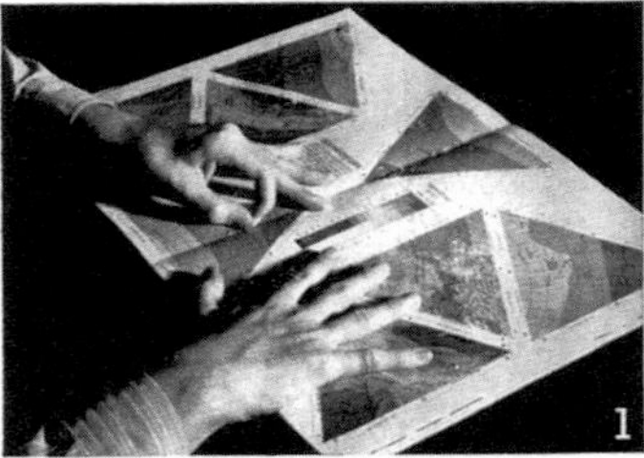

1

BENDING OF STAPLES is first step in removing map from copy of LIFE. Bent back, staples hold copy intact.

2

SCORING OF MARGINS of colored face of segment with clip or dull knife facilitates folding of flaps (*right*).

3

FOLDING OF FLAPS should follow margin of map precisely. Flaps of segments to be joined are keyed by letters.

4A

HOUSEHOLD PASTE or mucilage is best means for fastening flaps. It should be spread thinly to avoid warping.

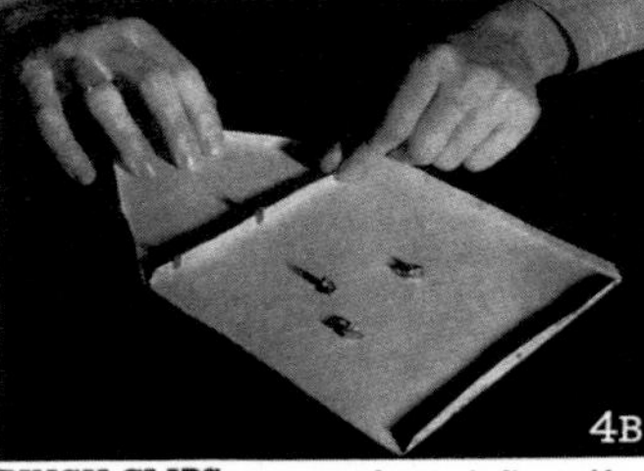

4B

PINCH CLIPS, easy to apply, permit disassembly of globe. If clips are used last segment must be taped or glued in.

4C

CELLOPHANE TAPE is substitute for paste and clips. It must be applied inside and out to keep edges together.

5

POLAR SQUARE and triangles should first be assembled into unit. Care should be taken to keep edges in register.

6

EQUATORIAL SQUARES are then joined to polar square-triangle assembly. Key letters simplify matching.

7

MOST DIFFICULT is this step in which square is joined to triangle. Polar square should be held flat on table.

8

SQUARES AND TRIANGLES are now fastened and structure is self-supporting. Paper has surprising rigidity.

9

SECOND POLAR ASSEMBLY is mounted. Paste should be allowed to dry a little before the flaps are joined.

10

LAST TRIANGLE is left unfastened until other flaps are secured. It can then be set by pressure from the outside.

—OR YOU CAN USE SEGMENTS FOR A FLAT, MOVABLE MAP (SEE PAGE 53)

MIDWAY

0.1% of world's land area . . . . 58,500 sq. mi.
6.1% of world's water area . . . 8,572,500 sq. mi.
0.1% of world's population . . . 3,115,800 inhab.

SQUARE MILES	TERRITORY	POPULATION
38,600	JAPAN	3,084,400
12,300	U.S.S.R.	30,400
6,400	ALEUTIANS	800
28	MIDWAY	118
1	WAKE	50

U.S.A. AND NEIGHBORS

8.8% of world's land area . . 5,011,800 sq. mi.
2.6% of world's water area . . 3,619,200 sq. mi.
8.0% of world's population . 173,684,700 inhab.

SQUARE MILES	TERRITORY	POPULATION
3,011,700	U.S.A.	130,974,700
1,040,300	CANADA	9,893,800
721,600	MEXICO	19,453,700
189,600	CENTRAL AMERICA	8,113,800
41,200	CUBA	3,846,700
94,700	GREAT LAKES	

CHILE AND ARGENTINA

1.7% of world's land area . . . 982,000 sq. mi.
5.5% of world's water area . . 7,649,000 sq. mi.
0.55% of world's population . . 10,538,220 inhab.

SQUARE MILES	TERRITORY	POPULATION
752,600	ARGENTINA	7,766,200
225,900	CHILE	2,772,000

GOOD HOPE

0.6% of world's land area . . . 338,900 sq. mi.
5.9% of world's water area . . 8,292,100 sq. mi.
0.2% of world's population . . 3,974,700 inhab.

SQUARE MILES	TERRITORY	POPULATION
187,900	SOUTH AFRICA	155,800
150,900	SOUTH-WEST AFRICA	3,814,000

INDIAN OCEAN

0.08% of world's land area . . 47,000 sq. mi.
6.1% of world's water area . . 8,584,000 sq. mi.
0.01% of world's population . . 286,400 inhab.

SQUARE MILES	TERRITORY	POPULATION
42,350	AUSTRALIA	285,200
4,500	KERGUELEN ISLAND	NONE
100	HEARD ISLAND	NONE

NEW ZEALAND

0.4% of world's land area . . . . 175,800 sq. mi.
6.0% of world's water area . . 8,455,200 sq. mi.
0.01% of world's population . . 2,293,300 inhab.

SQUARE MILES	TERRITORY	POPULATION
103,900	NEW ZEALAND	1,573,800
27,000	AUSTRALIA	57,200
26,200	TASMANIA	241,600
7,000	NEW CALEDONIA GROUP	49,800
7,000	FIJIS	220,800
2,500	NEW HEBRIDES	30,000
1,300	SAMOAS	72,400
800	OTHER ISLANDS	49,600

EUROPE AND AFRICA

8.4% of world's land area . 4,821,500 sq. mi.
2.7% of world's water area . 3,809,500 sq. mi.
20.0% of world's population 411,483,000 inhab.

SQUARE MILES	TERRITORY	POPULATION
212,700	FRANCE	41,834,900
196,600	SPAIN	25,209,500
178,600	GERMANY	57,856,900
101,600	ITALY	45,209,300
82,800	BRITISH ISLES	44,055,300
35,600	PORTUGAL	6,360,300
157,200	TURKEY	9,394,400
2,200,000	FRENCH AFRICA	29,009,000
115,500	SPANISH AFRICA	1,475,100

ASIA

12.2% of world's land area . 6,926,800 sq. mi.
1.2% of world's water area . 1,704,200 sq. mi.
27.0% of world's population 572,916,800 inhab.

SQUARE MILES	TERRITORY	POPULATION
2,737,100	CHINA	219,008,400
1,675,800	U.S.S.R.	16,211,300
1,419,700	INDIA WITH BURMA	273,136,500
463,200	TIBET	1,500,000
375,200	MANCHUKUO	28,856,900
180,200	AFGHANISTAN	8,328,000
76,200	CASPIAN SEA	

257 ·Cut-out page center spread, *Life,* March 15, 1943, with eight triangular and four square elements.
·Reverse side of page with captions for the eight triangular elements indicating population count, area, etc.
·258/259: Inner side of the cut-out page
·260: Reverse side with captions for the six squares

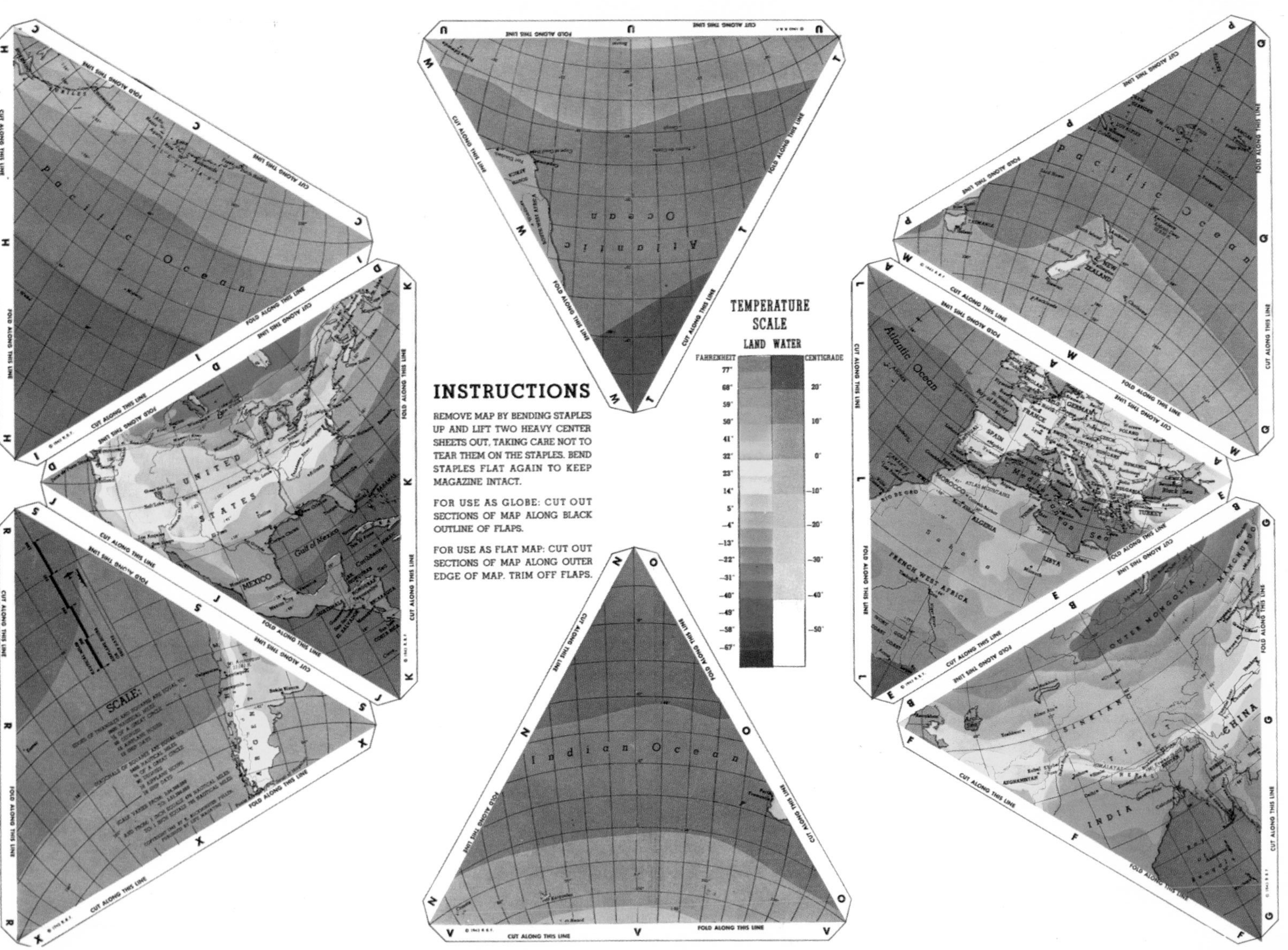
INSTRUCTIONS
REMOVE MAP BY BENDING STAPLES UP AND LIFT TWO HEAVY CENTER SHEETS OUT, TAKING CARE NOT TO TEAR THEM ON THE STAPLES. BEND STAPLES FLAT AGAIN TO KEEP MAGAZINE INTACT.
FOR USE AS GLOBE: CUT OUT SECTIONS OF MAP ALONG BLACK OUTLINE OF FLAPS.
FOR USE AS FLAT MAP: CUT OUT SECTIONS OF MAP ALONG OUTER EDGE OF MAP. TRIM OFF FLAPS.
TEMPERATURE SCALE
LAND WATER
FAHRENHEIT
77° 68° 59° 50° 41° 32° 23° 14° 5° −4° −13° −22° −31° −40° −49° −58° −67°
CENTIGRADE
20° 10° 0° −10° −20° −30° −40° −50°
CUT ALONG THIS LINE
FOLD ALONG THIS LINE
Pacific Ocean
Atlantic Ocean
Indian Ocean
UNITED STATES
Gulf of Mexico
MEXICO
NEW ZEALAND
FRANCE
SPAIN
GERMANY
TURKEY
Black Sea
MOROCCO
ALGERIA
LIBYA
FRENCH WEST AFRICA
ATLAS MOUNTAINS
OUTER MONGOLIA
SINKIANG
TIBET
CHINA
INDIA
MANCHUKUO
SCALE:

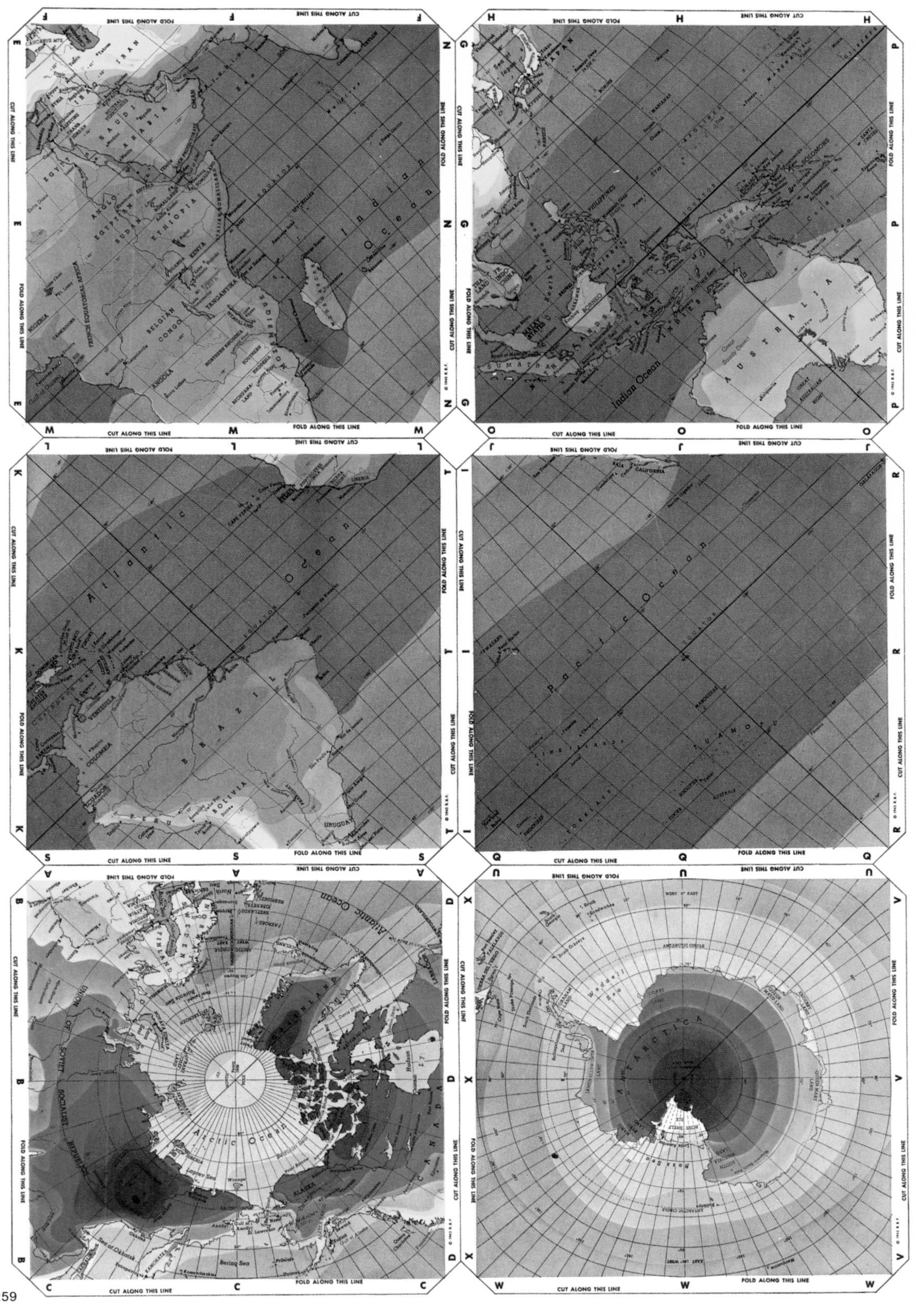
CUT ALONG THIS LINE
FOLD ALONG THIS LINE
Indian Ocean
SAUDI ARABIA
ETHIOPIA
BELGIAN CONGO
ANGOLA
PHILIPPINES
NEW GUINEA
AUSTRALIA
Indian Ocean
Atlantic Ocean
BRAZIL
VENEZUELA
COLOMBIA
PERU
BOLIVIA
URUGUAY
Pacific Ocean
TUAMOTU
Arctic Ocean
CANADA
ALASKA
Bering Sea
Sea of Okhotsk
Weddell Sea
Ross Sea
ANTARCTICA
ANTARCTIC CIRCLE

AFRICA AND ASIA

19.3% of world's land area . 11,053,500 sq. mi.
7.3% of world's water area . 10,219,500 sq. mi.
12.0% of world's population . 255,883,900 inhab.

SQUARE MILES	TERRITORY	POPULATION
2,581,300	BRITISH AFRICA	45,675,800
284,600	UNION OF SOUTH AFRICA	5,775,900
379,100	EGYPT	15,742,000
1,186,500	FRENCH AFRICA	9,964,700
911,500	BELGIAN AFRICA	10,356,200
777,300	PORTUGUESE AFRICA	3,484,000
350,000	ETHIOPIA	12,100,000
241,000	MADAGASCAR	3,797,900
1,000,000	ARABIA	7,155,300
645,000	INDIA	79,701,300
137,200	TURKEY	8,475,500
10,500	PALESTINE	1,568,700

SOUTH AMERICA

11.7% of world's land area . 6,706,000 sq. mi.
10.4% of world's water area . 14,567,000 sq. mi.
4.0% of world's population . 94,897,000 inhab.

SQUARE MILES	TERRITORY	POPULATION
3,275,000	BRAZIL	41,356,600
537,800	BOLIVIA	3,426,300
532,000	PERU	7,023,100
448,000	COLOMBIA	9,523,200
352,200	VENEZUELA	3,492,700
318,800	ARGENTINA	5,551,900
275,900	ECUADOR	3,200,000
208,800	GUIANAS	555,400
174,900	PARAGUAY	1,014,800
72,200	URUGUAY	2,146,500
59,300	CHILE	1,904,400
20,000	CARIBBEAN ISLANDS	8,646,000
392,000	AFRICA	6,154,500

NORTH POLE

17.2% of world's land area . . 9,799,100 sq. mi.
8.4% of world's water area . 11,473,900 sq. mi.
8.9% of world's population . 191,995,700 inhab.

SQUARE MILES	TERRITORY	POPULATION
4,019,900	U.S.S.R.	123,831,000
2,654,600	CANADA	1,526,000
736,500	GREENLAND	18,000
586,400	ALASKA	71,800
450,000	SCANDINAVIA	12,891,000
39,700	ICELAND	108,800
24,300	SPITSBERGEN	2,200

WARNING

DO NOT REMOVE THESE PAGES UNTIL YOU TURN

TO CENTER SPREAD AND READ THE INSTRUCTIONS

PACIFIC WAR AREA

9.6% of world's land area . . 5,477,600 sq. mi.
11.3% of world's water area . 15,795,400 sq. mi.
19.0% of world's population 402,462,800 inhab.

SQUARE MILES	TERRITORY	POPULATION
2,905,400	AUSTRALIA	6,287,400
735,200	NETHERLANDS EAST INDIES	49,343,700
382,200	CHINA	204,089,500
116,400	JAPAN	66,358,900
117,800	THAILAND	7,078,700
71,300	KOREA	22,531,000
114,400	PHILIPPINES	16,971,100
49,800	MALAY PENINSULA	5,494,300
800	JAPANESE MANDATES	102,500

PACIFIC OCEAN

0.1% of world's land area . . 63,700 sq. mi.
15.1% of world's water area . 21,209,300 sq. mi.
0.06% of world's population . 1,196,900 inhab.

SQUARE MILES	TERRITORY	POPULATION
10,700	U.S. PACIFIC COAST	694,500
42,400	LOWER CALIFORNIA	20,000
6,000	HAWAIIAN ISLANDS	423,300
2,900	GALAPAGOS	1,000
600	SOCIETY ISLANDS	21,300
400	MARQUESAS	2,400
2	PITCAIRN	140
700	OTHER ISLANDS	34,200

SOUTH POLE

8.7% of world's land area . . 5,020,000 sq. mi.
11.3% of world's water area . 16,253,000 sq. mi.
0.0004% of world's population 8,400 inhab.

SQUARE MILES	TERRITORY	POPULATION
5,000,000	ANTARCTICA	NONE
13,000	TIERRA DEL FUEGO	2,500
4,600	FALKLAND ISLANDS	2,400
1,000	SOUTH GEORGIA	400

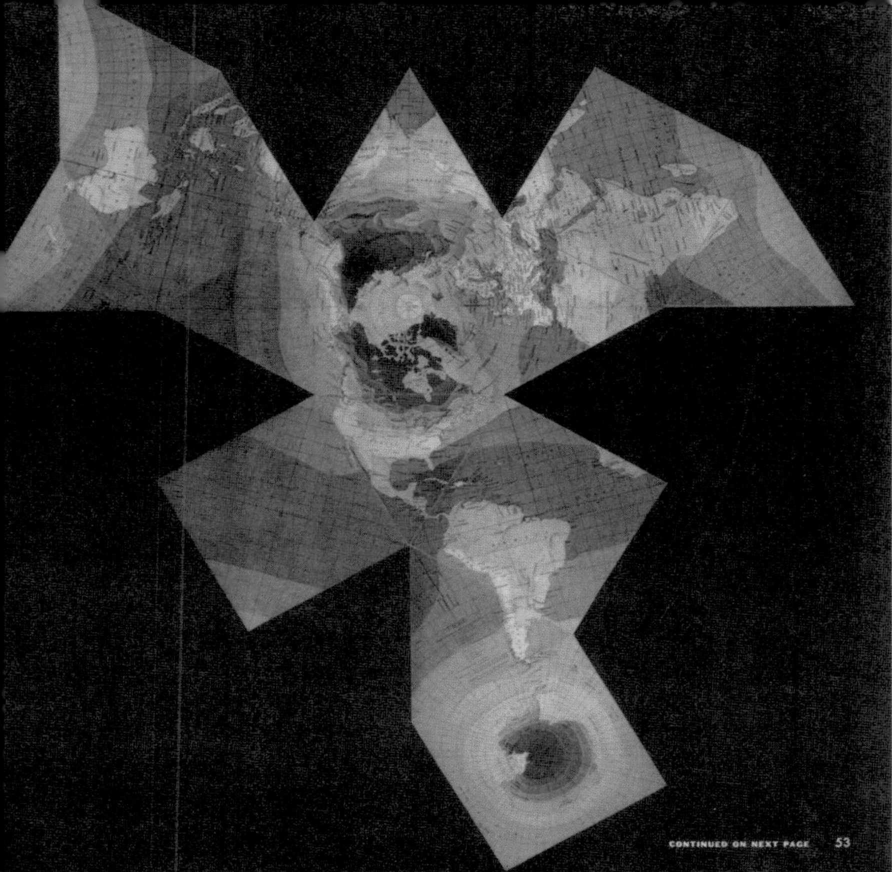

Flat map shows world in many perspectives

In its role as a flat map, the Dymaxion World holds fascination as endless as the possible combinations of its 14 segments. (...) Matched together, as below, the tiles can be arranged and rearranged to animate the facts of geography and clarify many of its obscurities. The layout may be centered on any world power, and it will suggest at once the geographical considerations that dictate its strategy and ambitions. Thus the motivations of Jap imperialism and Argentine isolationism can be spread out for inspection, and the recent revolution in the U.S. world-view can be set down in graphic narrative.

North Pole Layout of segments shows the world in new focus of air power. The U.S., Canada and Russia control almost the whole coastline of the Arctic Ocean. One of the first to recognize the strategic importance of the Arctic region, Mr. Fuller designed his first polar projection as an end paper for his Nine Chains to the Moon. (Lippincott, 1938)

Mercator World. Here the tiles are laid in a pattern that approaches the familiar appearance of the Mercator projection. The equator is a continuous line, orienting the world east to west. Not shown on the true Mercator are the poles, which appear here. The Mercator is still the best and standard base map of navigation, but its perspective is that of the 16th, not the 20th Century.

DYMAXION WORLD

(continued)

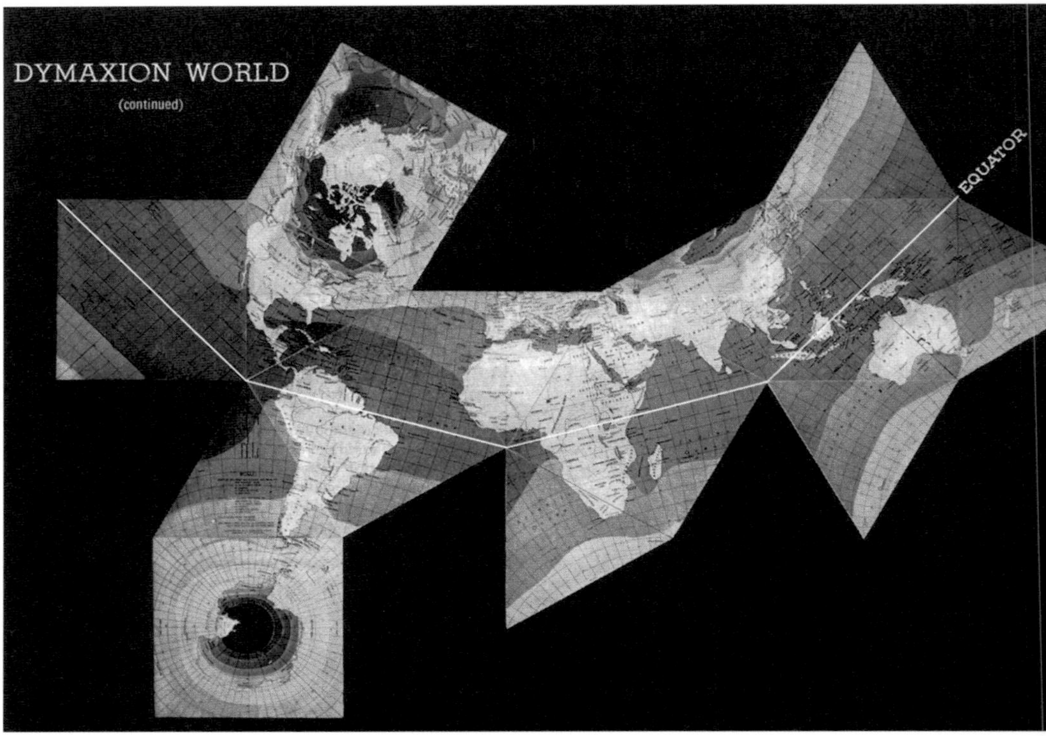

EQUATOR

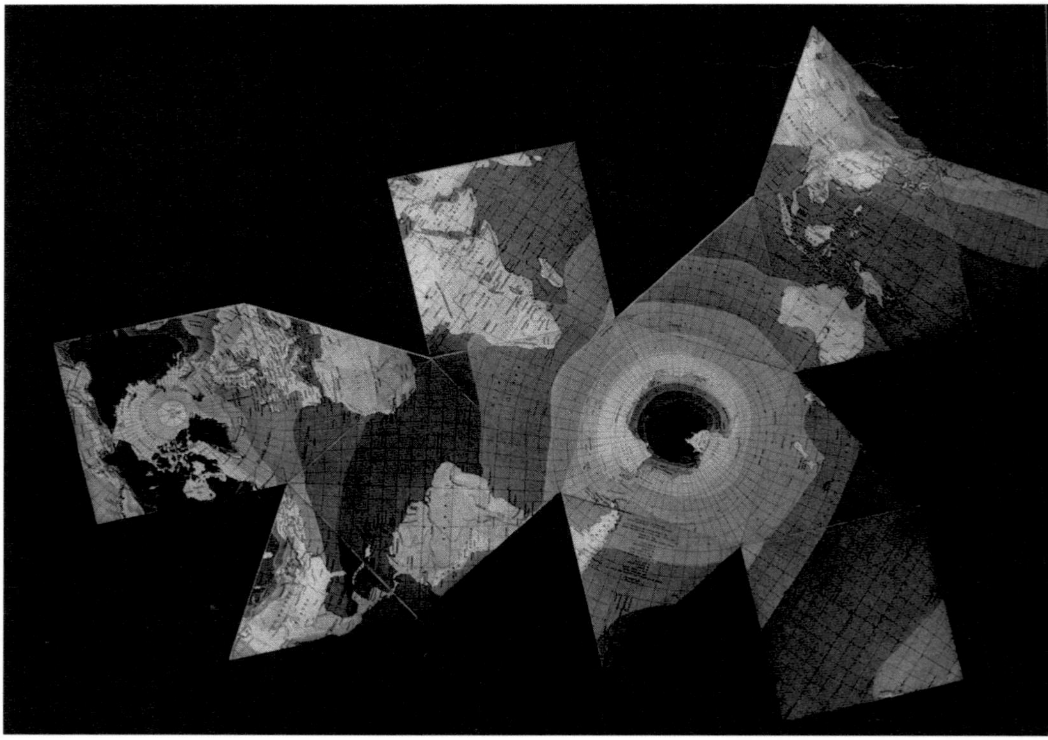

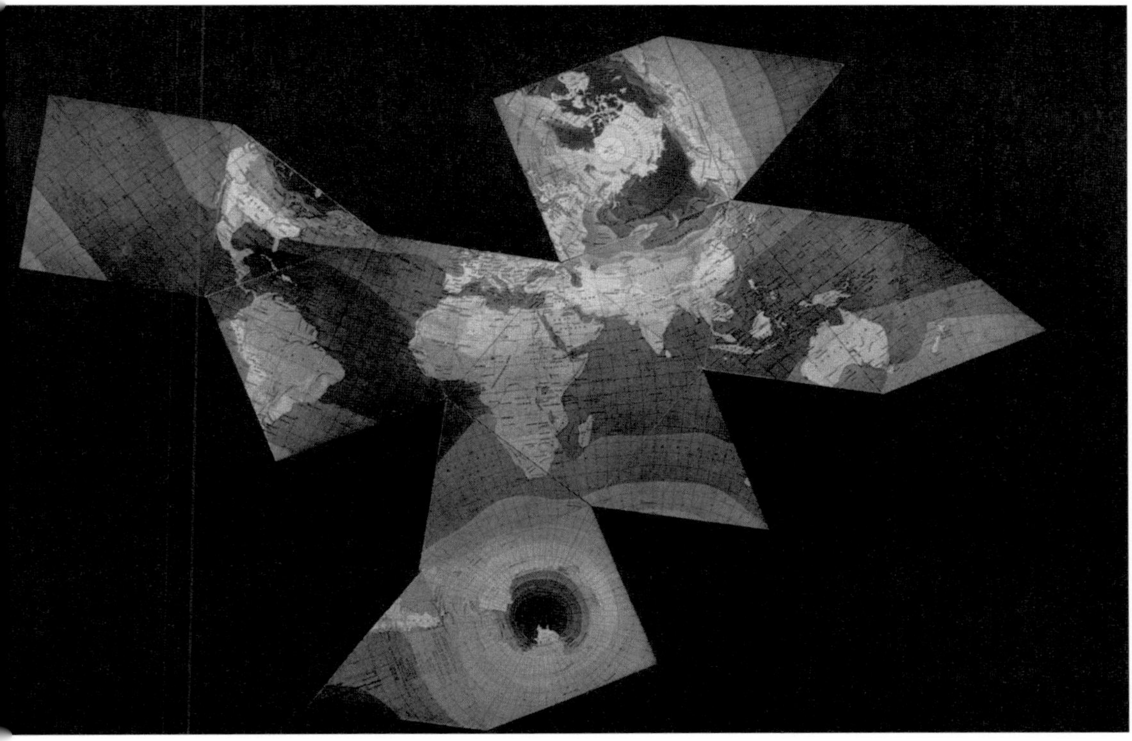

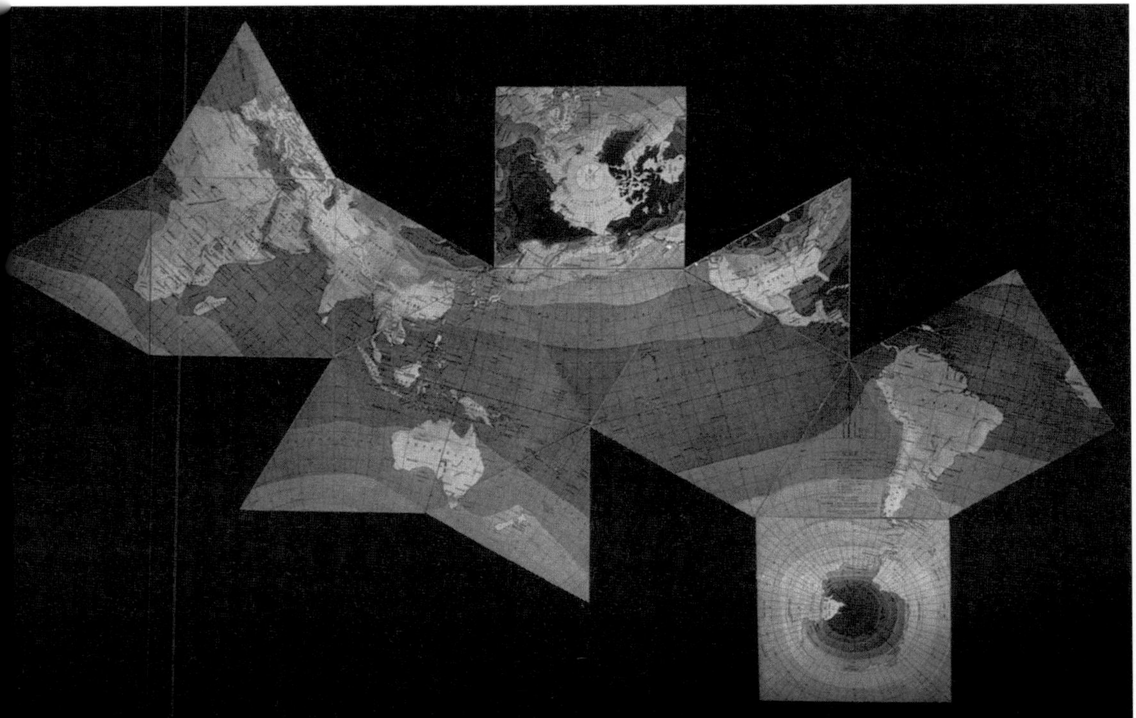

 · above left: "Mercator's World"
· below left: "British Empire"

· above right: "Hitler's Heartland Concept"
· below right: "Japanese Empire"

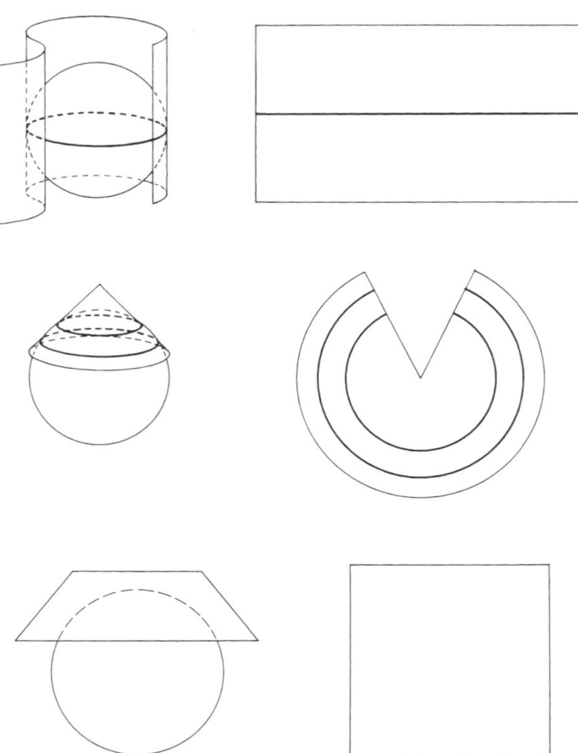

This Dymaxion map approximation of the Mercator projection brings the character of the Renaissance world into bold relief. Most striking is the vast expanse of ocean, the world's highway. Though a few hardy explorers hunted a Northwest Passage, the course of empire and trade was southward along the coast of South America and eastward around Good Hope into the Orient. It was in this period that the Dutch cartographers and Papal fiat divided the world in hemispheres.

British Empire: The British Empire was built by generations of mariners in the Royal Navy and merchant marine who knew not only ships but oceans. This layout of the Dymaxion map is not so much a picture of what they accomplished as it is a demonstration of how they did it. They found the secret to power where no one else thought to look – on the bottom of the world.

Due primarily to the rotation of the earth, the prevailing winds and ocean currents around the Antarctic region move from west to east, clockwise on the map. Sailing southward around Good Hope, British ships moved into the winds and currents as onto a turntable. From the turntable the waters opened out to the Indian peninsula, the Indies, the Pacific islands and the western coast of the Americas. Suez, after 1869, moved the lifeline northward through the Mediterranean Sea.

Heartland. Central Eurasia is the Heartland, the prize for which the German Army has once more marched against the world. It is a concept first expounded before World War I

·Schematic drawing of the Mercator projection (above), the cone projection (middle), the Azimuthal Projection (below)

by a British, Sir Halford Mackinder, who envisioned the decline of seapower before terrain-covering mechanized land armies. For land-bound Germany it was a ready-made program for World War II. The Nazis inflated it into something called "geopolitics," and set out to make it a terrible reality.

The world of geopolitics, as shown here, consists of one big continent on which hang the peninsulas of Europe and Africa. Australia, North America and South America are unimportant islands. The oceans, separate basins on map, lose strategic significance. "He who controls the Heartland controls the world," because he has outflanked seapower by capturing all its bases.

Jap Empire. The ruthless logic of the Jap imperialism is exposed by this layout of the Dymaxion World map. The seapower-minded Japs want nothing less than the Pacific. Control of the ocean and its shoreline, they calculate, will give them control of the hinterlands. Their thinking strikes an obvious contrast to the landlubber geopolitics of their German allies (above).

The Japanese can boast that they are closer to their goal than the Germans. Already they are masters of the western shores of the Pacific, from their northern toe hold in the Aleutians to their outposts in the Solomons. Consolidated there, they would be ready for the attack on North America as predicted by Ambassador Joseph C. Grew. On our side of the Pacific they saw foundations of their dream laid by colonies and trade on the U.S., Mexican and South American coastline.

Text in *Life*, 1943

267 ·Mercator shadow projection system: true-to-scale … the poles, illustrated using human figure (based on ideas

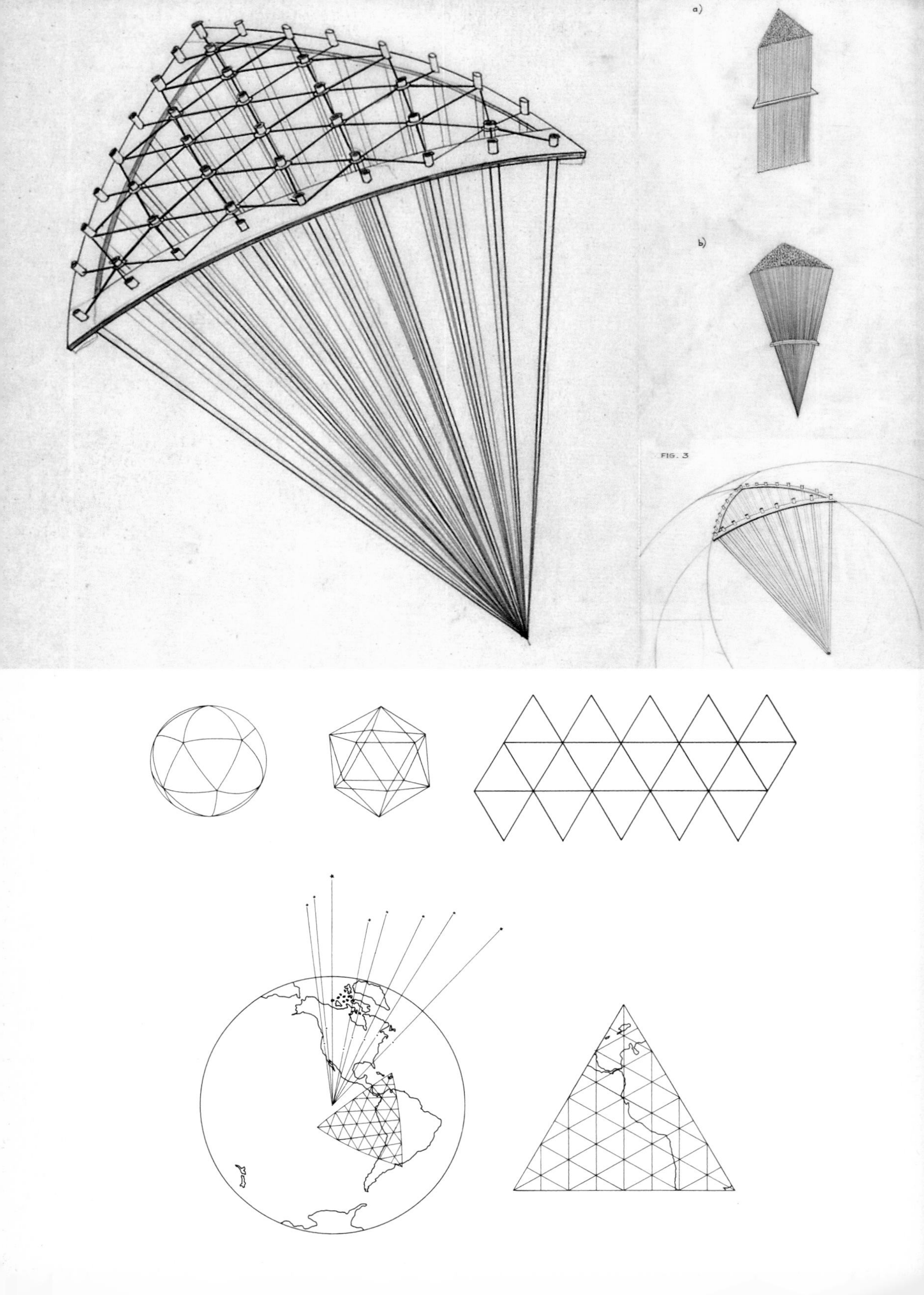
a)
b)
FIG. 3

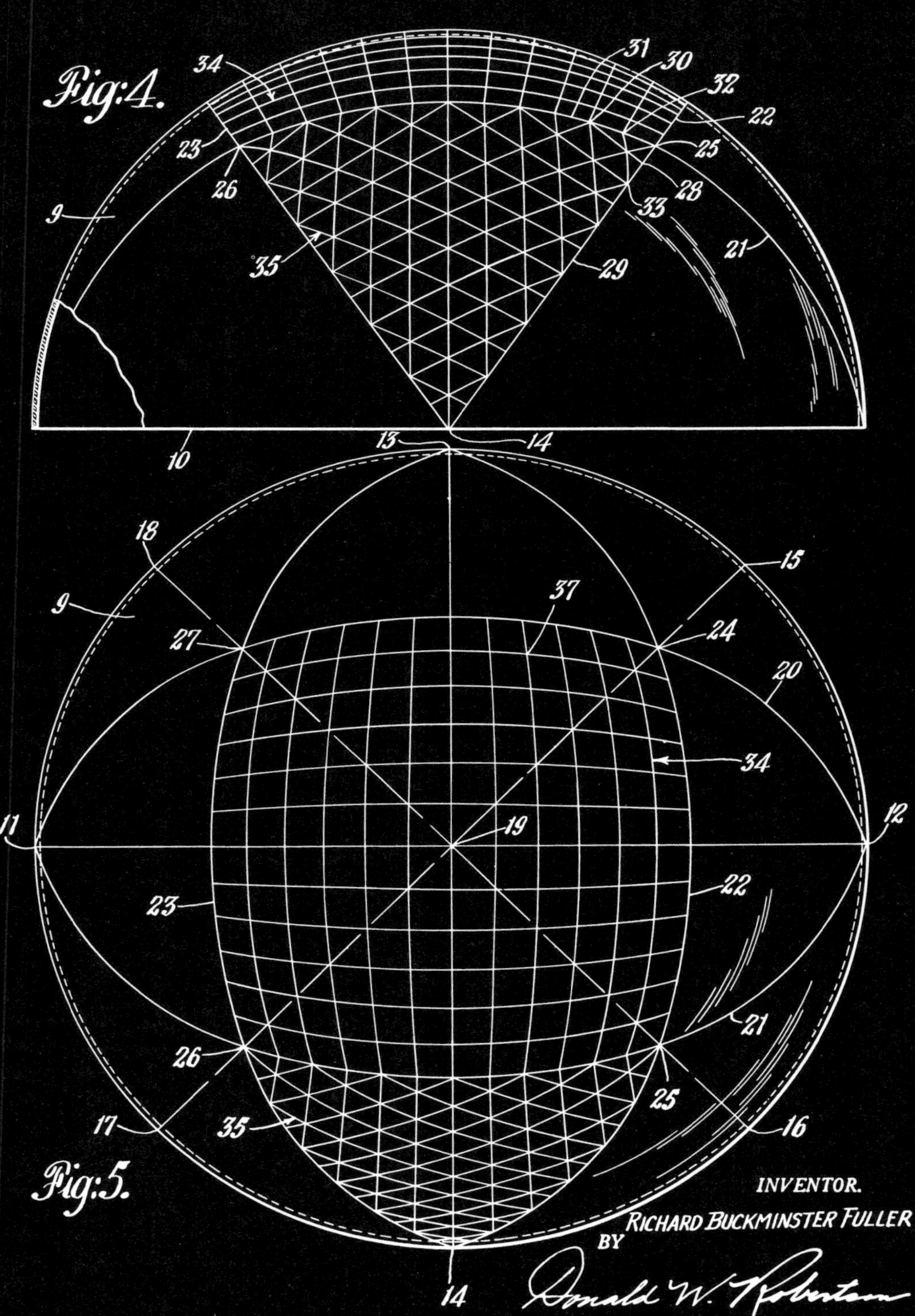

269 · Illustration from the Dymaxion Projection patent application: the discovery of the 3-way-grid, 1944

Construction of Dymaxion World Map

People are learning that "via the North Pole" is the shortest great-circle distance from America's midst to the center of population of the world. But when they were told during the war that Tarawa represented the first major gain in the direction of Tokyo, they were not well enough versed in their geography to realize that announcement that the Marines had taken the North Pole would have put the United States closer to Tokyo's center, and that the Marines were actually farther from Tokyo than Chicago is from London. And even those professional geographers and military tacticians who did know that these were the proportionate distances by great circle, have not, unless skilled in the dynamic sailorman's thinking, realized that in the terms of the air motions which twist the great circle courses all out of shape, the North Pole is a third nearer to Tokyo than Tawara, when full advantage is taken of atmospheric motions.

Because of this latter fact, it becomes obvious that the kind of map that static cartographers produce can only partially educate the people – that is, up to the realistic great-circle concept. All known traditional projection methods fall far short of providing a comprehensive, sectional cartographical device which may be mutably arranged in any direction in such a manner as to bring focus to bear on any of the dynamic inter-relationships of the world's surface affairs.

One such sailorman's mid-twentieth-century invention is the mutable map presented here. Appropriately, its faces represent the facets of the monometric construction shown in the accompanying drawing, which drawing represents the vector equilibrium of a sphere.

Unlike all known preceding projections, which represent transfers of spherical data to plane surfaces tangent to the sphere only at one point, as in the case of the azimuthal or gnomonic projection methods, or only along one central line, as in the case of the Mercator, or only along one

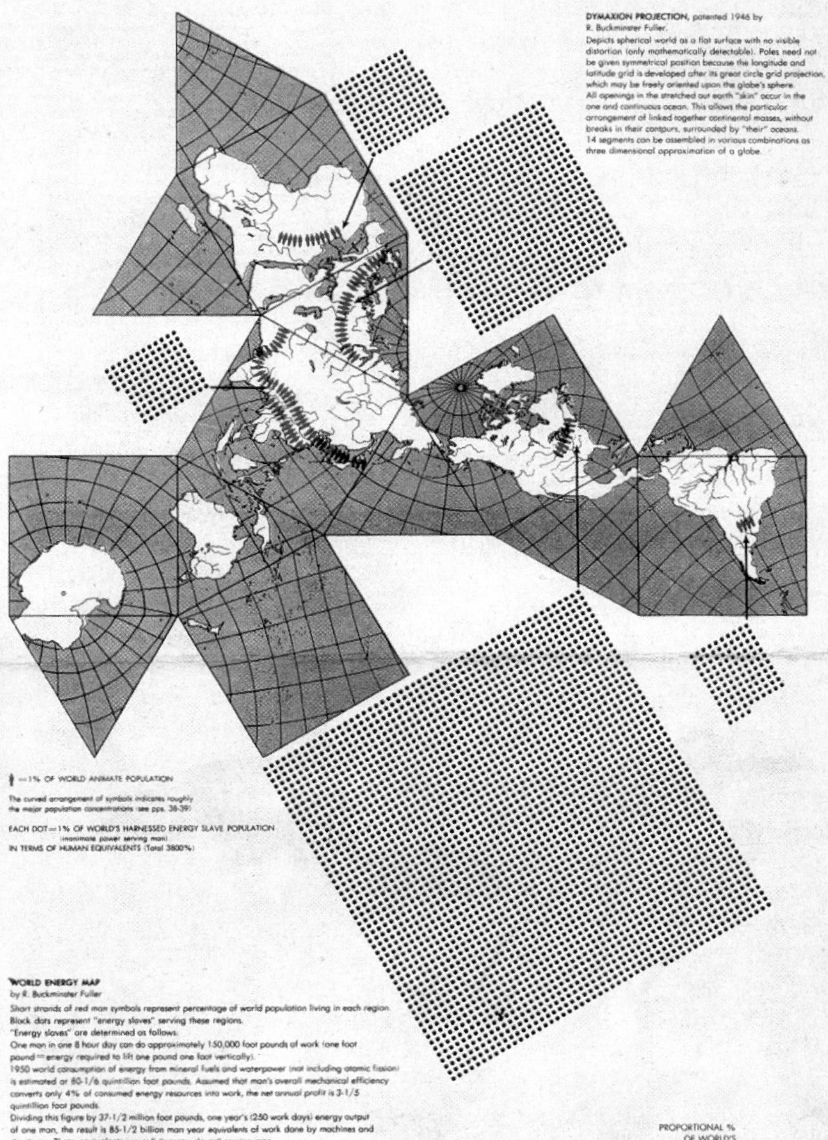

DYMAXION PROJECTION, patented 1946 by R. Buckminster Fuller.

Depicts spherical world as a flat surface with no visible distortion (only mathematically detectable). Poles need not be given symmetrical position because the longitude and latitude grid is developed after its great circle grid projection, which may be freely oriented upon the globe's sphere.
All openings in the stretched out earth "skin" occur in the one and continuous ocean. This allows the particular arrangement of linked together continental masses, without breaks in their contours, surrounded by "their" oceans.
14 segments can be assembled in various combinations as three dimensional approximation of a globe.

= 1% OF WORLD ANIMATE POPULATION

The curved arrangement of symbols indicates roughly the major population concentrations (see pps. 38-39)

EACH DOT = 1% OF WORLD'S HARNESSED ENERGY SLAVE POPULATION (inanimate power serving man) IN TERMS OF HUMAN EQUIVALENTS (Total 3800%)

WORLD ENERGY MAP
by R. Buckminster Fuller

Short strands of red man symbols represent percentage of world population living in each region.
Black dots represent "energy slaves" serving these regions.
"Energy slaves" are determined as follows:
One man in one 8 hour day can do approximately 150,000 foot pounds of work (one foot pound = energy required to lift one pound one foot vertically).
1950 world consumption of energy from mineral fuels and waterpower (not including atomic fission) is estimated at 80-1/6 quintillion foot pounds. Assumed that man's overall mechanical efficiency converts only 4% of consumed energy resources into work, the net annual profit is 3-1/5 quintillion foot pounds.
Dividing this figure by 37-1/2 million foot pounds, one year's (250 work days) energy output of one man, the result is 85-1/2 billion man year equivalents of work done by machines and structures. These equivalents we call "energy slaves" serving man.

$$\frac{\text{85-1/2 billion energy slaves}}{\text{2-1/4 billion world population}} = \text{38 energy slaves per capita}$$

Note

The atomic energy resource consumption during this period in various countries is not available but would probably tend to increase even further the present disparity of respective world energy advantages.
Also note that energy slaves are not confined to narrow range of physical conditions limiting man's activities for they can work "comfortably" anywhere between absolute zero and 5,000°F., at submicroscopic precision and at speed of 186,000 miles per second.

	A % OF WORLD POPULATION 1950	ENERGY SLAVES POPULATION 1950	% OF WORLD'S ENERGY SLAVES 1950	PROPORTIONAL % OF WORLD'S ENERGY SLAVES in terms of A and human equivalents (as shown on map) 1950	ENERGY SLAVES PER HUMANS PER AREA (in round numbers) 1950
ASIA	50	2,565,000,000	3	114	2
EUROPE	24	14,535,000,000	17	646	27
AFRICA AND MEDIT. WORLD	12	3,420,000,000	4	152	13
NORTH AMERICA	8	62,415,000,000	73	2774	347
SOUTH AMERICA	4	2,565,000,000	3	114	28
CENTRAL AMERICA	1	0	0	0	0
ALL OTHERS	1	0	0	0	0
	100%	85,500,000,000	100%	3800%	

or two segmental arc lines, as in the conics, this projection is one in which the coincidence with the projected sphere occurs all along the complete boundary of each section of the projection, thus retaining the unique cartographical feature of being the only projection in which uniform great-circle scale characterizes the logical terminal edge of each section to be projected, that scale being maintained intact after transfer from the sphere to the flat surface of the map.

Sailorwise, this new projection is made from the cosmic viewpoint – that is, the astronomical zenith and the center of the earth always remain, respectively, vertically above and below each and every point of the surface of the cartographic data. Not only is this true in its spherical arrangement but also in its planar projection into the sections of the comprehensive map.

Because the enclosing border scale cannot be elongated, distorted or contracted, and represents a great circle bent flat into a one-dimension line, the adjustment of the contained spherical surface segment to a plane surface segment must be satisfied by interior contraction of the data instead of by exterior stretching, as in all other methods of projection. Because of this feature the several pieces fit neatly together, being the mutual sides of adjacent polygons and being separated by the same great circle or straight line. Because the area of a circle increases as the square of its radius, the same error outwardly disposed must be distorted to four times greater extent than by inward disposition.

The segmentation of the earth's surface into eight triangles and six squares is not in any sense a matter of esthetic choice. It represents one known subdivision possible by means of this universal-projection viewpoint, for the radial and chordal lengths must be identical in order to allow this symmetrically hinged opening of the sphere.

Having six axes and four dimensions, its parts may be rearranged to unpeel the globe continuously in all directions. Gnomonic projections through the surface facets of any of the regular polyhedrons will serve to provide a variety of sectional world surface maps. Striking an optimum between angular contraction and numbers and sizes of pieces, an icosahedron is the least distorted, for these projection purposes, of any of the regular solids. However, the spherical vertexes of the icosahedron's twenty triangles must be reduced from 72° to plane triangular vertexes of 60°, a reduction of twenty per cent, which percentage times the number of pieces gives the total distortion. On the other hand, the Dymaxion's fourteen pieces accomplish translation with a distortion of only sixteen per cent, the Dymaxion's spherical triangles being only 70° and the vertexes of the square bearing the same percentage relationship between the spherical and the plane figure.

The Dymaxion projection method of transferring spherical data to the plane surface is extremely simple. Because of its universal viewpoint, it need deal only with the surface of the sphere and the plane surface of the map. A lattice of four great circles is formed about the sphere, each intersecting the other in such a manner as to subdivide each circle into six symmetrical arcs of 60°. This lattice provides fourteen spherical surface areas, eight of them triangular, six of them quadrangular. These spherical triangles and squares are equilateral. The surfaces of these triangles and squares are then interwoven with great circle grids, the triangles by a three-way grid of great circles and the squares by a two-way grid of great circles. These grid lines spring from uniform scale modular subdivisions of the 60° arc sections in as fine a degree as is desired.

The geographical data coinciding with these grids is then transferred to eight equilateral plane triangles and six plane surface squares. For purposes of this translation, the plane triangles and squares have been prepared as follows: Their surfaces are subdivided by respective three-way and two-way grids of straight lines. These straight lines spring from modular subdivisions of their boundaries which correspond in scale and number to the subdivisions of the original spherical arc segments of 60°. The spherical geographical data is then posted to corresponding positions in the appropriate plane grid spaces. The spherical great circle grids are thus treated as constituting straight lines in the plane geometrical surface. The principle of treating great circles and straight lines as constituting one and the same thing effects the distribution of the angular contraction in a concentric disposition on the plane sections.

The key to understanding why this method accomplishes its translation with a minimum of distortion is that it treats the 180° spherical gores in two ways. The irreconcilable conditions of convergence and parallelism, characterizing the terminals and mid-part of the 180° gores, are treated separately and symmetrically as triangles and squares respectively. All other projections impose the advantage of one feature against the advantage of the other by trying to solve both convergence and parallelism by one grid. These resolved gore parts of the Dymaxion map, by treating these conditions separately, allow four-dimensional unwrapping of the sphere.

Fluid Geography (1946), in Ideas and Integrities (1963), 137

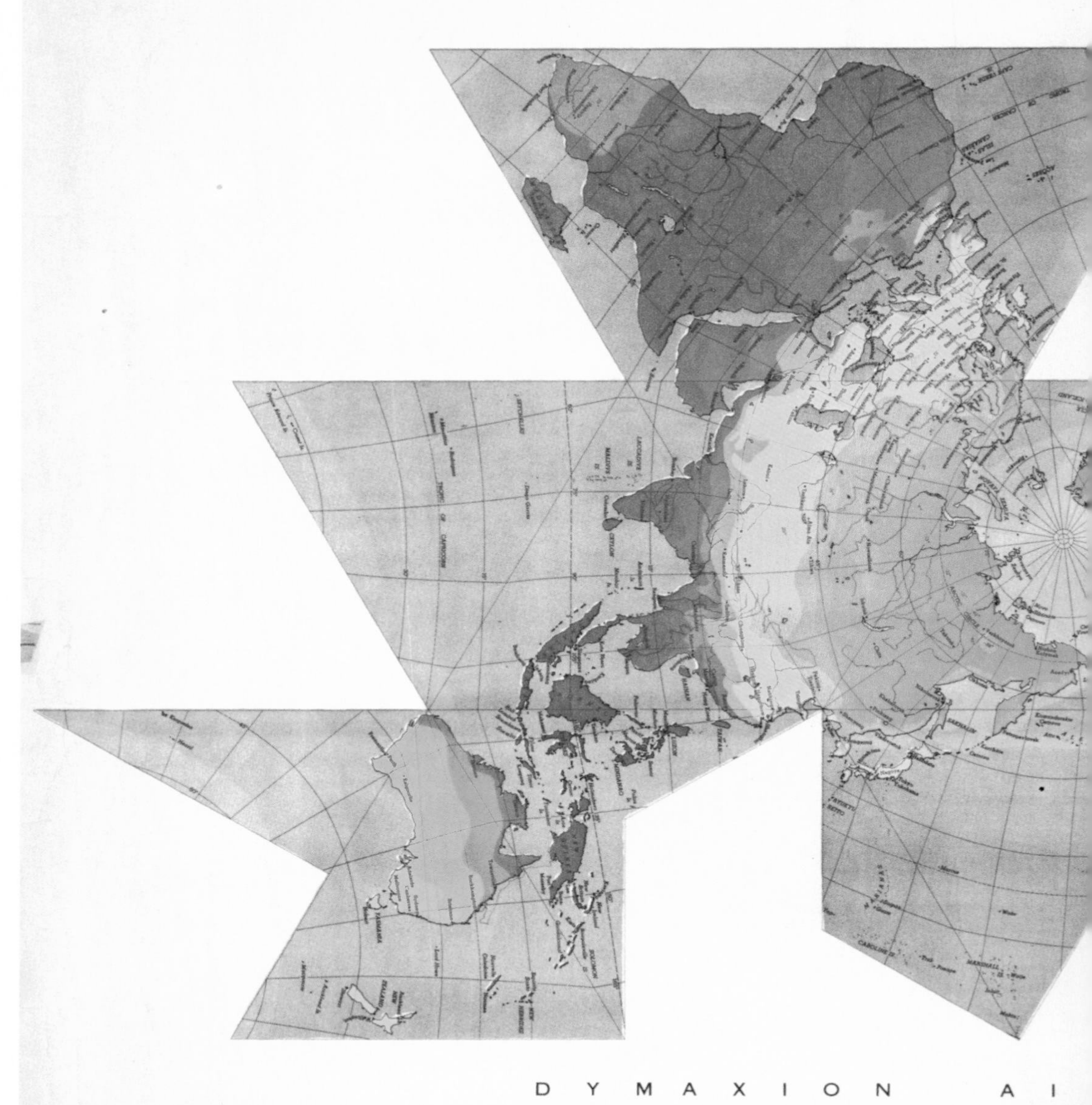

D Y M A X I O N A I

N. C. STATE COLLEGE SCHOOL OF

R. BUCKMINSTER FULLER &

PUBLISHED BY STUDENT PUBLIC

COPYRIGHTED 1952

TEMPERATURE	LAND / WATER	−58° F −50° C	−49° F	−40° F −40° C	−31° F	−22° F −30° C	−13° F	−4° F −20° C	5° F

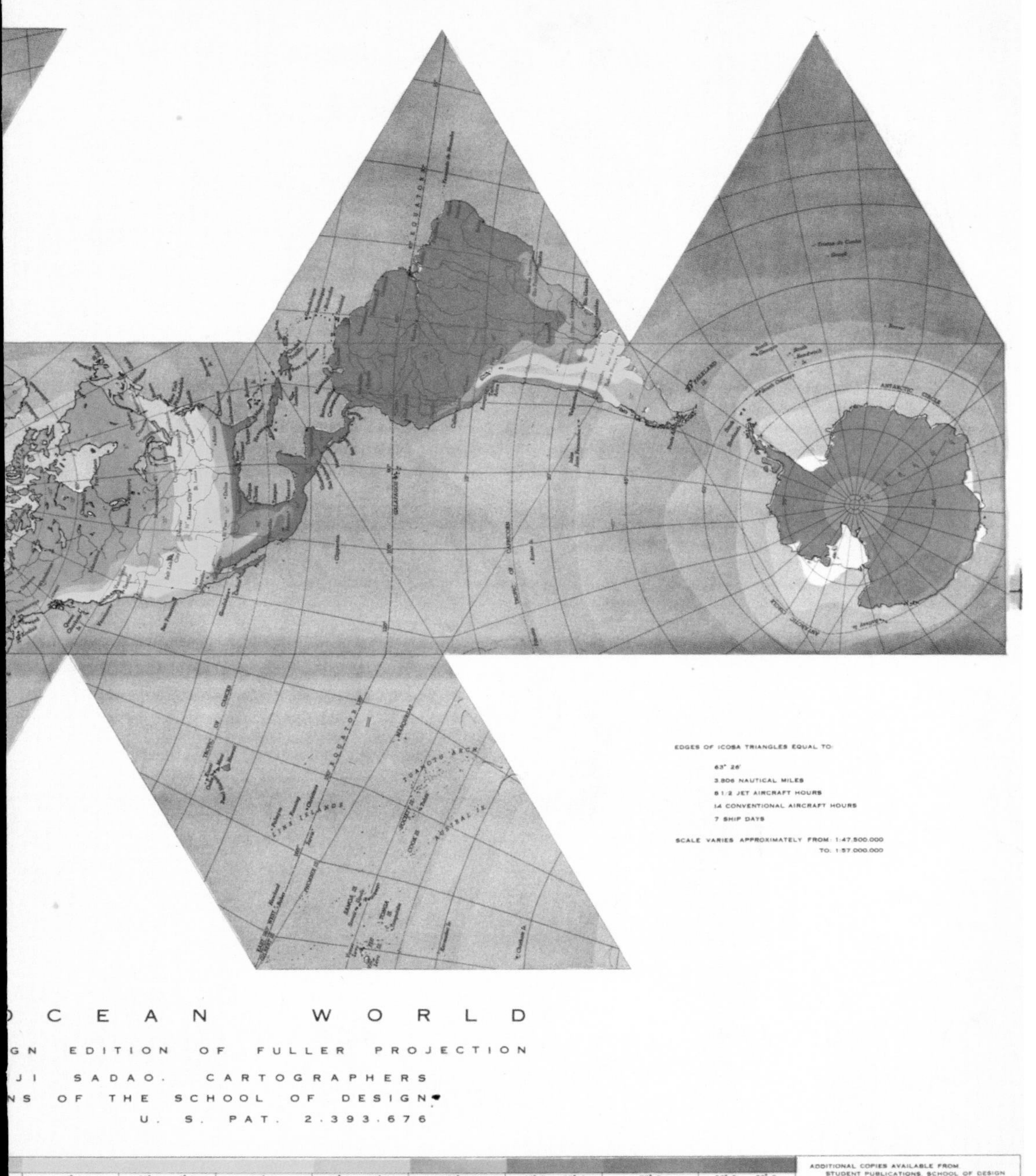
EDGES OF ICOSA TRIANGLES EQUAL TO:
63° 26'
3.806 NAUTICAL MILES
8 1/2 JET AIRCRAFT HOURS
14 CONVENTIONAL AIRCRAFT HOURS
7 SHIP DAYS
SCALE VARIES APPROXIMATELY FROM: 1:47.500.000
TO: 1:57.000.000
OCEAN WORLD
GN EDITION OF FULLER PROJECTION
JI SADAO. CARTOGRAPHERS
NS OF THE SCHOOL OF DESIGN
U. S. PAT. 2.393.676
23° F
32° F 0° C
41° F
50° F 10° C
59° F
68° F 20° C
77° F
86° F 30° C
ADDITIONAL COPIES AVAILABLE FROM
STUDENT PUBLICATIONS, SCHOOL OF DESIGN
BOX 5273
RALEIGH, NORTH CAROLINA

Since about 1943, Fuller was working increasingly toward his own form of research on nature. Self-taught and without much respect for the specialized research of academia, he developed his own questions about the material world. He saw himself as a follower of Plato, Aristotle, Kepler, Euler, and Newton. He sought a "geometry as Nature uses it" and discovered the mathematical formula for the closest packing of spheres. His discovery of the "jitterbug" transformation that same year, 1948, was epoch-making. It states that regular geometric bodies do not stand statically next to one another but they are subjected to various phases – tetrahedron, octahedron, icosahedron, cuboctahedron – of a process of mutual transformation. Insights like these seem not to have been imparted or put in perspective for some time afterward. In some cases, confirmation of their significance came only after Fuller's death.

FULLER RESEARCH

Closet Packing of Spehres

Comprising a finite system, seven wires, when compacted into a parallel bunch, always twist to form a cable of hexagonal cross section, comprised of one nuclear wire at the center, symmetrically surrounded by the other six wires... and, also demonstrating a finite system in universal geometry, 12 spheres of equal diameter will always collect in tight assembly around one sphere of the same diameter, with each sphere tangent to all of its respectively adjacent spheres. Pursuing this concept of a finite system in universal geometry still further, in order to relate it to the significance of the 92 elements, we apply another layer of spheres of identical diameters to those of the original 13, thus completely surrounding the original 13 in the most compact, natural positions. The spheres of this outer layer are also tangent respectively to all adjacent spheres and therefore tightly and symmetrically packed. This second layer totals 42 spheres. If we apply still another layer of equal-dimension spheres, we find that they, too, compact symmetrically and tangentially. The number of spheres in this outer layer is 92.

If we add on more layers, we find that they always compact symmetrically and tangentially, and that this process of enclosure may seemingly be repeated to infinity. However, each layer is in itself a finite or complete and symmetrical enclosure of spheres. These enclosing layers of spheres constitute finite systems, and together with the layers they enclose and the final single sphere center (or nucleus), constitute a concentric, finite system.

As many layers are added, however, it is found that a symmetrical pattern of concentric systems begins to repeat its dimensions in all directions. That is, the systems of three layers around one sphere, with 92 spheres in the outer layer, repeat themselves infinitely in such a

·277: Geometry Laboratory: RBF with assistant fashioning a model. Studio in the Chicago Institute of Design, 1948

·Glass sphere model demonstrating the closest packing of spheres, basic forms of energetic-synergetic geometry: triangle,

WHOLE NUMBER ...

BASIC RELATIONSHIP

COMPREHENSIVE SYS...

PARTICLE GROWTHS,

= RATE OF GROWTH OF HEXAGON OF BALLS IN ONE LAYER

$$= 6 \times \frac{N^2+N}{2} + 1$$

$$= 3(N^2+N)+1$$

WHERE CENTER ROW, OR BALL = 0

THIS SAYS THAT TWO ADJACENT LAYERS OF HEXAGONAL BALL GROUPS ONE LAYER OVER THE OTHER ...

			81920	$16^3 \times 20$
			67500	$15^3 \times 20$
			54880	$N^3 \times 20$
			43940	$13^3 \times 20$
			34560	$12^3 \times 20$
			26620	$11^3 \times 20$
			20000	$10^3 \times 20$
			14580	$9^3 \times 20$
			10240	$8^3 \times 20$
			6860	$7^3 \times 20$
			4320	$6^3 \times 20$
			2500	$5^3 \times 20$
			1280	$4^3 \times 20$
			540	$3^3 \times 20$
			160	$2^3 \times 20$
			20 × 160	$1^3 \times 20$
	1	8		$[(0 \times 12)+2] \times 10$
26	7		140	$+1 = [(1 \times 12)+2] \times 10$
	19	56	380	$+2 = [(3 \times 12)+2] \times 10$
98	37		740	$+3 = [(6 \times 12)+2] \times 10$
	61	152	1220	$+4 = [(10 \times 12)+2] \times 10$
218	91		1820	$+5 = [15 \times 12+2] \times 10$
	127	296	2540	$+6 = [21 \times 12+2] \times 10$
386	169		3380	$+7 = [28 \times 12+2] \times 10$
	217	488	4340	$+8 = [36 \times 12+2] \times 10$
602	271		5420	$+9 = [45 \times 12+2] \times 10$
	331	728	6620	$+10 = [55 \times 12+2] \times 10$
866	397		7940	$+11 = [66 \times 12+2] \times 10$
	469	1016	9380	$+12 = [78 \times 12+2] \times 10$
1178	547		10940	$+13 = [91 \times 12+2] \times 10$
	631	1352	12620	$+14 = [105 \times 12+2] \times 10$
	721		14420	$+15 = [120 \times 12+2] \times 10$

this no = this no when ÷ 20

...UTH i.e. EACH BALL $= \frac{\sqrt{R}}{10}$ EACH DYMAXION $= \frac{\sqrt{R}}{6}$ = AS $\sqrt{R} = 1$ ...

way that the successive layers outside of the 92-sphere layer begin to penetrate the adjacent systems, and that only the concentric system of spheres within and including the layer of 92 are unique and individual systems.

It will be found that the number of spheres in any complete layer around any nuclear group of layers always terminates with the digit 2; first layer 12, second layer 42, third layer 92, fourth layer 162, fifth layer 252, sixth layer 362, and so on to infinity. The digit 2 is always preceded by a number which corresponds to the "square" of the number of layers surrounding a nucleus. The third layer's number of 92 is comprised then of the number 3 multiplied by itself ("squared") with digit 2 as suffix. This third layer is the outermost of the symmetrically unique nuclear system patterns, and may be identified with the 92 unique atomic systems, and with the 92nd element, uranium. The geometrical form thus most compactly developed is not that of a composite sphere, but is always a polyhedron of 14 faces, comprised of six squares and eight triangles, and 12 points extending in tangential radius from the original 12 spheres surrounding the nucleus.

These successive layers, which permeate each other in all directions, may be identified with energy waves radiant in all directions from a nucleus. The volumetric measurement of any one wave is (variously) determined by the "square" of the radius multiplied (constantly) by the nuclear 10, plus (constantly) the nuclear biological unity: 2. Ergo Einstein's formula is E (energy) equals M (for the particular mass whose relative value is being measured) multiplied by – or comparisoned to – C^2 (where C is the speed of radiation in any one direction – radially – squared).

Earth Inc. Appendix 1 (1948)

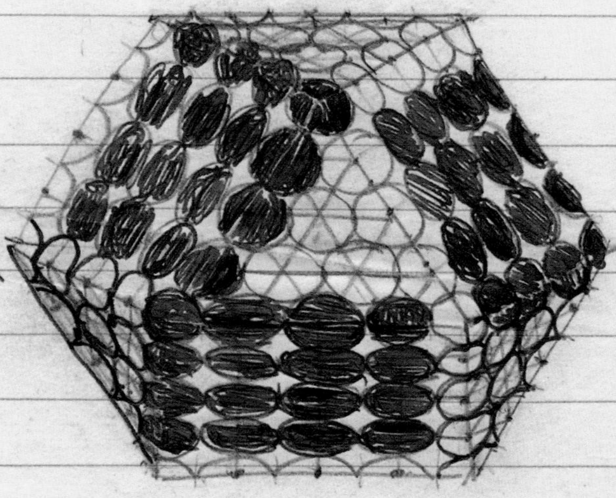

See Good drawing in color next dated — 1/10/48

3 3/8

27/8 > 13.5/8 > 6.75/8

Jan 9th 48

7/8 > 27/16 > 27/32

23/16 > 11.5/16 > 5.25/16

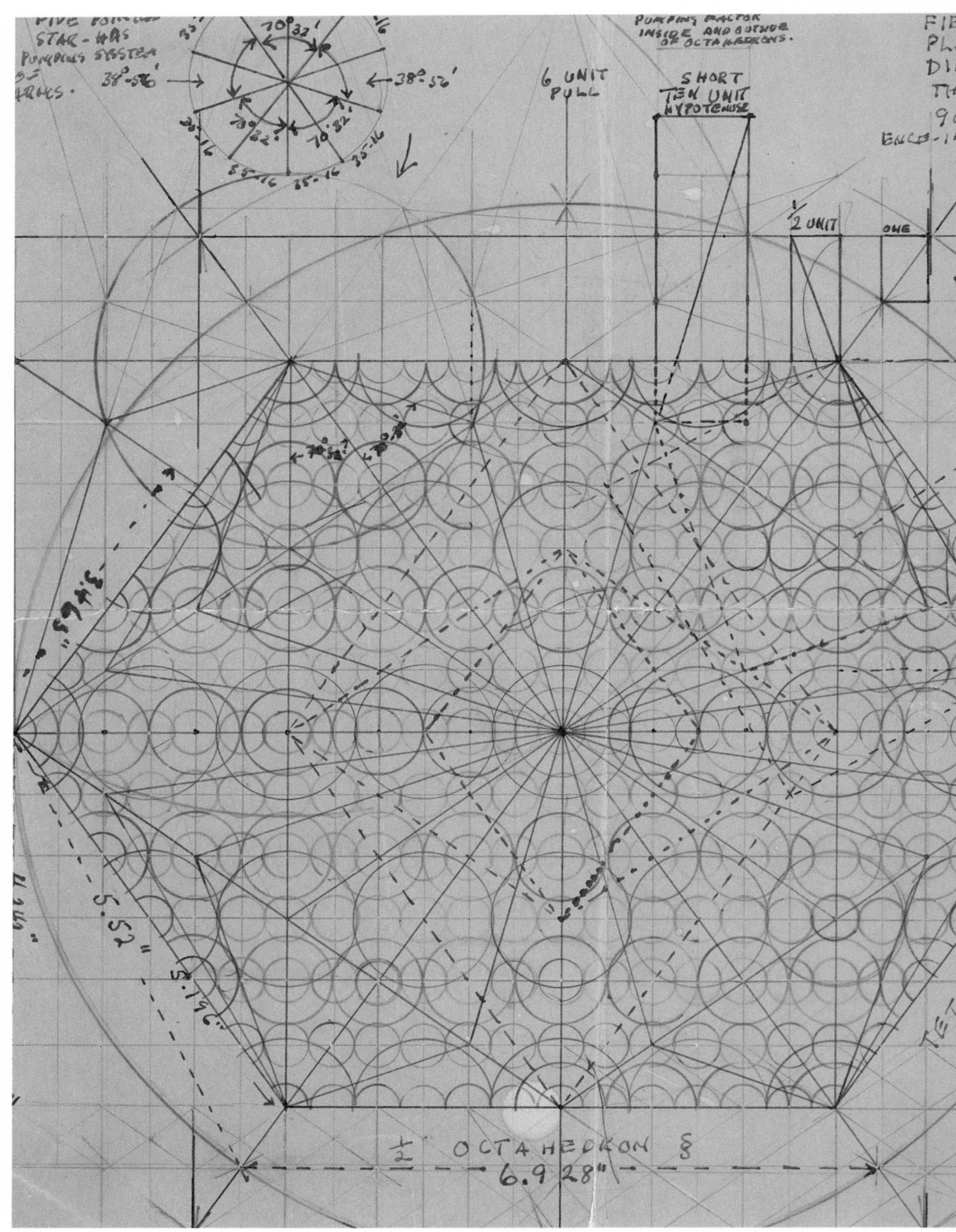
STAR - HAS
PUMPING SYSTEM
38°-56'
70° 32'
← 38°-56'
70° 32'
6 UNIT
PULL
PUMPING FACTOR
INSIDE AND OUTSIDE
OF OCTAHEDRONS.
SHORT
TEN UNIT
HYPOTENUSE
½ UNIT
ONE
3.463"
5.52"
5.196"
½ OCTAHEDRON §
6.928"

DYMAXION KINETIC VECTOR STRUCTURE

R Buckminster Fuller
Dec 15th 1947.
6 Burns St.
Forest Hills.
N.Y.C.

S DOUBLED IN
GREAT CIRCLE S)
ONS OF ITS COORDINATES
ITS AXIS IS ROTATED
ITS VOLUMETRIC INFLU-
ASED EIGHT FOLD
TRUCTURE ALLOWS IN
MANNER FOR
CESSIONAL ROTATION
TO APPLIED LINE OF

ONSHIP ... THESE ANGLES HYPOT ... PROVIDES OCTAVE EXPANSION OR CONTRACTION

NOTE THAT THE DIAMOND INVERSION INCREASES ULTIMATELY TO REACH NUCLEUS OF ADJACENT SYSTEM. BUT PROVIDES NO GRAVITATIONAL PATH TO CENTER OF PRIMARY NUCLEUS.

TWO UNIT WAVE LENGTH

NOTE THAT SAME 1:1.414214 QUADRANGLES MAY BE FORMED ON ANY OF THESE HYPOTENUSES CONSIDERED AS LEGS BY JOINING CENTERS OF SPHERES THUS PROVIDING MEANS OF ROTATION

THE LONG HORIZONTAL 24 BALL AXIS OF SPHERES IN CLOSE OR TANGENT ALIGNMENT IS DIAMETER OF DYMAXION POTENTIAL VECTOR STRUCTURE. WHOSE RADIUS WAS UNITY (HERE REPRESENTED BY 12 6-ON 4) FORMULA - DVF = VECTOR RADI OF DYMAXION (BOTH POTENTIAL AND KINETIC AND $\left(\frac{DVB}{12}\right)^2 + \left(\frac{DVB}{12} \times 1.414214\right)^2 =$ WAVE LENGTH

FOUR UNIT HYPOT

NOTE THAT WHILE THE HYPOTENUSE GROUPINGS ARE .5 - 1 - 2 - 4 - 10 - 20 THAT THE ALIGNMENT GROUPINGS ARE ALL IN 3 - 6 - 12 s

LONG TEN UNIT HYPOTENUSE

NOTE CONSTANT RECURRENCE O THE 10 AND 20 FACTOR IN ATOMIC STRUCTURE ... STRUCTURE $\frac{1}{\text{MASS OF PROTON}} = 20 \times 92$ SPEED OF ELECTRON $\frac{1}{10}$ C

12 UNIT PUSH

CONSTANT PROPORTION

A : B = B : 2A = 2A : 2B = 2B : 4A = 3A : 3B = 3B :

A B
A A B
A A B B
A A A A B B

SECTION THROUGH EQUATOR OF ANY OF SIX IDENTICAL AXIAL ROTATIONS OF DYMAXION SHOWING METHOD OF OCCORENCE OF SPHERES ON CENTERS OF GRAVITY AND AT JUNCTURES OF A. B. C. + D PARTICLES - IN A TIGHT TANGENTIA ASSEMBLAGE OF SPHERES IN ALL DIRECTIONS.

WAVE LENGTH DERIVED FROM UNIT MASS OF A.B.C.+D PARTI TO ... CON

SHORT TWENTY UNIT HYPOT. 4A WIDE 5B HIGH

1"

6.928"
5.52"
5.196"
4.242"
3.463"
1.225 x WAVE LENGTH 1.225"
3.675"

NOTE THAT THERE ONLY TWO COMPLETE SETS OF THREE CONCENTRIC RINGS BESIDE THE CENTER

L BY PNEUM... PUMP
...ULIC CENTRAL AND
...OUGH ...IDE TUBES

STATIC DYMAXION BECOMES SPHER... ICOSAHEDRON BY ROTATION OF TETS WH... DETACH MASS RELATIONS WITH A...

THAT IS ·

TET = $\frac{1}{20}$ OF COMPREHENSIV... WHILE A & B EACH EQUAL $\frac{1}{...}$

~~THEY MAI~~

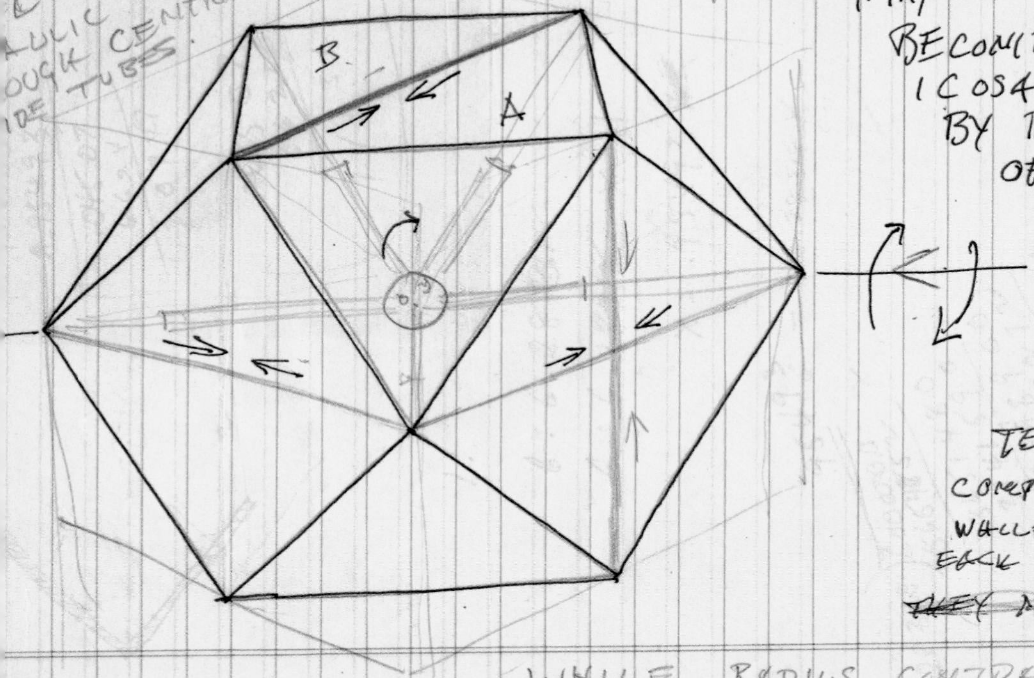

B

A

WHILE RADIUS CONTRACTS IN PROPORTION FROM π TO 3 ∴ GT CIRCL...

BLACK LINES DO NOT CHANGE IN LENGTH.

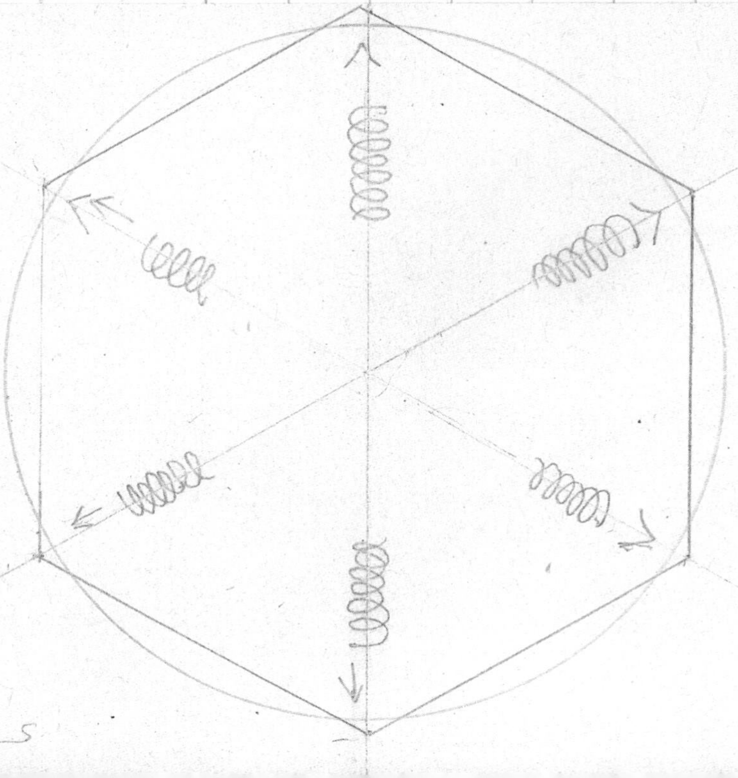

3/10/48

B Fuller

28
22
11
91
12
5
5

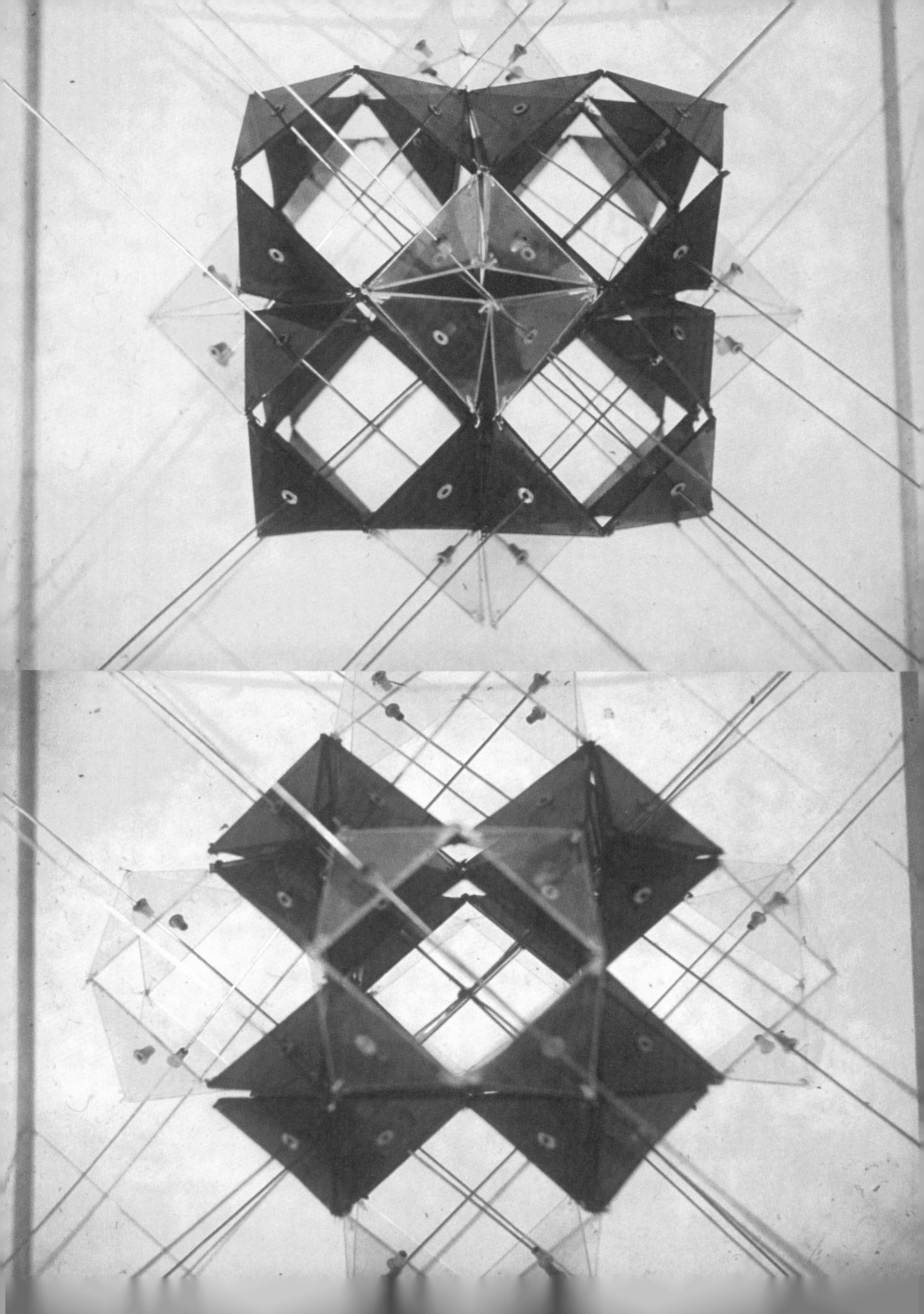

Fuller Research, commentary

The basing of architecture on time in the 4D concept was a challenge to standard geometric modeling. Observations since his youth added up to give rise to the suspicion in Fuller that the development of natural forms operates with coordinates other than the Cartesian system. He was on the quest for a geometry "as Nature uses herself." For a long time, the theme remained in the background, and in the polyhedron studies and large circle designs of the world map first came to a breakthrough: that the half-regular cuboctahedron, utilized for the map projection as a globe divider, has geometric properties that make the figure usable as a vector matrix. The cuboctahedron, which Fuller named "the Dymaxion," results as an invariable configuration from the "thickest packing of spheres" around a nucleus. Fuller examines this method, known since the time of Kepler, with a view toward nuclear physics and chemistry, and in April 1948, he was successful in modeling quantum effects that brought out Platonic bodies in a new light. Fuller interpreted them as phases of a dynamic transformation process that manifested itself in a spiral-shaped contraction. The system oscillates, expanding and contracting over tetrahedrons, octahedrons, icosahedrons to again end with the cuboctahedron. This central discovery, that implied a new science of forms, was given the name "Jitterbug" by Fuller – after a popular dance of the 1940s. His model of a "quantum machine" formed the conceptional background of most of his discoveries.

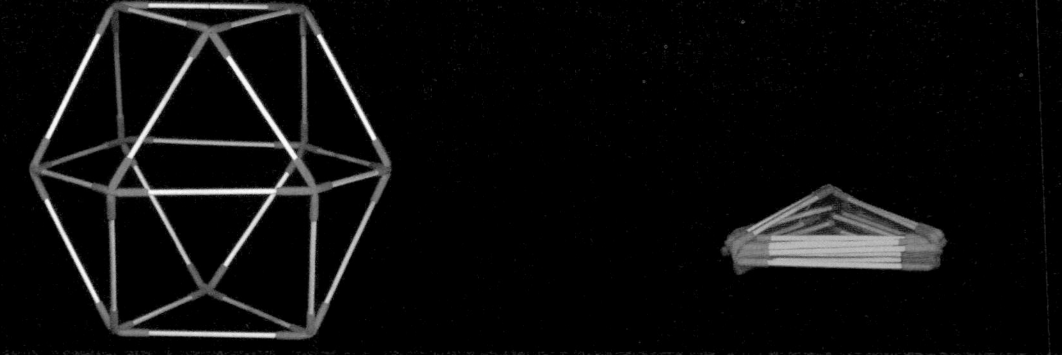

 ·*Jitterbug* Transformation: the phase of maximal expansion

·Color sketch of Jitterbug Transformation: with the rotation of

AS 8 TRIANGLES ROTATE
TO SATISFY DIAGONAL
STRESS IN (OCTA) SQUARES.
(RADIUS CONTRACTS) + THEY ROTATE POINT TO POINT
AS GEARS
PUT
ANTIPODAL
PAIRS
ROTATE
IN
SAME
DIRECTION.
∴ SPIN
INDUCES
CONTRACTION
4 DIMENSIONALLY

8 TRIANGLES
ROTA

SQUARES
BECOME
(A) DIAMONDS
AND THEIR
TWO TRIANGLES
EACH
TOTALING
2×6 = 12
ARE THEN
IDENTICAL
TO DYMAXIONS,
EIGHT, MAKING
TOTAL OF 20 ∴
THE ICOSAHEDRON

CONTRACTING FURTHER
FROM ICOSAHEDRON
AS DIAGONALS
OF DIAMONDS
ELONGATE

THE DIAMONDS
CONTRACT TO ZERO
AND ONLY 8 TRIANGLES
REMAIN ∴ = OCTA -
BUT OCTAS TENSION
BONDS ARE
DOUBLE

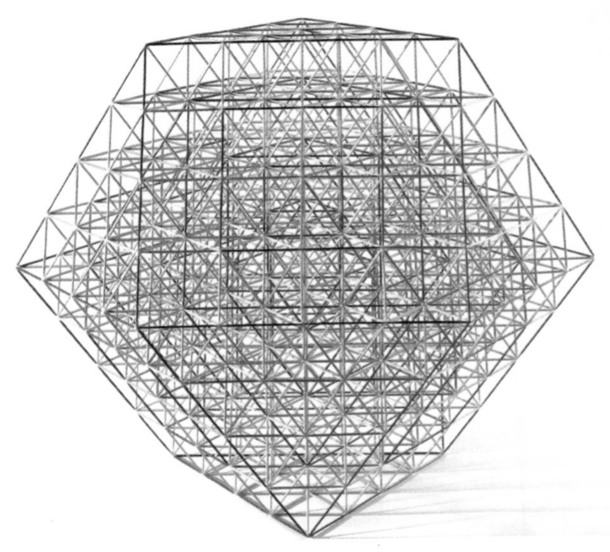

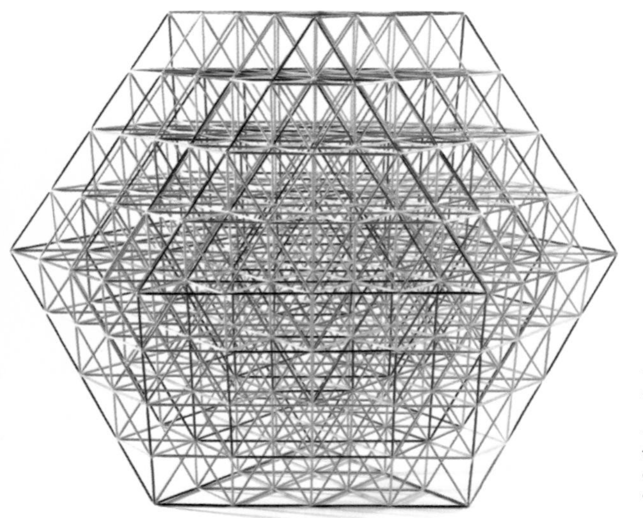

·Two views of the isotropic vector matrix, model constructed out of toothpicks, produced around 1970

·Manuscript sheet from April 28, 1948: Discovery of the formula for the law of transformation of the vector equilibrium, derived from the closest packing of spheres; discovery of the transformation sequence of the *Jitterbug*.

EUREKA — EUREKA —
THIS IS WHAT ARCHIMEDES SOUGHT AND PYTHAGORAS AND KEPLER AND NEWTON —
4/23/1948 B. Fuller
AND Eureka again !!!

We set out to find such a new comprehensive system 20 years ago. and there it is

In terms of this discovered relationship of growth of balls $= 10R^2 + 2$ and growth of Dymaxion $= 60R^2 + 2$ and volume of Dymaxion construction where vol $= R^3$ in terms of modulus of R in wave length — from nucleus and side of square $= \sqrt{2}$ we now see that it is possible to convert all the "absolute" system reference units from their inception in reference to linear 3 dim. of centimeter and cube and its volume and its gram and reference the whole comprehensive system of measurement to basic wave length from nucleus and to whole numbers of spheres analogous to numbers of molecules or atoms. as they grow and therefore to whole new number in place of Avogadro number which a simple relationship of whole or integral number grows corresponding to spheres and therefore all the mass. velocity heat, energy, radiation and gravity and radiation laws (R^2) come into same integral expression.

We discover that "cannot" laws of lifting weight "for work" should be restated as "expressing" ocurrence or acceleration — etc.

We find pressure law of $\frac{2}{3}$ energy of motion of all molecules in vol. to be integral number function —

C. H. Fuller [illegible] Apr. 25/48

Examined April [illegible] 1948 [illegible]

 ·292/293: Technical drawing of the 31 great circles of the spherical icosahedron.

·RBF in studio, Black Mountain College 1948; in background, large-circle hemisphere as model for the first geodesic dome

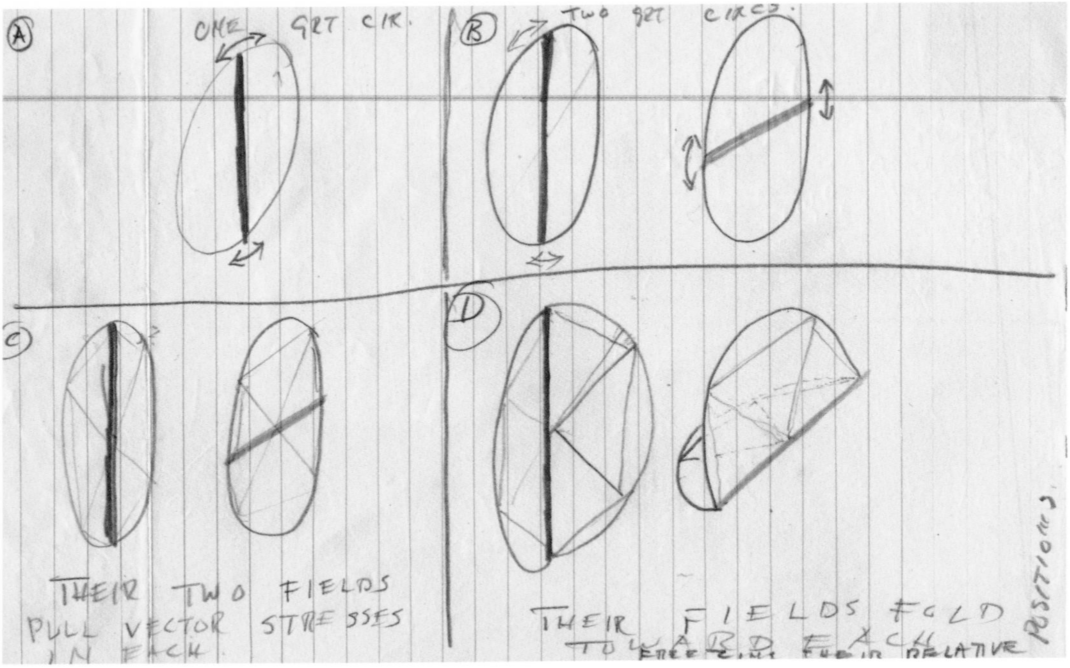

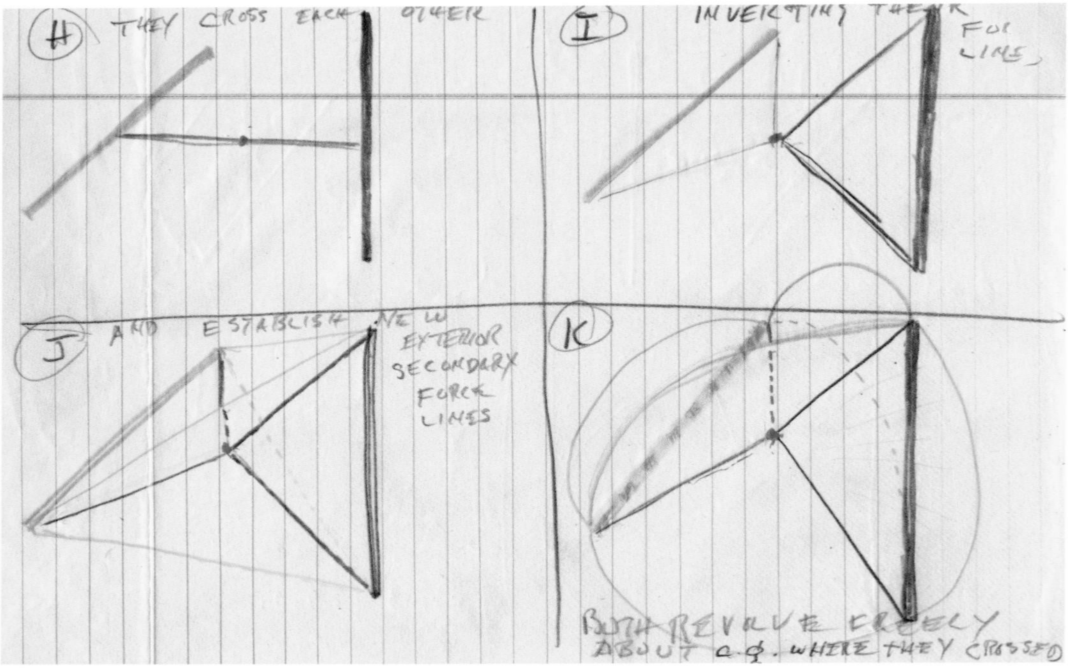

·"Wedding" and penetration of two rotating rods in the folding of their great circle. Sketch. 1948

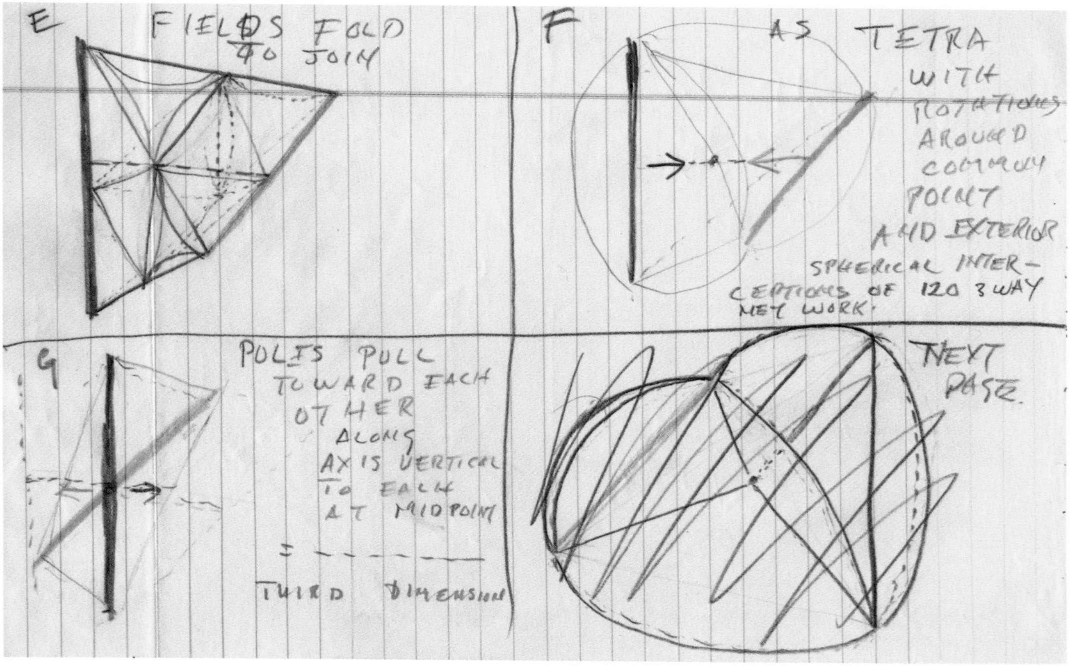

E FIELDS FOLD TO JOIN

F AS TETRA WITH ROTATIONS AROUND COMMON POINT AND EXTERIOR SPHERICAL INTER-CEPTIONS OF 120 3WAY NET WORK.

G POLES PULL TOWARD EACH OTHER ALONG AXIS VERTICAL TO EACH AT MIDPOINT

= ------

THIRD DIMENSION

NEXT PAGE.

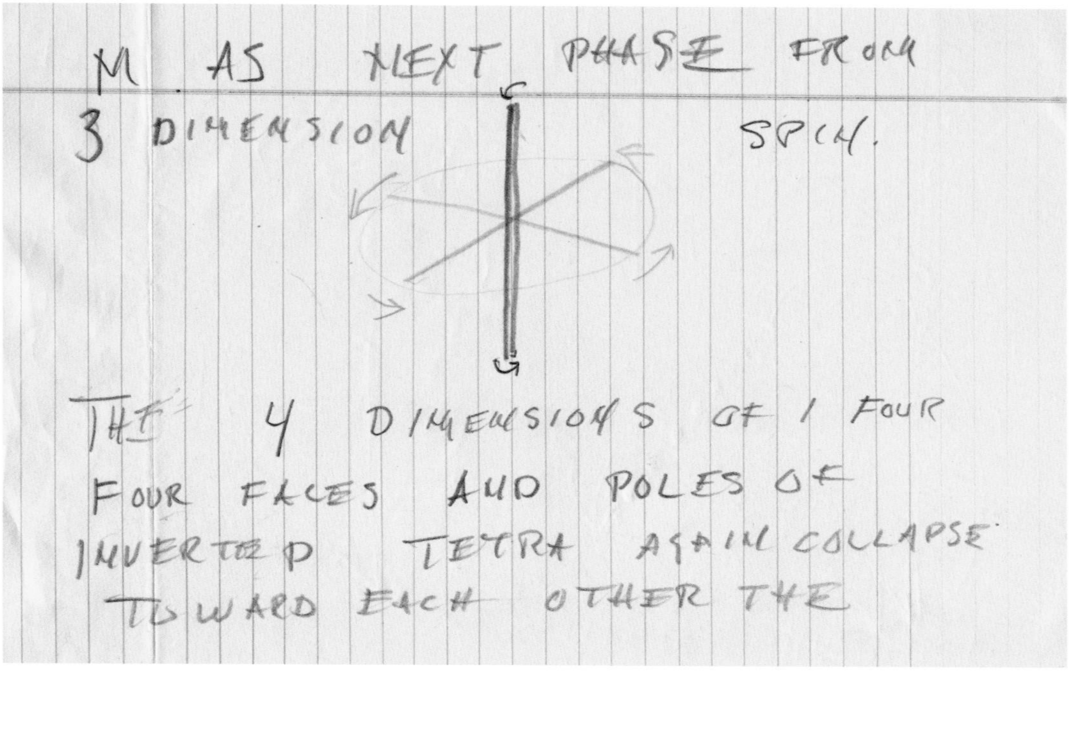

M AS NEXT PHASE FROM 3 DIMENSION SPIN.

THE 4 DIMENSIONS OF 1 FOUR FOUR FACES AND POLES OF INVERTED TETRA AGAIN COLLAPSE TOWARD EACH OTHER THE

EXTERIOR FIELDS

CONVERGE AT COMMON c ♂

AND PROVIDE (DYMAXION

COMPREHENSIVE OF 7 DIM.

3 ORIGINAL PLUS NEW FOUR

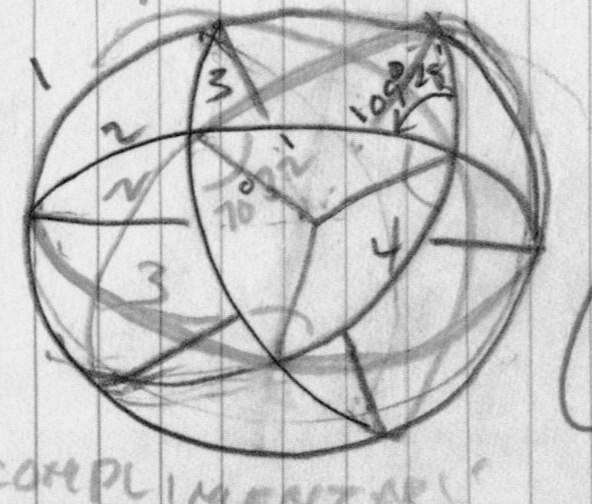

AND WITH INTERNAL
109°28' ANGLES OF SPHERICA
TETRA NOW EXTERNALIZE
(TURNED INSIDE OUT
AGAIN TO EXTERIOR
VERTEXES OF SPHERICAL SQUARES

TH COMPLIMENTARY
0°32' EXTERIOR ∠-S-△S

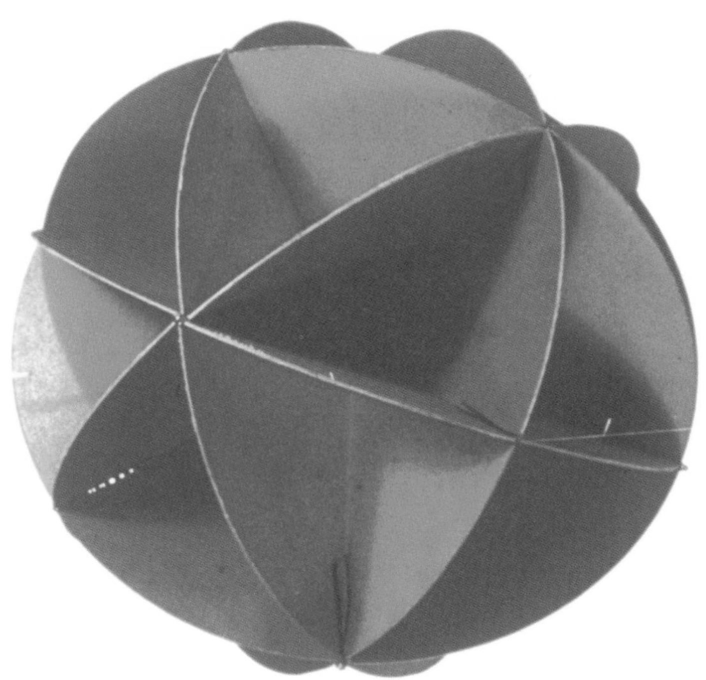

·Model of the 31 great circles of the spherical icosahedron. Sketch of great circle construction in the spherical

·Study model of great circle foldings of various spherical polyhedrons (also the following double-page)

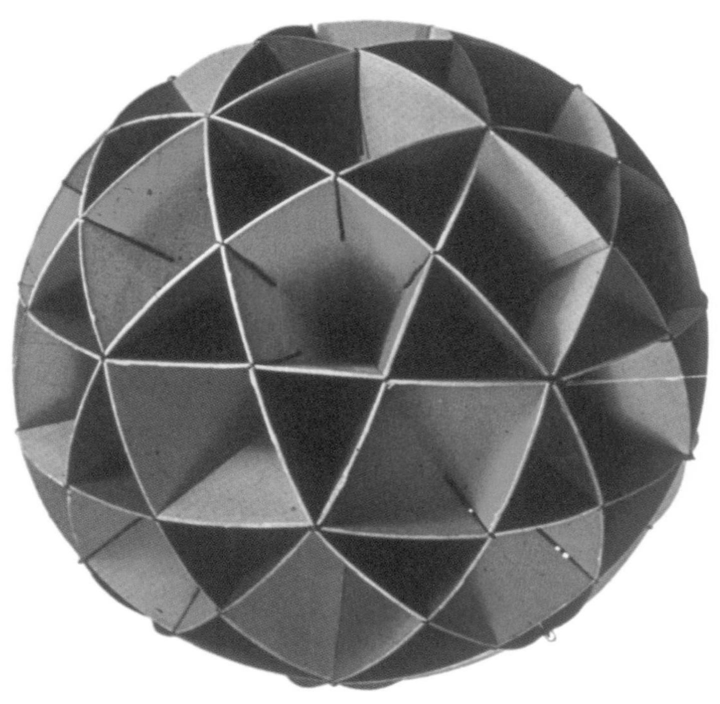

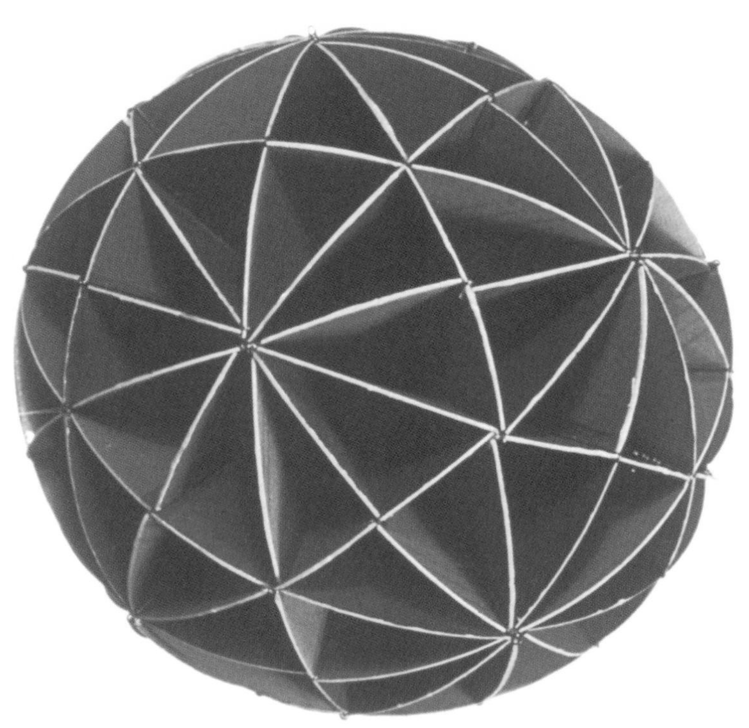

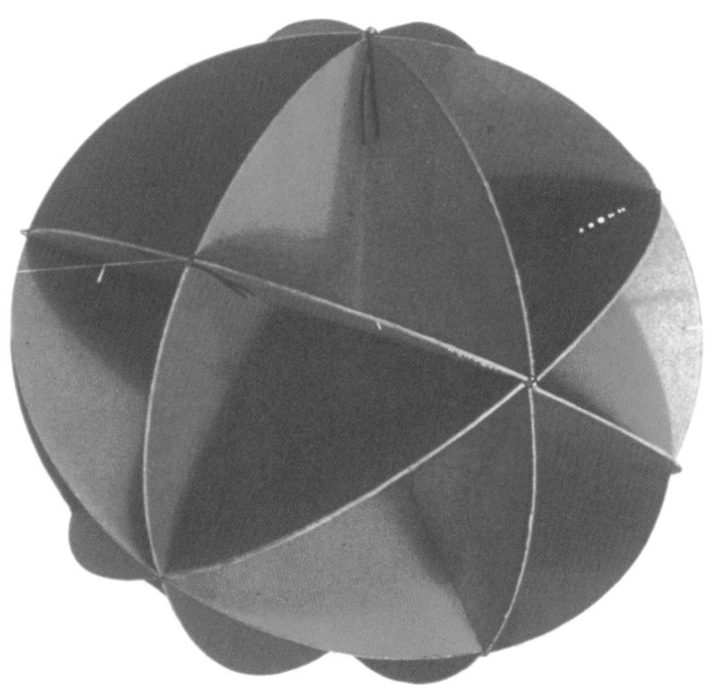

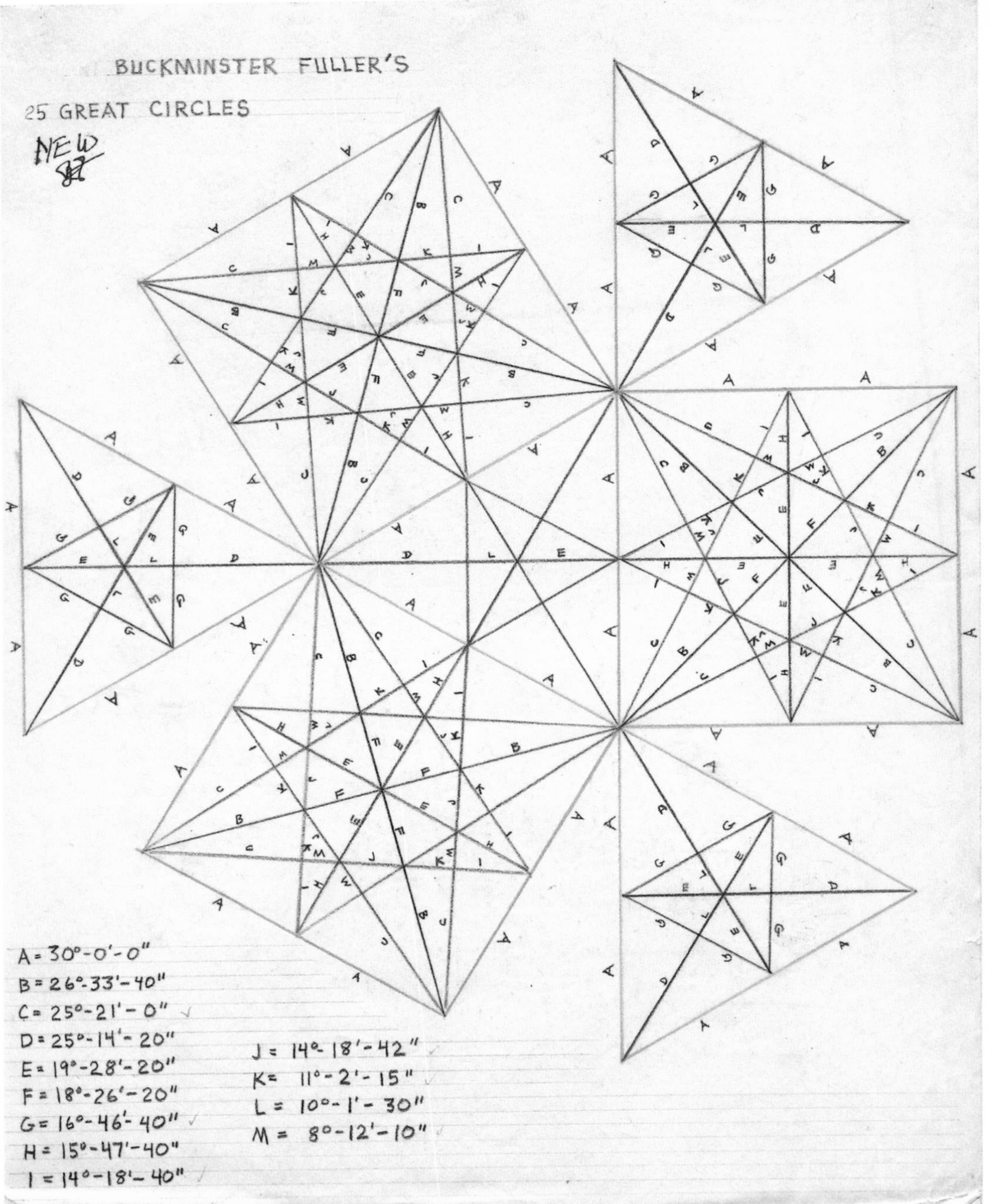

·Construction of the 25 great circles of the

·Study of the construction of the *Geodesic Grid* from great

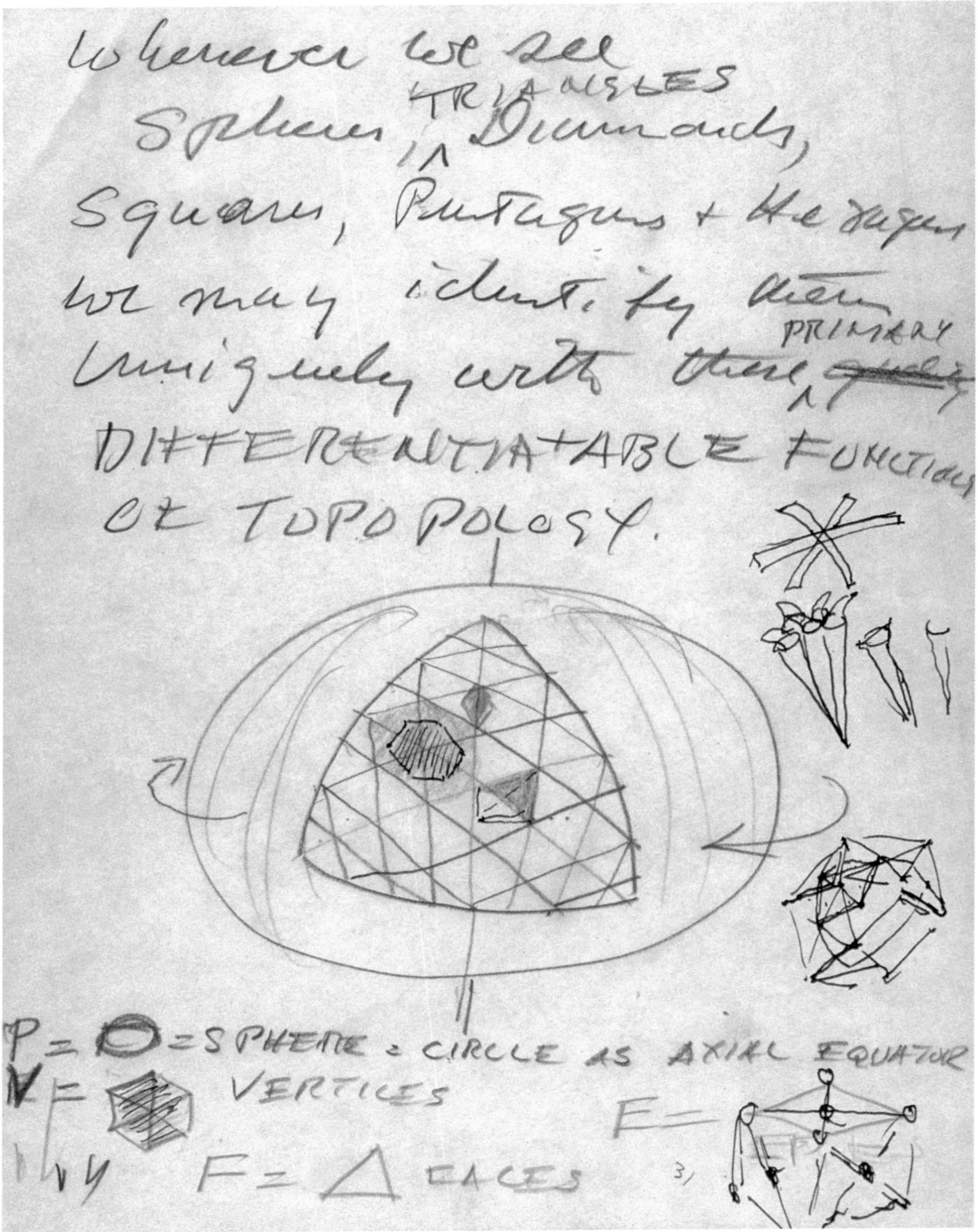

Whenever we see
Spheres, TRIANGLES Diamonds,
Squares, Pentagons + Hexagons
we may identify them
uniquely with these PRIMARY
DIFFERENTIATABLE FUNCTIONS
OF TOPOPOLOGY.

P = O = SPHERE = CIRCLE AS AXIAL EQUATOR
V = VERTICES
F = △ FACES
F =

Generation of the various facets of the *Geodesic Grid*

ICOSA SPUN TO 15 GT. CIR.
BECOMES INDIVISIBLE 2
TETRA (POS + NEG) DODECA +
2 OCTA (POS + NEG.)

STRUCTURALLY REALIZING
EMPHASIZED
VERTECES BECOME
SURFACES COMPLETIONS AS

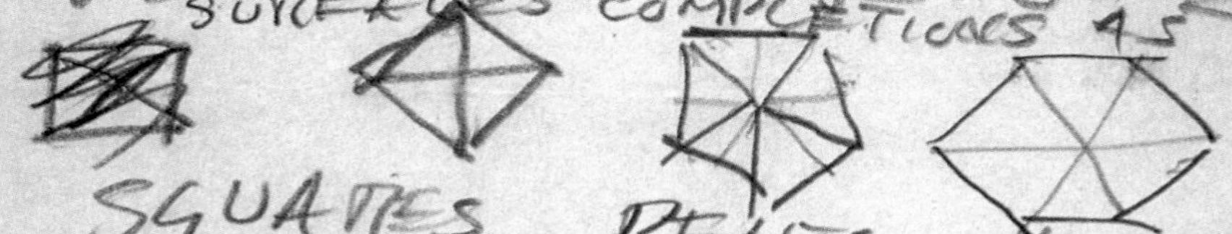

SQUARES, PENTS + HEXAS

EMPHASIZED
EDGES (EXTERIOR) BECOME

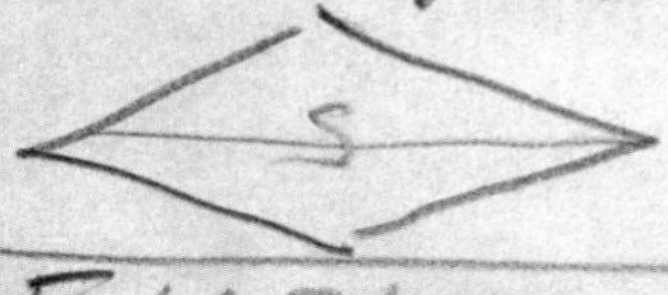

S

EMPHASISED FACES
BECOME

S

EMPHASIZED POLES BECOME 
EITHER
SPHERES OF GYRATION
OR DIAMETRIC ~~RADII~~ TOTALITY.

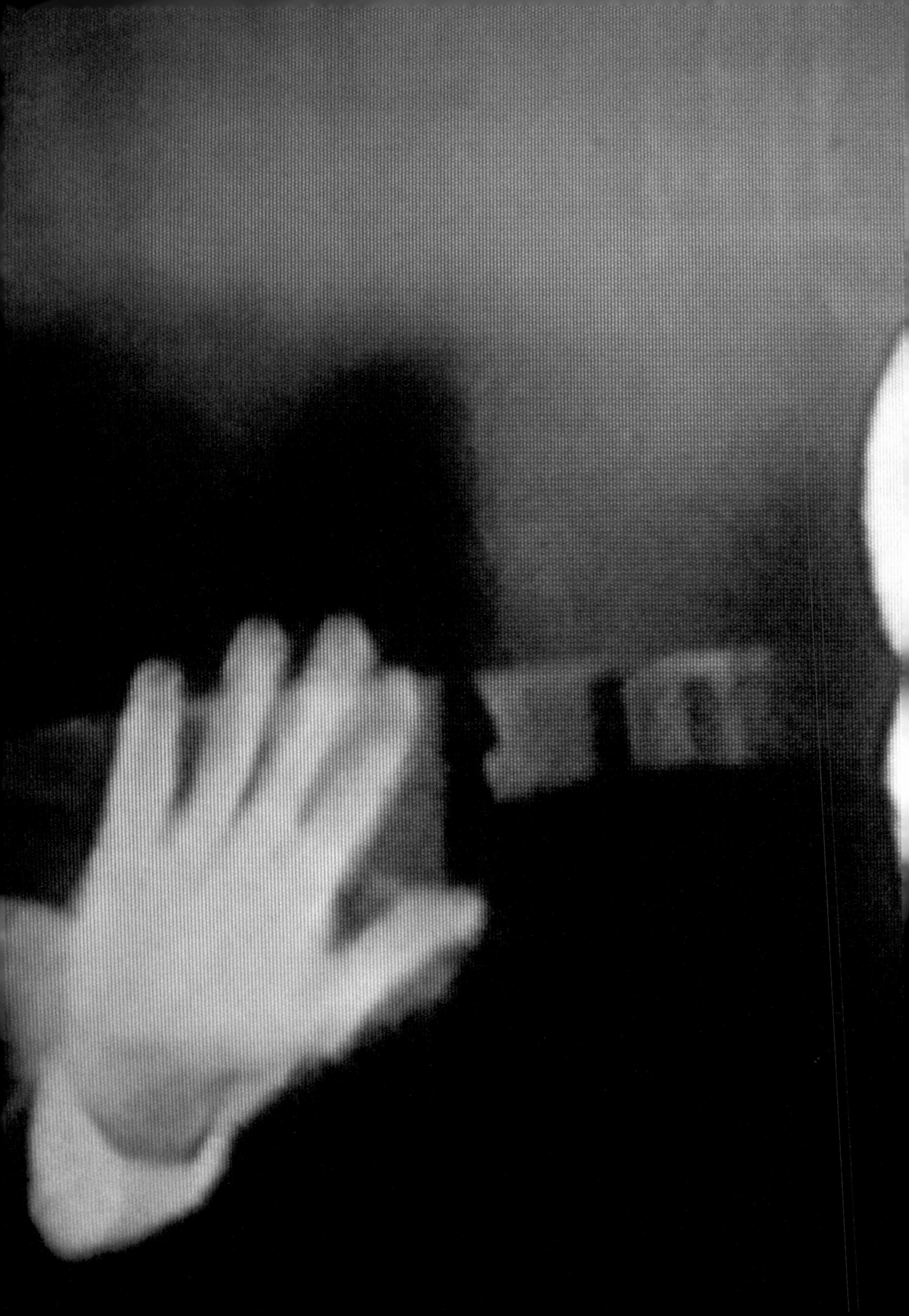

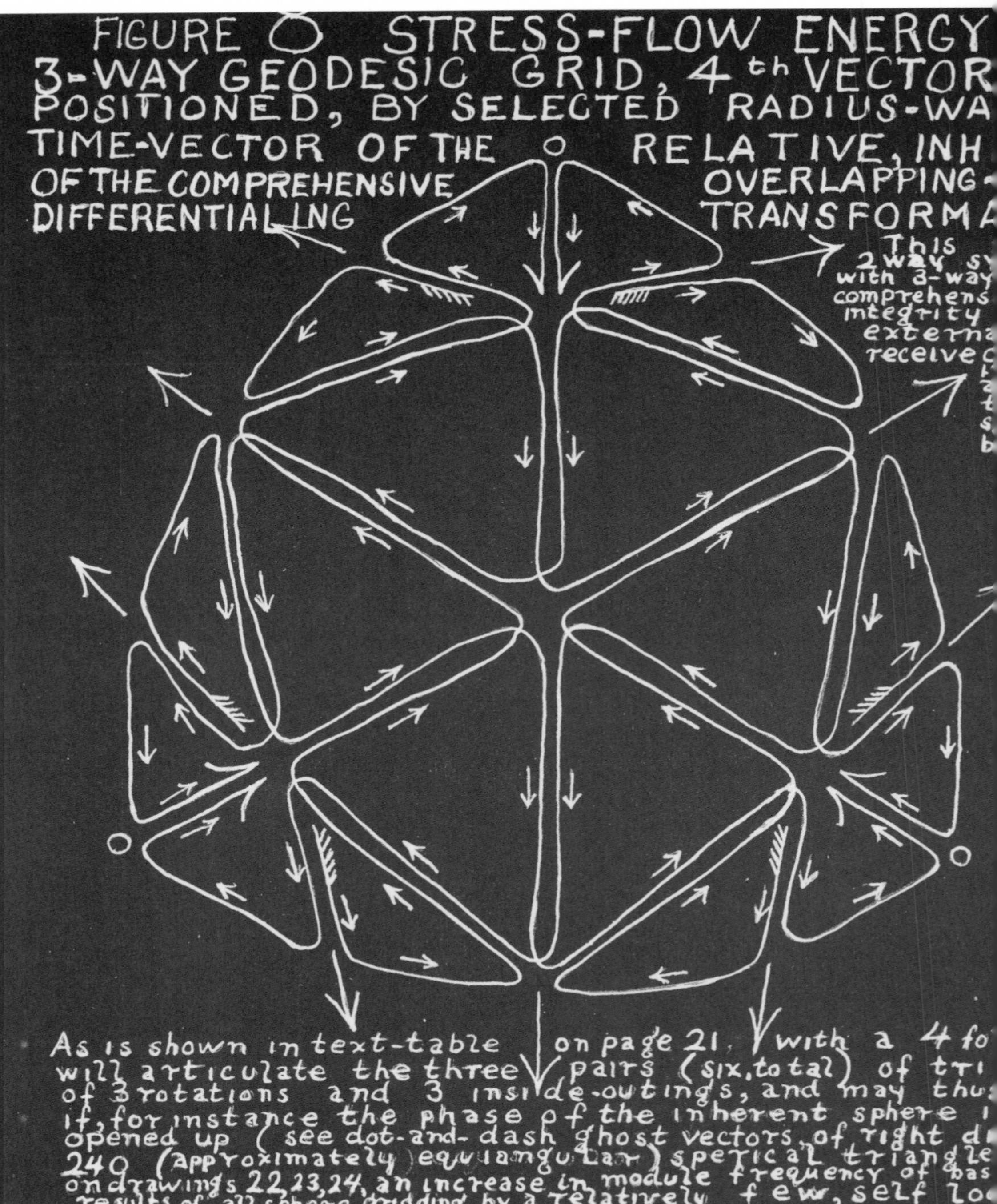

· Drawings from the manuscript *Noah's Ark 2,* 1950: illustration of complementary flows of stress and energy in geodesic grids; compare to basis for construction of *Necklace Dome* and *Geoscope*

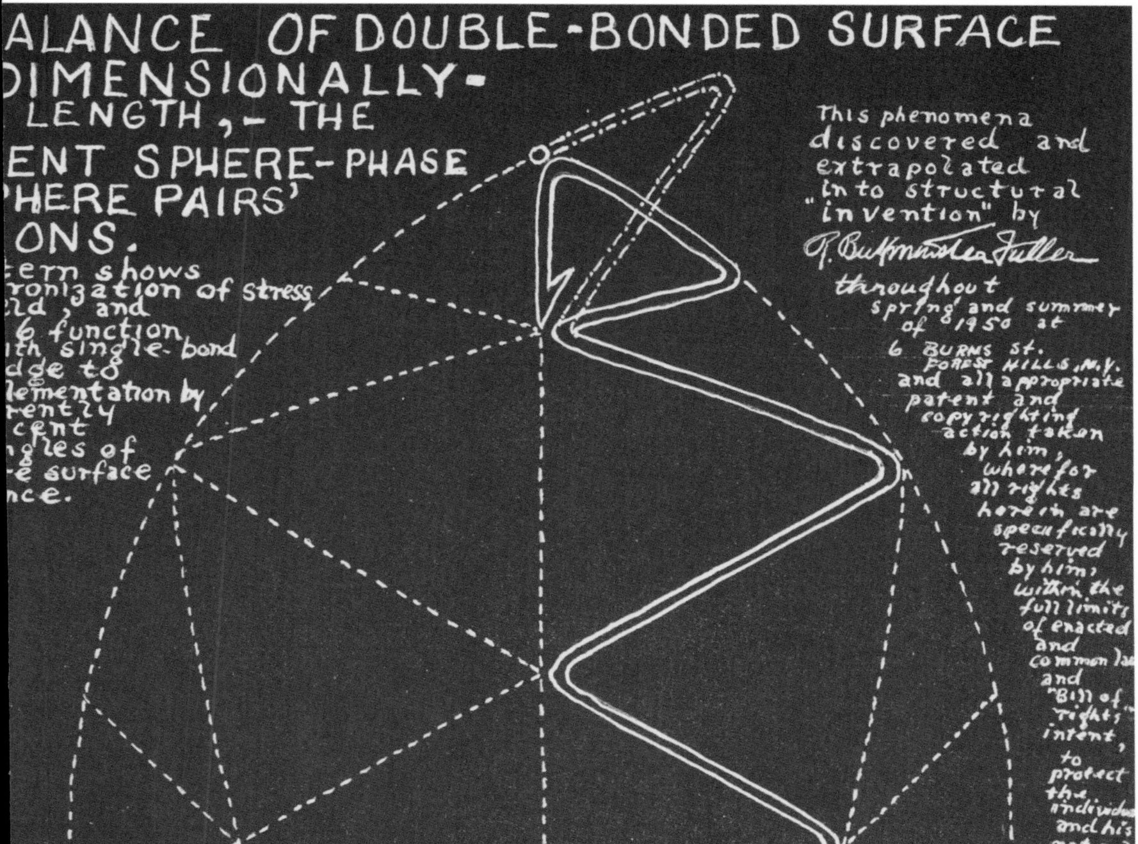

ALANCE OF DOUBLE-BONDED SURFACE
DIMENSIONALLY-
LENGTH, - THE
ENT SPHERE-PHASE
PHERE PAIRS'
ONS.

tern shows
ronization of stress
ld, and
6 function
th single-bond
dge to
lementation by
rently
cent
ngles of
e surface
nce.

This phenomena discovered and extrapolated into structural "invention" by R. Buckminster Fuller throughout spring and summer of 1950 at 6 Burns St. Forest Hills, N.Y. and all appropriate patent and copyrighting action taken by him, wherefor all rights herein are specifically reserved by him, within the full limits of enacted and common law and "Bill of rights" intent, to protect the individual and his natural custodianship of functions physical and abstract that his individual life may be freely enhancing of the common weal and common wealth of integrity of growth.

basic module frequency this one ZIGZAG component
ometric aspect-emphasis of function interrelationships,
mplete the spherical triangles here shown. And
cosahedron the ZIGZAG component may be
am above) they will threeway grid the sphere with
As described in text, especially on pages 20 and 21, and
ge subdivisions may be infinitely reintroduced, with similar
g double-bond, rate of change-vector-ZIGZAG-DIFFERENTIALS

Noah's Ark 2

The research into the fundamentals of geometry during the years 1947 to 1949 can be seen as an ambitious attempt to translate phenomena from the world of solid bodies and static forces into phenomena with a regenerative pattern and energetic dynamics. The goal was to produce a deeper understanding of the behavior of dynamic systems.

In 1950, in a study titled *Noah's Ark 2* (see vol. 2), he summarized his thoughts on the development of geodesic structures by analyzing model experiments. The study contains the basic principles for the geodesic dome constructions that were to follow. It explored the variety of possible geodesic grids and of groups of fields in triangular networks, which opened up a rich typology of geodesic load-bearing structures.

Fuller devoted special attention to the analysis of the distribution of local stresses along the overall structure of the network, a feature that is characteristic of geodesic domes and that accounts for their capacity. Fuller explained it in terms of the interaction of complementary components. Their overall performance is more effective than the sum of the capacities of the individual parts. To describe this increase in capacity, he introduced the term "synergetic" in *Noah's Ark 2.*

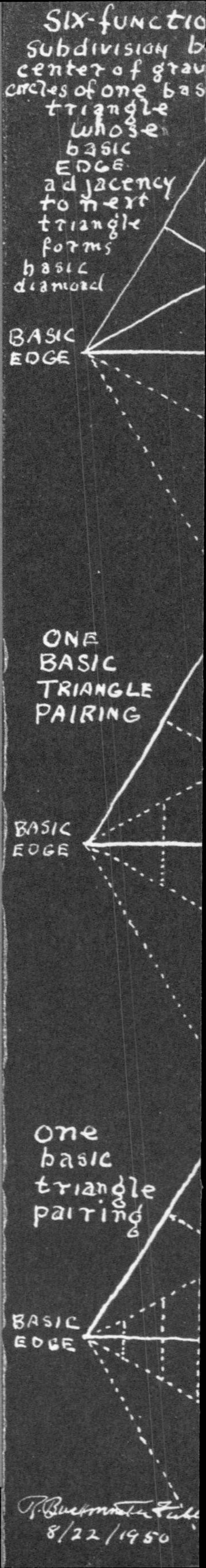

·Relationship of ZigZags to Basic Module Frequency, examination of the divisibility of the sphere in regard to a system of construction, in *Noah's Ark 2,* 1950 (detail)

·312/313: Programmatic drawing of an operational geometry. undated

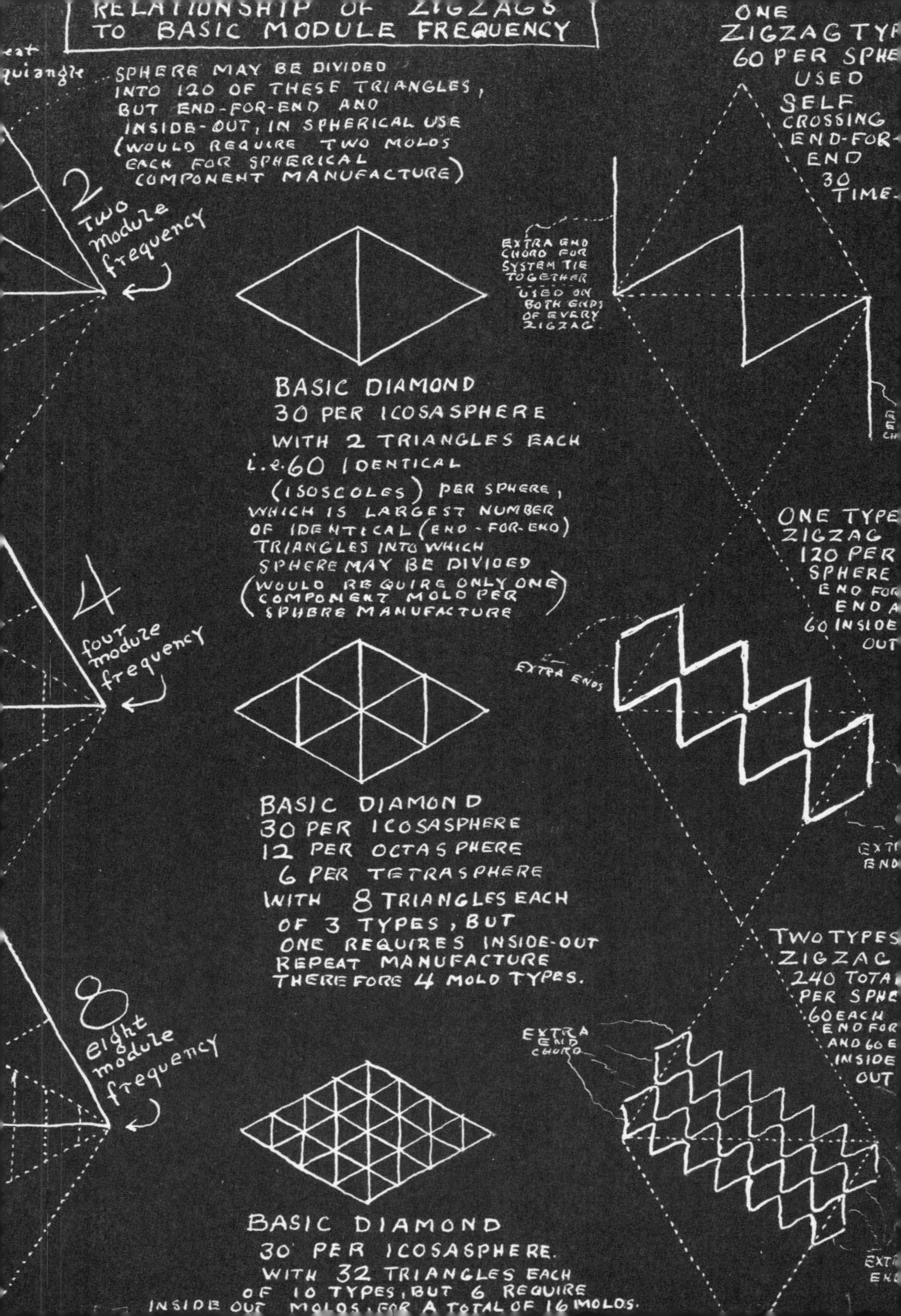

RELATIONSHIP OF ZIGZAGS
TO BASIC MODULE FREQUENCY

eat
quiangle

SPHERE MAY BE DIVIDED INTO 120 OF THESE TRIANGLES, BUT END-FOR-END AND INSIDE-OUT, IN SPHERICAL USE (WOULD REQUIRE TWO MOLDS EACH FOR SPHERICAL COMPONENT MANUFACTURE)

2
Two module frequency

BASIC DIAMOND
30 PER ICOSASPHERE
WITH 2 TRIANGLES EACH
i.e. 60 IDENTICAL (ISOSCOLES) PER SPHERE, WHICH IS LARGEST NUMBER OF IDENTICAL (END-FOR-END) TRIANGLES INTO WHICH SPHERE MAY BE DIVIDED (WOULD REQUIRE ONLY ONE COMPONENT MOLD PER SPHERE MANUFACTURE)

ONE
ZIGZAG TYP
60 PER SPHE
USED
SELF
CROSSING
END-FOR-
END
30
TIME

EXTRA END CHORD FOR SYSTEM TIE TOGETHER USED ON BOTH ENDS OF EVERY ZIGZAG

4
four module frequency

BASIC DIAMOND
30 PER ICOSASPHERE
12 PER OCTASPHERE
6 PER TETRASPHERE
WITH 8 TRIANGLES EACH OF 3 TYPES, BUT ONE REQUIRES INSIDE-OUT REPEAT MANUFACTURE THEREFORE 4 MOLD TYPES.

ONE TYPE
ZIGZAG
120 PER
SPHERE
END FOR
END A
60 INSIDE
OUT

EXTRA ENDS

EXTR
END

8
eight module frequency

BASIC DIAMOND
30 PER ICOSASPHERE.
WITH 32 TRIANGLES EACH OF 10 TYPES, BUT 6 REQUIRE INSIDE OUT MOLDS, FOR A TOTAL OF 16 MOLDS.

TWO TYPES
ZIGZAG
240 TOTA
PER SPHE
60 EACH
END FOR
AND 60 E
INSIDE
OUT

EXTRA END CHORD

EXTR
END

TWO VECTORS
PASS THROUGH
SIMULTA

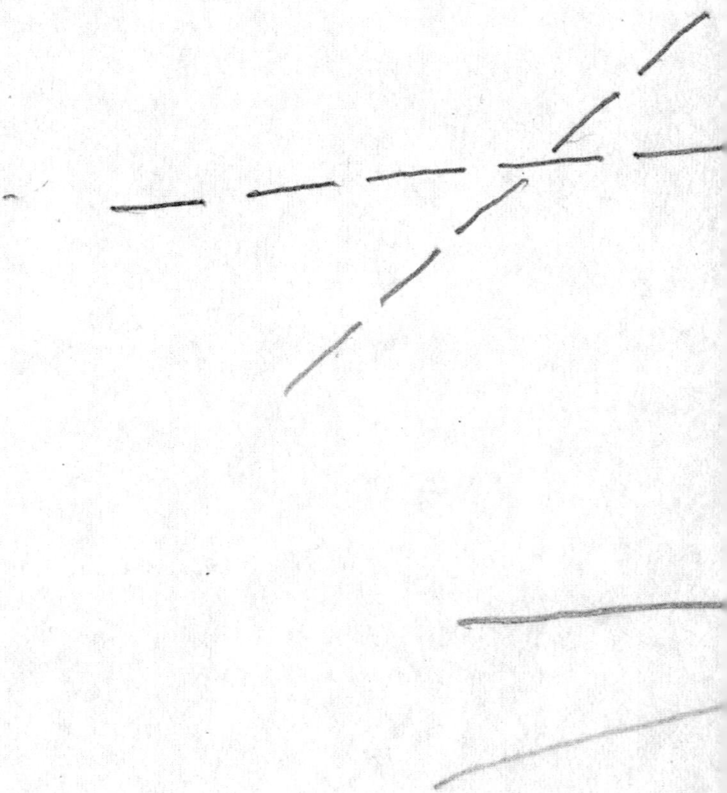

add turbining sphere

DO NOT
SAME POINT
EOUSLY.

- 30 struts (~~e~~edges of icosa

Immediately following the geometric discoveries of the spring of 1948, Fuller took teaching positions at the Institute of Design in Chicago and at Black Mountain College. His research activity merged seamlessly into his modeling-based work with students. It was the start of an unparalleled level of teaching activity at a growing number of art and design schools, universities, and colleges. Fuller had a talent for inspiring his students to achieve genuine innovations and inventions. Without the innovative contributions of numerous students and colleagues – students like Kenneth Snelson, Jeffrey Lindsay, Ted Pope, Don Richter, and Shoji Sadao and colleagues like Duncan Stuart, James Fitzgibbon, T.C. Howard, and many others – Fuller's success of the fifties would have been all but unthinkable. He established life-long friendships in the unique atmosphere of Black Mountain College: Fuller shared with John Cage, Merce Cunningham, and Josef and Anni Albers a profound intellectual accord.

ARCHITECTURE OUT OF THE LABORATORY

It was more or less coincidence that Fuller came to Black Mountain College, a tiny art school in North Carolina, where Josef and Anni Albers were carrying on the Bauhaus tradition. At the last minute, Fuller stepped in for Bertrand Goldberg, whose partner in the Chicago architectural firm, Leland Atwood, had suggested Fuller as a substitute. Atwood had helped Fuller in the 4D House project.

Fuller came with his aluminum mobile home trailer, packed full with his mathematical models. His laboratory was now mobile, his research nomadic. According to reports from participants, including Elaine de Kooning, Kenneth Snelson, Richard Lippold, and Merce Cunningham, Fuller's first, three-hour-long lecture must have had an electrifying effect on the audience. "Bucky," Elaine de Kooning recalled, "whirled off into his talk, using bobby pins, clothespins, all sorts of units from the five-and-ten-cent store to make geometric, mobile constructions, collapsing an ingeniously fashioned icosahedron by twisting it and doubling and tripling the modules down to a tetrahedron; talking about the obsolescence of the square, the cube, the numbers two and ten (throwing in a short history of ciphering and why it was punishable by death in the Dark Ages); extolling the numbers nine and three, the circle, the triangle, the tetrahedron, and the sphere; dazzling us with his complex theories of ecology, engineering, and technology. Then he began making diagrams on a blackboard. He drew a square, connecting two corners with a diagonal line. 'Ah', he said affectionately, 'here's' our old friend, the hypotenuse.'"
Elaine de Kooning quoted in Mary Emma Harris, *The Arts at Black Mountain College* (Cambridge, Mass., 1987), p. 151

In addition to his models, Fuller came to Black Mountain packed full of ideas and projects. Shortly before his departure, he had sketched out a project on 15 June: the construction of a transparent geodesic dome that would enable its occupant to locate his or her correct position in the universe. This was clearly the origin of the idea that was connected to the construction of geodesic domes. Fuller called it "Your Private Sky" (see Geoscope, p. 344). At Black Mountain College, he planned to execute the small great-circle model on a larger scale, utilizing lightweight metal Ventian blind sashes to create the first dome.

 ·315: The Montreal dome, December 1950, the first geodesic dome construction realized by Fuller

·Black Mountain College, RBF surrounded by study-models from his teaching and research, 1949

Merce Cunningham

The lecture at Black Mountain College that Bucky gave was about his work, but also about his way of thinking of the world even at that time, as a single entity ... I remember that marvelous idea of his about how people traveled either with the wind or against the wind. And the first part of the lecture were the ideas that he had been concerned with over his life. And the second part ... He began to demonstrate with these small models which were made of paper ... they were quite small, on the table ... he would pick up one of these, I remember, and turn them around slightly to show us how one thing could work against another in that way he had about building ... Yes, he built one of those geodesic domes, and my recollection about it is that he asked for a specific kind of material but the width was less. So he was heard very quietly to make this remark as they were trying to put it up: Well it won't work but we'll go right ahead and do it anyway. And it did fall down. But it was such an experience for any of us who watched.

– Interview with choreographer Merce Cunningham, in *Thinking Out Loud: Buckminster Fuller,* a documentary movie in the series *American Masters,* directed by Simon and Goodman. New York, 1995

“The whole world has to be turned into music or into a Fuller university.”

John Cage

 ·318/319: Workshop at Black Mountain College, summer school 1948: constructing a geodesic dome out of the metal stripping used for Venetian blinds

·RBF▲ with friends, Black Mountain College, 1948, identified: front left, Nancy Newhall; front right, Anni Albers and to her right, Max Wilhelm Dehn

One high point in the program of the Summer Institute of 1948 was the performance of Erik Satie's *Le piège de Meduse* under the direction of John Cage. Cage gave a lecture defending Satie and organized a Satie festival using amateur performers. Fuller and Elaine de Kooning took the main roles; Merce Cunningham choreographed it and played a mechanical monkey; Arthur Penn directed it. The reworking of the piece was noteworthy: It was a magnificent time…

Arthur Penn, quoted in Mary Emma Harris, *The Arts at Black Mountain College.* Cambridge, Ma. 1987

·321: Merce Cunningham at Black Mountain College, 1948

·RBF as Baron Méduse, with Merce Cunningham and with Elaine de Kooning, in the piece *Le piège de Méduse* from Erik Satie, Black Mountain College, 1949

for Bucky at 85 with love

highest purpose Be
dUst

Camera

now we Know
what coMes
to It
NOT moving
he perceived thoSe
highesT
solvE

duRation did it
and on the same Floor next door

measUrements 2
with Less dust

smaLL
dEtails
fReed music from its rules

he said i Breathe

work withoUt work

aCt
liKe nature
if we Must
otherwIse 3
earth iN
Spring
as Though
following his advicE
he had gone undeRground

Feeling

Knew what was Up

Life
on a cLoud

lovE
befoRe hand

John Cage

(part of Themes and Variations: Theme #VIII)

 ·John Cage: Poem for RBF, on the occasion of his 85th birthday, July 1980

·John Cage at Black Mountain College, around 1950

Necklace Dome

It was a big task to get from great-circle models of globes to a practical supporting structure. The first step in the process was creating the models that Fuller called necklace dome structures. The great circles were produced from straight lines, namely, tubes with a cable running through the hollow cavity that tied them together and gave the structure its name. The individual great circles had joints, and the structure as a whole was meant to fold out and back together. The points of intersection were covered with plate-like fittings. One of the main points of Fuller's theory said that as the diameter of the dome increased, the force on the individual strut decreased, and the synergetic principle of the diffusion of force became more and more effective. The increasing size of the models and experimental domes, which were always subjected to weight-stress tests, was an important criterion in this phase of successive exploration of the characteristics and possibilities. The structure in the upper right of the illustration (at Black Mountain College) had a suspended platform and was subjected to irregular impacts to test its response to an artificial earthquake.

The discovery of the tensegrity principle at Black Mountain College was due to the magical atmosphere of learning of this school and its teacher Fuller. The wire-spoke wheel had always been important to Fuller: "I saw that the exterior of the equatorial compressional island rim-atoll of the wire wheel must be cross-sectionally in tension as also must be its hub-island's girth" (*Synergetics,* 353); the step from monopolar to multipolar tensegrity was the discovery by his young student Kenneth Snelson. (See Tensegrity, p. 392 ff.)

Work on early geodesic structures, the discovery of the tensegrity principle and the problem of a support structure's foldability (1947 to 1950)

Multipolar tensegrity; this principle was discovered by RBF's student Kenneth Snelson. RBF with students and his Airstream Trailer, with rolled-up great circle sphere. Necklace-Dome geodesic structure that folds together.

Standard of Living Package

Fuller assigned this design project at the Institute of Design in Chicago in 1948: To make the complete furnishings for a household of six people that could be packed into a container and transported with a trailer. The students developed a box that was eight yards long and two yards high and wide, with collapsible walls that were fastened by hinges. When folded down, the walls also served as the floor for the furniture and household items, as well as for the corresponding zones of the floor plan: bedroom, living room, two baths. The overall useable area was ninety square yards. The exercise was influenced by the technique developed during the war for transporting delicate airplane parts in trucks that were packed full; it is also another attempt to deal with closest packing (see p. 278 ff.).

This exercise was also related to the design for the Sky Break, an easily constructed house whose climatic skin is a geodesic cover, designed according to the necklace principle, that stretches over the unfolded – in this case two-story – living space and storage compartments. The model shown here was built at Black Mountain College in 1949.

STANDARD OF LIVING PACKAGE

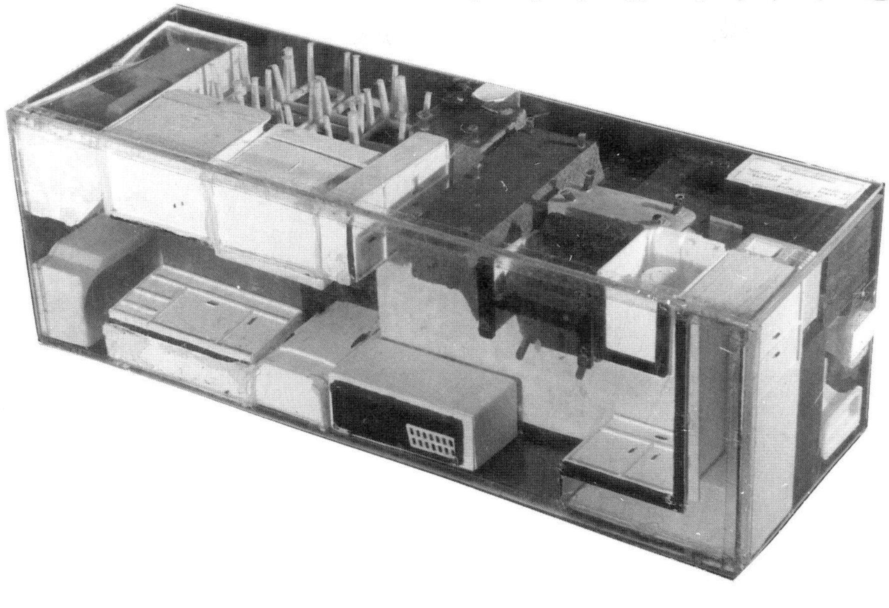

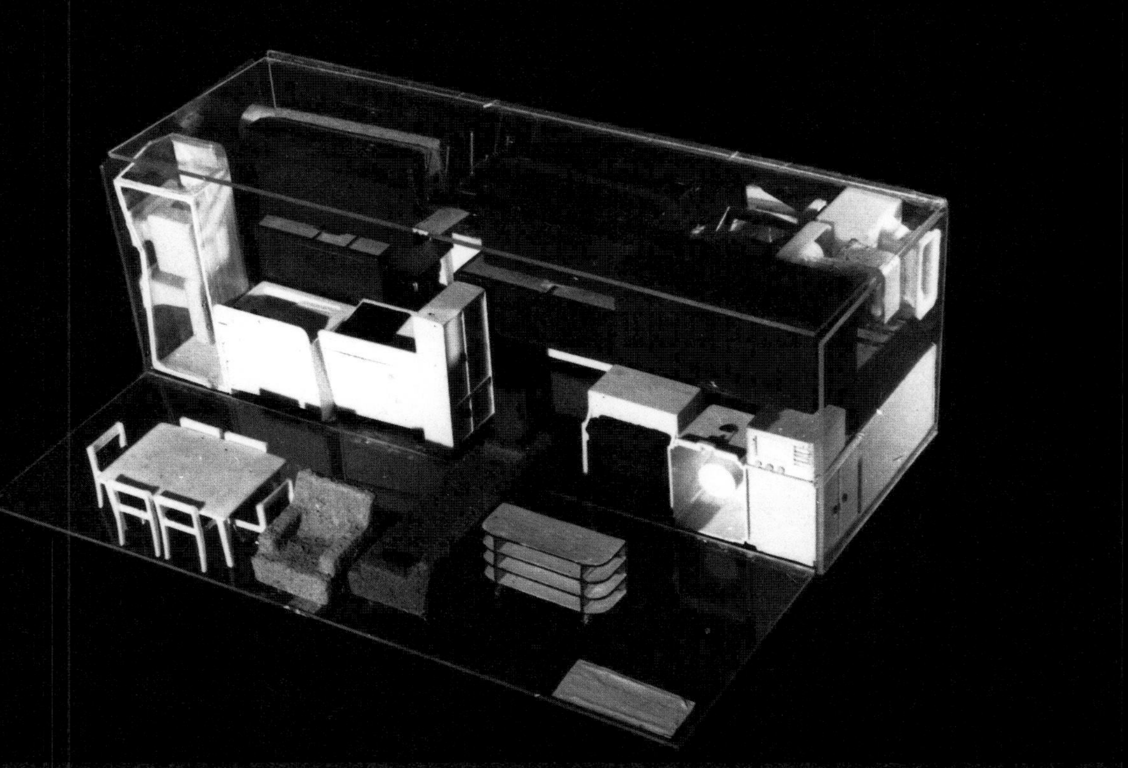

·328/329: *Necklace-Dome* with outer plastic skin
·*Skybreak:* project for a dwelling unit out of a container box: modular construction of the dwelling, with transparent screen over top that folds up (*Necklace-Dome*)
·The Standard of Living Package: packed together and partly folded up

332 ·Dome in Aspen, wooden slats and outer plastic skin, 1953 ·View from interior toward the ventilation opening in the zenith. (below) Charles Eames filming in Aspen, 1953, RBF on right

Geodesic structures

In the fall of 1947 I assembled a four-foot diameter three-way grid geodesic structure at my Forest Hills New York apartment, and that winter and spring I made two more four footers. In the summer of 1948 the first generalized prototype model of a 50-ft. Geodesic structure was assembled at Black Mountain, North Carolina. In December 1949 a 14-ft. necklace Geodesic was assembled at 6 Kinzie St., Chicago, at the request of the Air Force, and in February 1950 it was installed in the Pentagon Building garden at Washington, D.C. In December of 1950 the prototype of a specialized geodesic structure 49 feet in diameter was built in Montreal. I designed it to be an Arctic installation. The components of the structure were tubular aluminum struts weighing about one pound each. The structure was so light that we did not need a mast to lift it. Instead it was lifted locally in order to add more struts to the bottom. When the structure was completed we looked up at the blue sky through this thing and began to realize that something very pleasantly exciting was happening to us. We knew that it was light, knew that it was strong, but we did not know that it was going to do just that to a blue sky. Those are the very typical sensations we get when we tend to solve only the scientific side of the problem. The qualities of economy that are synergetically resultant in the end do something to us in the way of challenging our sensibility to new sensorial limits. Looking over against the birch trees, the slenderness ratios of these very high strength trees and of the Geodesic struts seemed to be very much akin.

Architecture Out of the Laboratory (1955), 17

 ·Geodesic dome near Montreal, December 1950, erected by Fuller Research Foundation, Canadian Division

335 · Before the textile skin is put in place: the metal rods form the pressure-subjected domed vaulting, the tautly-strung wires form the tension-subjected outer shell

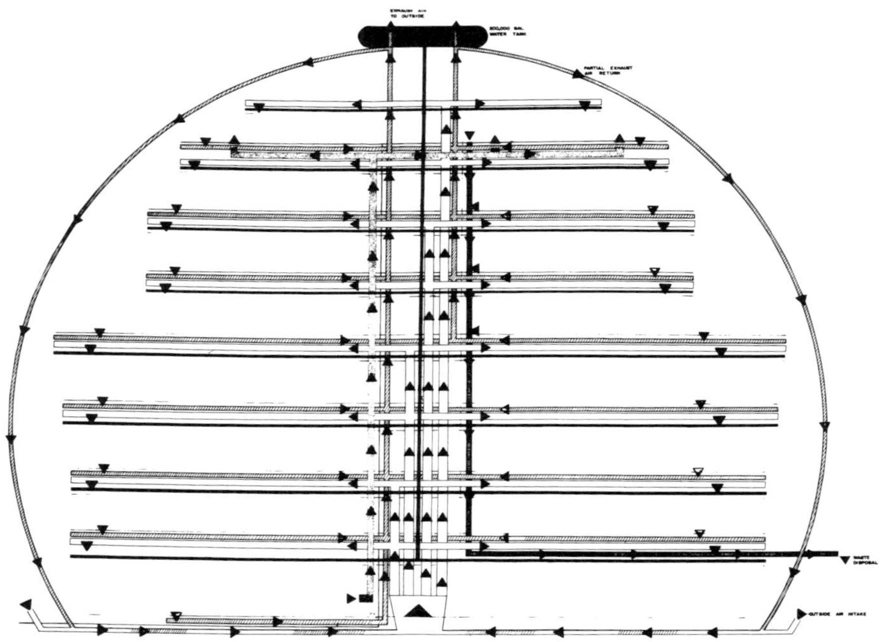

90% Automatic Cotton Mill
North Carolina State College 1951

How an entirely new factory architecture may precipitate the mechanical evolution.

A new cotton mill as a specific prototype problem generic to the greater imminent problems of the widely deployed semi-autonomous elements of an invulnerable industrialization.

(...) The first assumptions of cotton manufacture were that cotton mills must be installed beside mill dams because their machinery would "always" be powered by water wheels, requiring translations of power by belting pulleys and meticulous paralleling of shafting in the most geometrically economical patterns, consistent with all the functions of picking, carding, combing, twining, spinning and weaving. Obviously logical for the preservation of the expensive machinery was its housing: One hundred percent compressively conceived stone masonry for vertical components and heavily sectioned wooden platforms for the horizontal components. The natural clustering of the necessary mechanical stages of the manufacturing called for a geometry of four or five tiers of flooring.

(...) The whole early concept of factories assumed flooring requirements able to carry 200 lbs. to the square foot at any square foot, because which foot might be eventually employed, could not be foretold. The revolution of new cotton mill design, characterizing the switch from New England to the Southern States for economic reasons occurred after electric motors were introduced and steel structuring was available, and made possible by air conditioning and large spans of steel trusses, etc. However (...), the machinery itself, though now installed on one floor, represented widely separated operations (...). Due to their enormous foundation layout

·A project for a 90% automated cotton mill, team design effort at North Carolina State University, 1951; floor levels as Octet-Grids

·90% automated cotton mill, diagram of model

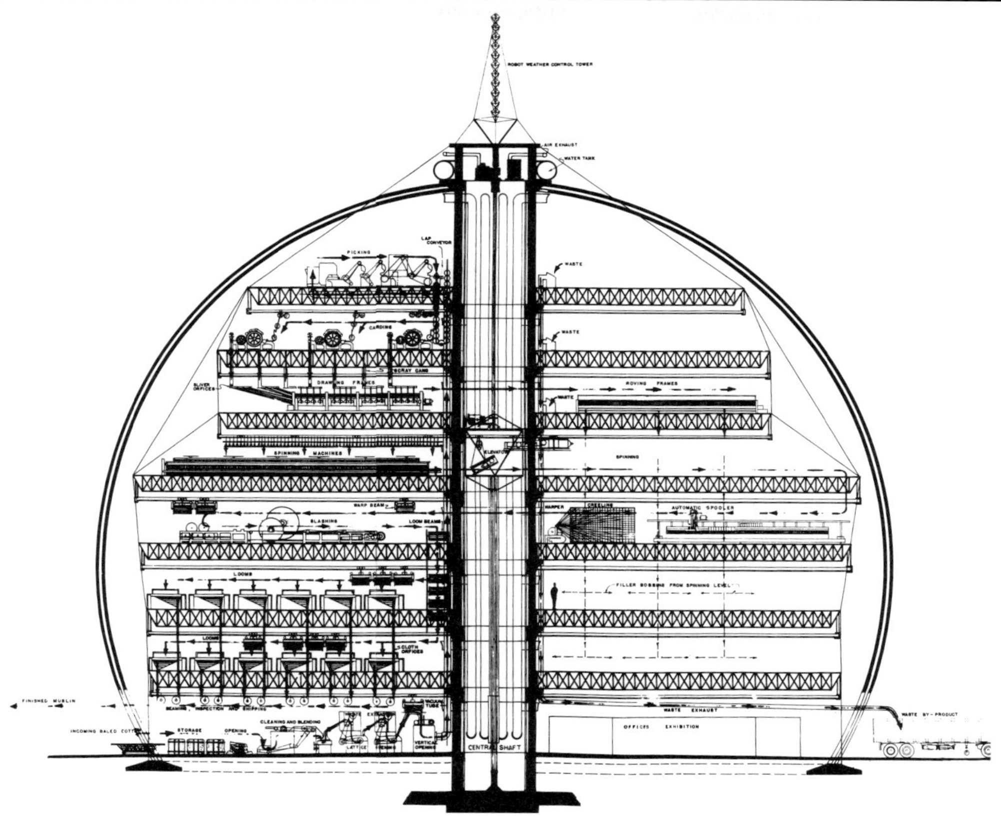

and assumptions of "anywhere-everywhere"-loading, the cost of buildings has continuously risen and along with it their maintenance cost.

Two industries with more recent starts than that of cotton manufacture and therefore with less accumulated tradition of conceptioning and operating strategy have shown far greater susceptibility to scientific treatment and therefore to acceleration in evolution of both general and particular physical mutations, i.e. the chemical manufacturing industry and electrical power generating and distributing industry. (...)

No. 1. The chemical manufacturing industry soon took scaling ladder advantage of its own generally advancing chemical knowledge of natural association-dissociation principles by developing non-deteriorating all-weather and continuously operating machinery, and like a chicken, burst out of its own classical arch corniced brick egg to stretch out nakedly and majestically in the local day and night sky with its omnidirectional system of receiving, self-meteringly valved, fractionating, accumulating, regeneratively circuiting and distributing facilities.

No. 2. The relatively very new electrical generating industry and its distribution system (and its by-product coke and coke-derivative chemical processing) have been repeating the out-of-egg process and are now better than half emerged. Its greater transformers and other major apparatus are now in open-air switchyards.

A totally new start in cotton mill conception is now proposed in which only the essential articulating and reciprocating mechanical components of the various stages of manufacture are separated out from their pre-conceived separate chassis, which now obsolete chassis assumed their parallel and "standing" placement on floors, relative to shafting.

Instead, it is proposed that a unitary, three-quarter spherical air conditioned enclosure be provided, with a secondary three-quarter sphere interior geodesic structuring entirely indepen-

dent of the enclosing structure, which shall serve only to support the articulating mechanisms to be suspended in preferred four-dimensional arrangement within. Thus a true flow pattern, similar to the digestive, shunting, secretive, and regenerative pattern of the human anatomy, will digest and process the cotton, taking advantage of ample height and gravity to drop-lead, space-accumulate and meter the progressive bars, slivers and threads and thus eliminate the primary "re-introduction" chores of workers, and the enormous and heavy floors around which roll the relatively light loads. The savings in amounts of metal required to permanently truss-sling the apparatus into the preferred flow patterning positions will be of dramatically significant magnitudes, and the heavy flooring itself will be eliminated. The free space positioning of the apparatus will be such as to permit its interpenetration by swiftly conceived mobile staging and to bring the few expert attendants into intimate advantage relative to any stoppage in the automatic processing. The whole process may be kept in constant review from central space advantage. The preferred aerodynamics of the three-quarter sphere in its economy of laminar circulation will provide highly efficient air conditioning and illumination, etc., and a minimum of concentrated foundation to support the base of the geodesic structuring.

This seminar will involve research, on the part of all its architectural students, within the N.C. State Engineering, Textile, and Business Administration Schools and expert consultation in establishing the entirely new fundamental assumptions and a trial-balance-design must be taken involving a working familiarity with all the functions. Within thirty days a general assembly, and primary sub-assembly, set of drawings must be developed, clearly revealing the fundamental scheme and cogently demonstrating the net economic gains in pertinent industrial logistics in metals, energy and time investments of original installation and subsequent operation, etc. To realize so comprehensive a program in so short a period, the participating students will

 ·Look at the construction of the lower-levels and overall view

have to establish procedural team discipline in which they will sub-divide the special tasks, but will avoid isolation from comprehensive responsibility by frequent convening of teams, reporting to the whole group and group discussions and tactical decisions.

It is hoped that the time budget will provide enough margin to undertake a few schematic sketches on the part of each student in which he himself applies this principle of automatic and integrated mechanics suspension to other fields of manufacture. It will be clear, as the problem develops, that this omni-directional, multi-dimensional spherical patterning introduces relationships and energy efficiencies that are not only novel but to be contrasted to the present 1-, 2-, and 3-dimensional geometry limitation of intermittent batch and production lines.

Richard Buckminster Fuller, Visiting Seminar Director

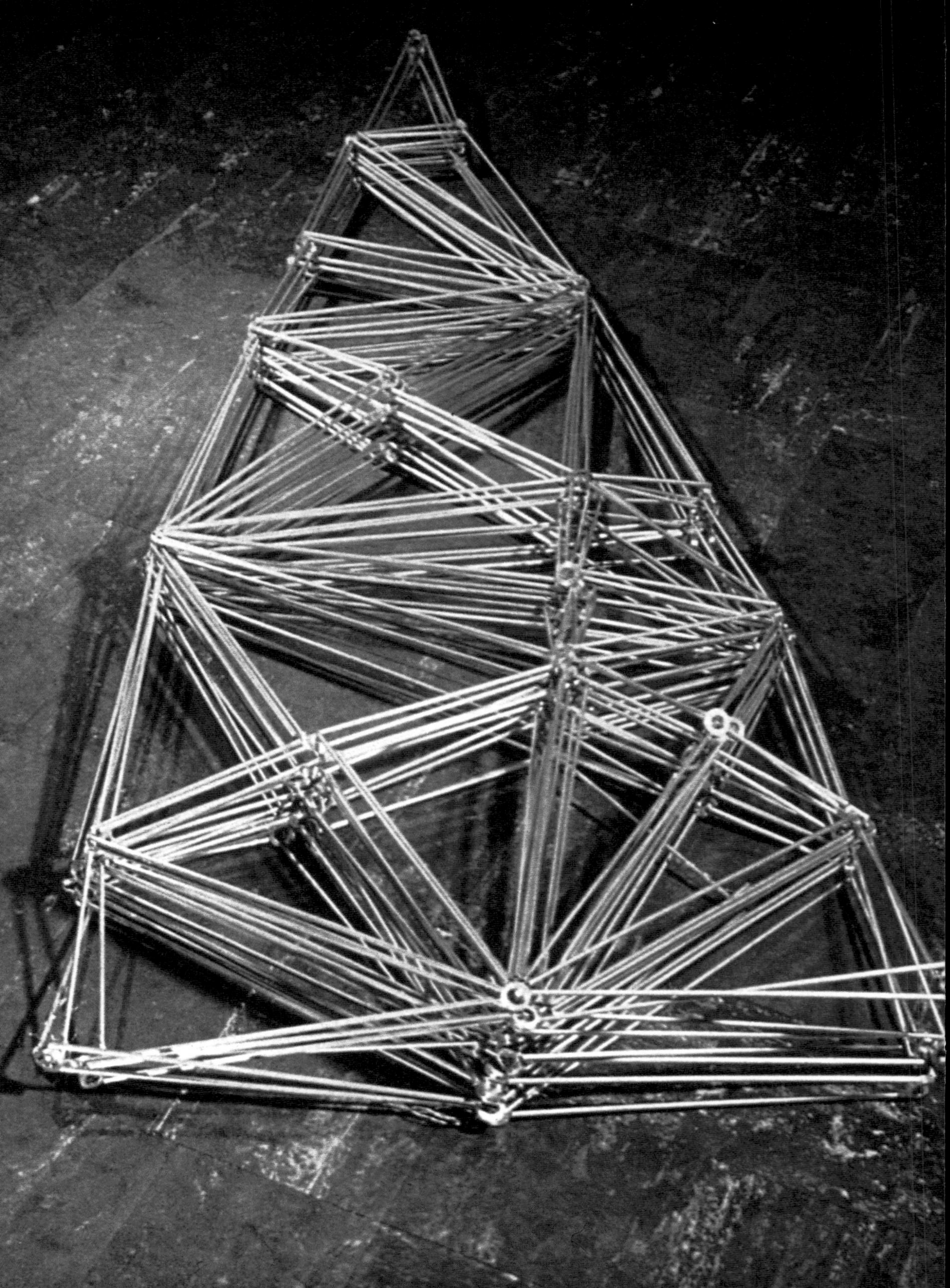

Flying Seedpod

This structure was made at Washington University, St. Louis, in 1953 under my invention and design instruction. You may possibly be looking at the prototype of the structural principles that we may use in sending history's first (little) scientific dwelling to the moon. As you see all the structural members are tightly bundled together in parallel so that they may be transported in minimum volume within a rocket capsule. The parallel-strut bundles of light-weight magnesium alloy consist of sets of three, fastened together at one end with ball joints clustering them in tripods. Each set is like a camera tripod with three, tubular magnesium, legs. Each set has ball joints at the tripod head. All the tripod's feet are also fastened together with ball joints in clusters of five and six tubular-tripod feet per ball joint. We have a little mast coming out of the top of each tripod. This mast is pushed out automatically by a piston in a cylinder mounted in each tripod head. We put 200 pounds of gas pressure inside the cylinder and the gas pressure, (when triggered by a lanyard), will push the masts of all the tripods outwardly from each of the tripod heads. The pushed-out masts each have three tension members leading to their respective tripod's feet. As the masts are pushed out by the 200-pound pressure the tension members pull the legs of the tripods outwardly from one another. The tripods all open wide with their ball joint feet fastened together in hexagons and pentagons. There is a triangular net of aircraft cable in a regular, geodesic, 'star', spherical grid that restrains the tripod legs from moving any further outwardly from one another than is necessary to form a symmetrical geodesic dome or sphere.

World Design Science Decade: 2 (Mexico Lecture, 1964), 51

343 ·*Flying Seedpod:* project at Washington University, St Louis, 1953: geodesic structure able to fold up

·Phases of the assembly process

·Detail: junction of the compression rod and tension

In 1952, students at Cornell University under Fuller's direction built a seven-meter sphere for a planetarium called a Geoscope. From inside the sphere, it was possible to observe the moon and the stars. The point behind it was, however, only partially the view looking away from Earth toward the stars; ultimately, the point was to look at Earth itself. The name Geoscope had precisely this double meaning. Years before the first photographs of our planet from space, the installation sketched a view of the globe "seen from outside"; it was meant to unsettle the cherished image, used even by confessed academics, of the sun rising and setting. In general, the Geoscope can be seen as an appeal to a perception of reality that was critical and intellectually independent.

GEOSCOPE

Your sky may be surfaced inside with sections of global map with zenith of 'sky' and oriented accurately to the north. Thus the inhabitant may 'see' his geography correctly. Can have his swimming pool at center a hemisphere with his antipodal data.
Can follow constellations and star paths and tell date and hour by their location.
Sun clock in day
Star clock at night.

Elucidation of earlier idea – listed with polarized sky rotating screens. Idea of June 15th 1948 mulled over for some this. – B. Fuller

Geoscope: point of view

"The point of view, through introspection, unlimited to the segmental area of our temporal eyes, is our abstract central position in the center of the universe, looking or building from inside out, as from the center of a great glass globe of the earth. Through this globe may be viewed the progression of relative positions to the starry universe, looking along the time lines in all directions. The separate paragraph thoughts are only connected by their common truths, which are the material crystalline spheres of sensible and reasonable fact, through which the radial time lines of individualism must inevitably pass in their outward progression towards the temporal infinity."

4D Timelock (1928/1972), 31

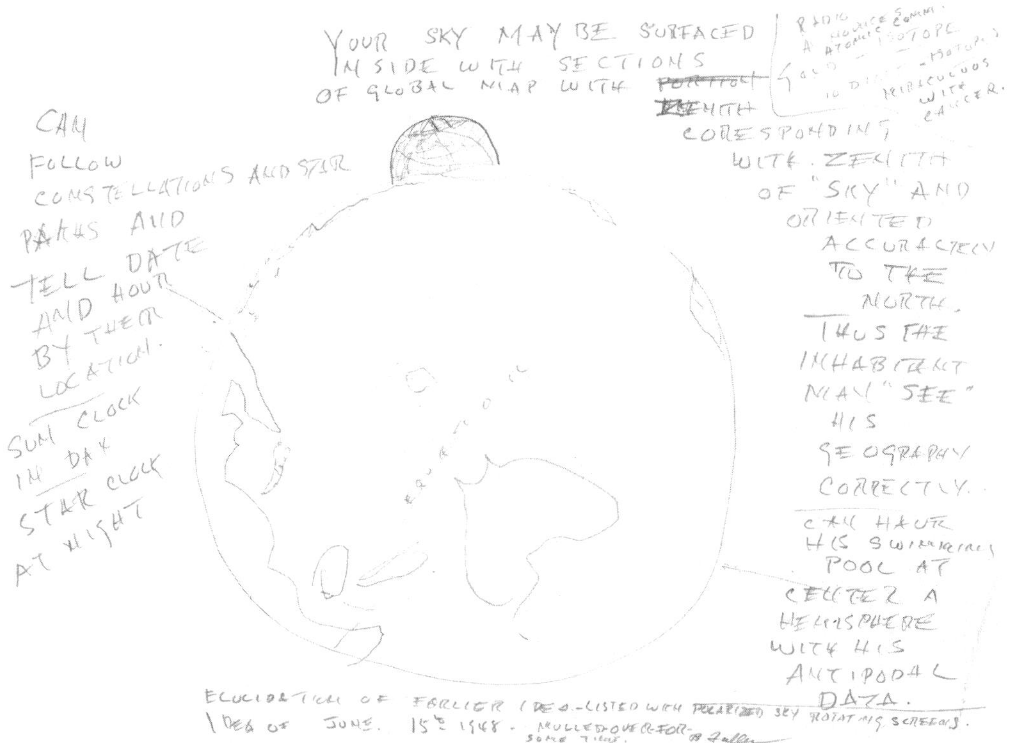

Glass Globe of the Earth

The earliest formulation of the Geoscope concept goes back to the year 1928, as a speculative concept in the 4D book. Fuller tries to clarify theoretically what design «from inside out» can mean. He handles the question in such a way that a determination of the human's position in the universe is made necessary. In the absence of a world design, likewise no house can be designed "from inside out." The "from inside out" in the model of the "expanding sphere" also implies, however, a founding of a time-based design methodology, already given expression to in the 4D program. The geodesic hull of the Geoscope supplies the adequate model of an expanding universe.

 ·*Geoscope,* model of the earth as transparent sphere using a geodesic Three-Way-Network. Constructed with students at Cornell University, Ithaca, N.Y., 1952

·Sketch from the manuscript *Your Private Sky,* 1948. The geodesic model of the earth was intended to become the basis for correct perception of the earth's rotation in relationship to the sun and stars

·348: Completed Geoscope, Cornell University. Geodesic structure (diameter 7m) with observation platform in center. The land masses of the continents are represented as cut-out pieces of transparent mesh wire-netting. Their positioning on the surface framework is true-to-reality, so that an observer can correctly perceive the real relationship between places on the earth and the stars.

·349: Geoscope Workshop: Pre-assembly and completion of the geodesic framework with rhomboid-shaped facets. Installed on the roof deck of one of the brick buildings on campus

·352/353: Geoscope land masses as a transparent screen.

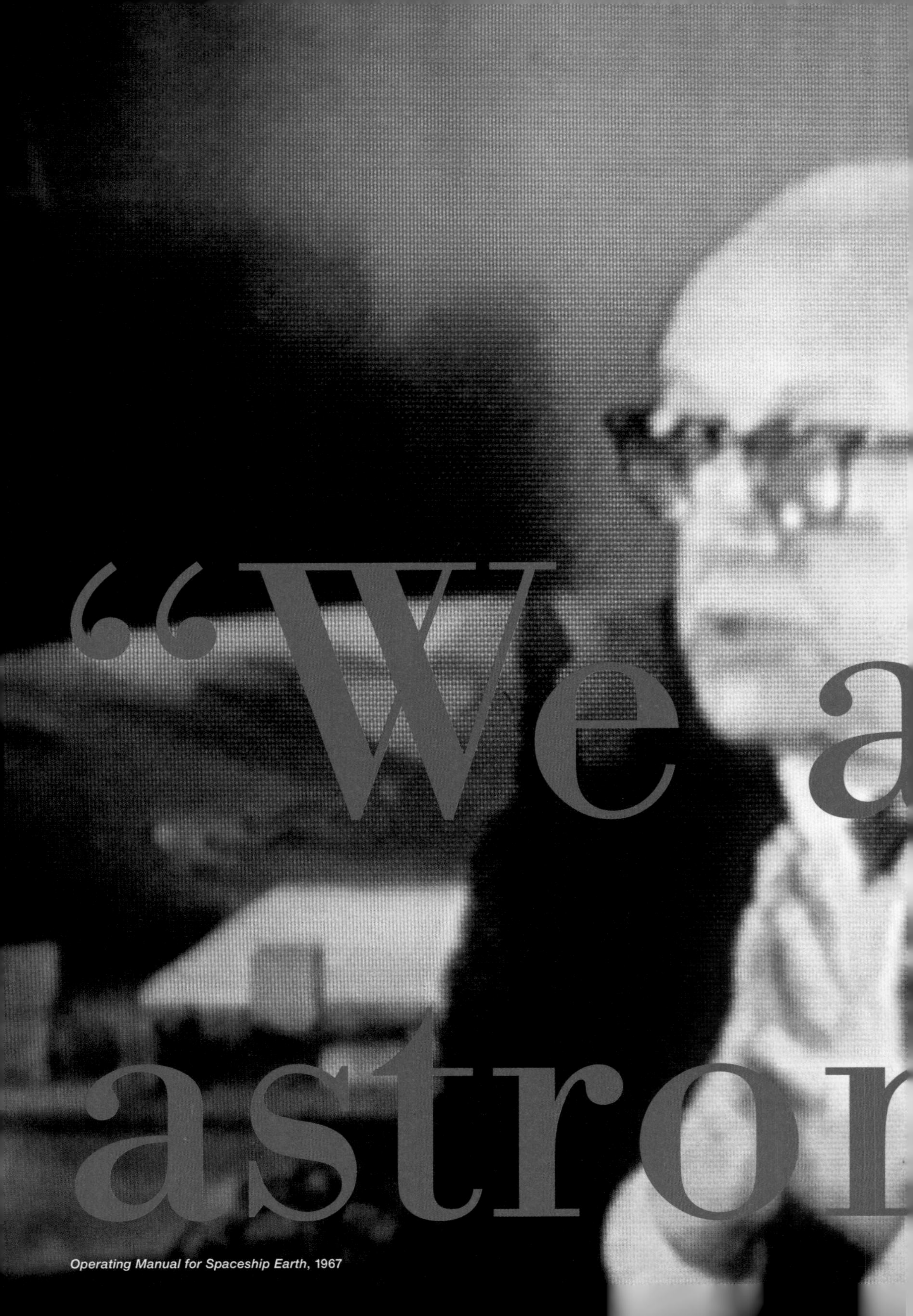

“We a

astron

Operating Manual for Spaceship Earth, 1967

re all

nauts"

Geodesic domes have become Fuller's trademark. Patenting them was the only financial success of his life. Since they were introduced, the domes have been signficant in two different ways. As technical artifacts, they aimed at maximum efficiency in the relationships of volume to weight, use of materials to useful surface, and assembly time to mobility. As sociocultural alternatives to typical rectangular architecture, the domes crystallized society's dreams of a life liberated from constraints and tutelage.

GEODESIC DOMES

·355: *Union Tank Car Company,* Baton Rouge, Louisiana: dome with span of 130m, rod structure with suspended

·*Fort Rotunda,* exhibition building, Dearborn, Michigan. Originally built for the 1933 Chicago World's Fair.

Ford Rotunda Dome

In February and March 1953 I was flying back and forth to Detroit working on the Ford Rotunda in Dearborn. I made the preliminary design of the dome with a crew from Yale and M.I.T. The testing of the Octet trussing for the dome was done at the aeronautical engineering laboratory at the University of Michigan. We put electric strain gauges on the truss members during static load testing of the central tetrahedron of the whole group. The Ford structure was designed using only six different lengths of members and all the holes were pre-punched to a tolerance of 1/1000 of an inch. There were 72 different orientations of end holes. When the structure was erected the holes lined up perfectly for a very right riveting job.

The first step was making up the triangles and then combining them into octahedrons. The octahedrons were fastened together and thus formed the grid structures. The aluminium spines of the primary structure were fastened together with four different types of hubcastings. Three spines put together made one large triangle. Inside this triangle we had ten octahedra which were placed on a flat plane and then lifted up. The octahedrons cluster together in an octahedra-tetrahedra-system, a system that is self-centering and self-aligning in such a way that you can lift up the whole group without putting a single fastening in. No triangle will fall out. However, they were firmly fastened together to reinforce the spines.

The second sequence of operations consisted of putting the gusset and shear plates into the Octet truss. The workmen installed them very rapidly. It was estimated that the total work on each truss would take two hours, but they were riveted together at the surprising rate of one every 20 minutes.

After the dome was designed we put up scaffolding which consisted of a 30 by 100 foot deck with a turntable on ball bearings and hydraulic lift so that the dome could be sent aloft and by doing so use my system of working from top to bottom. The Ford Dome workers never had to leave the deck to assemble the dome. When it was finished they were able to climb safely on the feather-weight parts. As the structure went up, looking at it from the court, it took on very interesting patterns. We saw a larger and larger dome turning and rising away.

The dome was to be covered with polyester fiberglas that was supposed to be prefabricated in a shop, but the manufacturer could not get the material in time. For this reason, all cementing work was done outdoors and proved to be unsatisfactory. When the dome was finished we put a vinyl skin over the whole thing and it seems to be doing all right.

It is significant that in this dome we obtained very reliable data on the structural performance of domes. [...] The dome was finished in 5 weeks, and as soon as the electricians put on the lights, we took some pictures, and it really looked exciting. The effect under the yellow lights was extraordinary.

Architecture Out of the Laboratory (1955), 19

 ·*Octet-Truss*, Octet Grid, geometry of several models

·The elements of the Octet Grid: octahedral and tetrahedral

Even before the patent for the principle of the geodesic dome had taken effect, Fuller was commissioned to design a dome for a popular tourist attraction, the Ford Rotunda. This building, a light-construction steel building by the architect Albert Kahn had been a decorative, colorful night attraction at the Chicago World's Fair in 1933 ("A Century of Progress"). For the fiftieth anniversary of the Ford Motor Company, Fuller was to design a cover for a circular interior courtyard of 28 meters in diameter. They chose Buckminster Fuller because the alteration could result in significant additional stress on the building's support structure. So, at the age of fifty-eight, Fuller received his first important commission, for which there was little time available for development. "Bucky Fuller Finds a Client" was the title of a small piece in "Architectural Forum" about it. Because of the large span, Fuller opted for a construction of spatial supports in a triangular arrangement: the octet truss (octahedral-tetrahedral structure). That was his name for a spatial framing with sides of equal length comprised of tetrahedrons and octahedrons in alternation that define a volume with the height of the tetrahedron and a normal projection that forms a large triangle by means of the inclusion of additional tetrahedral cells. Fuller saw the octet truss as intimately related to the problem of the "closest packing" of spheres. The octet truss has two parallel main surfaces, but its structure is fully triangulated, and thus it abandons the usual patterns of Euclidian space. In the interest of minimizing the weight, Fuller and his colleagues selected aluminum as the material and shaped it into rods with a triangular cross-section. (Louis I. Kahn, who met regularly with Fuller at that time, used the octet truss in the Yale Art Gallery in New Haven, Connecticut, as a cast concrete construction method, that is, not in Fuller's linear pattern but with "interpenetrating" concrete bands.) This was the only time that Fuller applied the octet truss, with its horizontal construction that is somewhat reminiscent of the molecular structure of graphite. Even at that time, however, Fuller

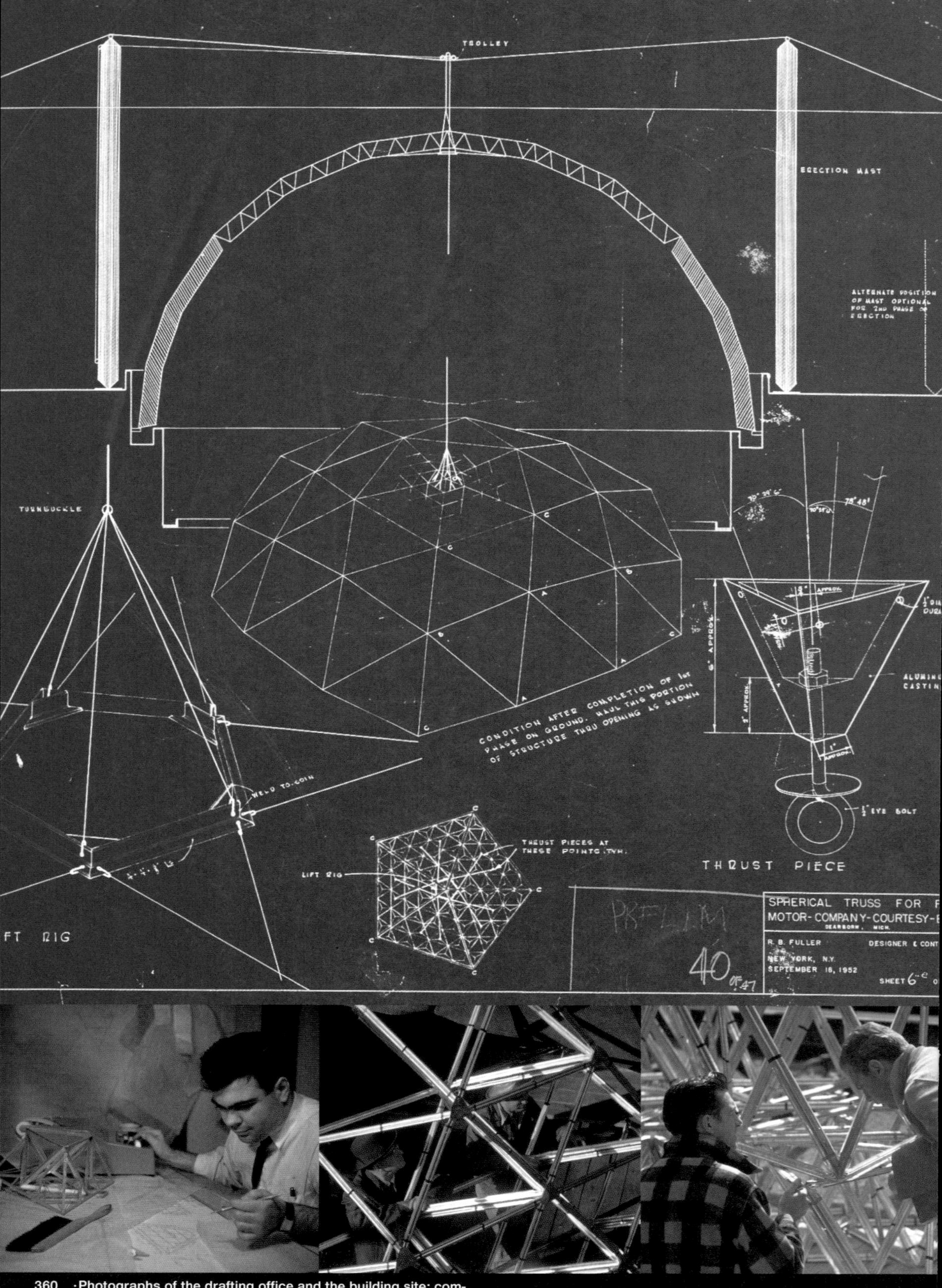

 ·Photographs of the drafting office and the building site: complex arithmetical calculations, high-precision production tended more to “spatial” structures, that is, structures that developed on all sides. In Dearborn, Michigan, the decision was probably driven by the intention to use a continuous skin of synthetic material, which was well suited to the flat facets of the octet truss. The dome was covered with a vinyl skin, because the polyester fiberglass that had originally been selected could not be produced to adequate standards.

In 1960, Fuller realized another dome for the Ford Motor Company, this time with a diameter of about thirty-five meters at the base and approximately one thousand square meters of floor space. It was thus almost twice the surface area, but it weighed only a quarter of the Dearborn dome, something that Fuller, ever the statistician, immediately emphasized, as it represented an improvement of eight times in just a few years.

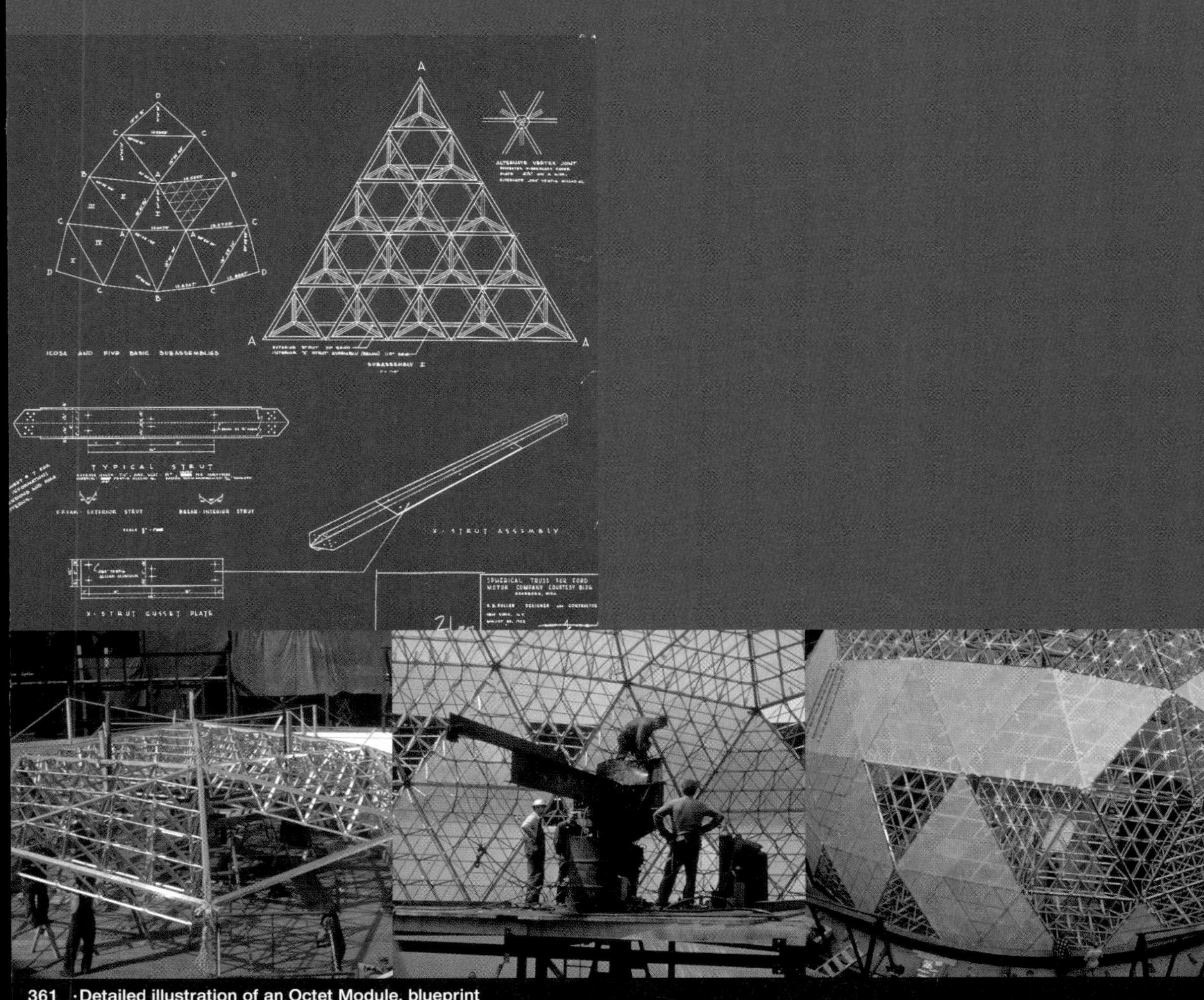

361 · Detailed illustration of an Octet Module, blueprint

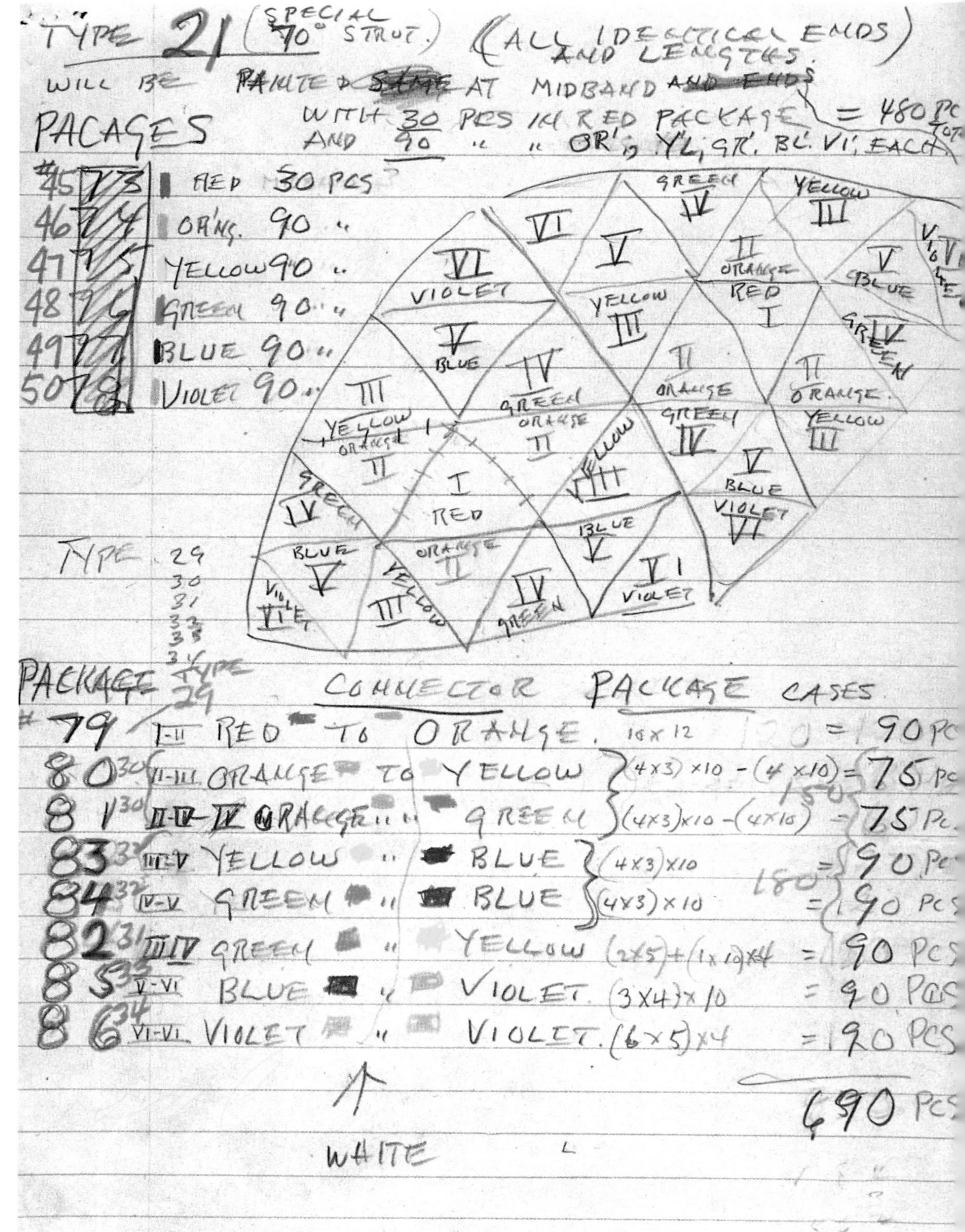
TYPE 21 (SPECIAL 90° STRUT.) (ALL IDENTICAL ENDS AND LENGTHS.)
WILL BE ~~PAINTED SAME~~ AT MIDBAND ~~AND ENDS~~
PACAGES WITH 30 PCS IN RED PACKAGE = 480 PC TOT.
AND 90 " " OR', YL, GR, BL, VI', EACH

#45 73 RED 30 PCS
46 74 OR'NG. 90 "
47 75 YELLOW 90 "
48 76 GREEN 90 "
49 77 BLUE 90 "
50 78 VIOLET 90 "

TYPE 29 30 31 32 33 34

PACKAGE TYPE 29 CONNECTOR PACKAGE CASES

#79	I-II RED	TO ORANGE.	16×12	120	= 90 PC
80 30	II-III ORANGE	TO YELLOW	(4×3)×10 − (4×10)		= 75 PCS
81 30	II-IV ORANGE "	GREEN	(4×3)×10 − (4×10)	150	= 75 PC.
83 32	III-V YELLOW "	BLUE	(4×3)×10		= 90 PC
84 32	IV-V GREEN "	BLUE	(4×3)×10	180	= 90 PCS
82 31	III-IV GREEN "	YELLOW	(2×5)+(1×10)×4		= 90 PCS
85 33	V-VI BLUE "	VIOLET.	(3×4)×10		= 90 PCS
86 34	VI-VI VIOLET "	VIOLET.	(6×5)×4		= 120 PCS

690 PCS

WHITE

363 ·RBF: Sheet of sketches and notes on the typology of Octet elements, undated (1953)
·364/365: Look from below at the Ford Dome.

·The Ford Dome after completion, on the 50th anniversary of the Ford Motor Company, 1953, with city lights of Detroit in the background

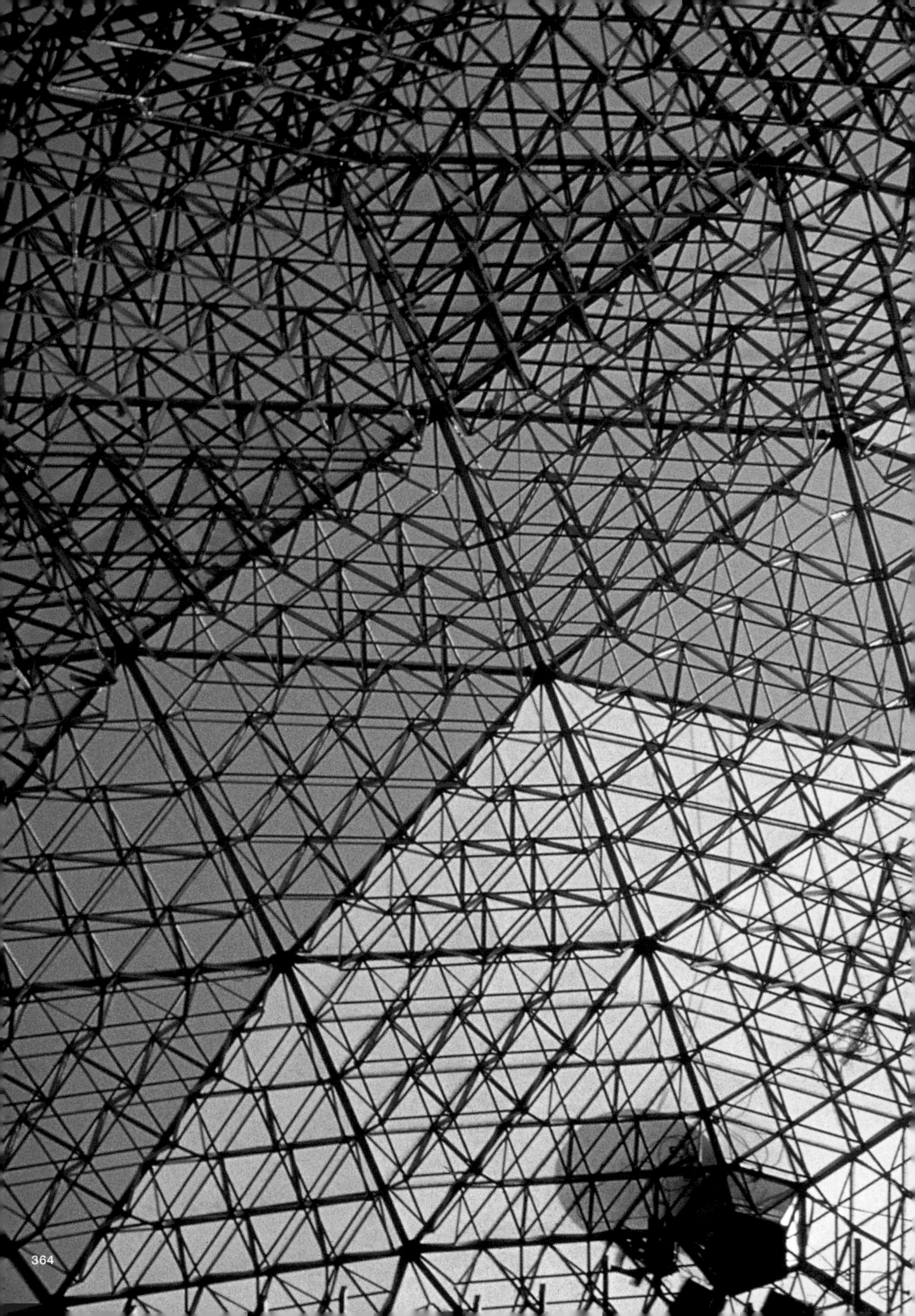

Fuller's Home Dome

"At the end of the day, Bucky and I drove the few blocks from his office to his geodesic dome house on South Forest Street for dinner with Anne. The house was a nice Microcosm of Fuller's universe: spherically coordinate, uncompromisingly simple in design, and at home in its environment. Its scale and weathered-wood framing were quite in harmony with the conventional houses with front porches and side yards that composed the rest of the elm-shaded neighborhood. As you enter the house the first impression is the absence of the familiar four-square cubical framework of rectangular floors and straight walls. The effect is total disorienting to our reflexive assumption that rooms should be shaped more or less like shoe boxes. The result is that Fuller as an architect has created an artifact – like all of his inventions from the Dymaxion Car to the vast dome at Montreal's Expo '67, an artifact intended to instruct. You cannot enter the house on South Forest without receiving a lesson on how we might organize our environment with spherical and hexagonal economics simply not available in a structure where all the rooms have to be cubes. The dome leads our eye in, out, and around – not up and down like the box. The interior walls of the house are a complex of prisms (...)"

E.A. Applewhite: Cosmic Fishing (1972), 13

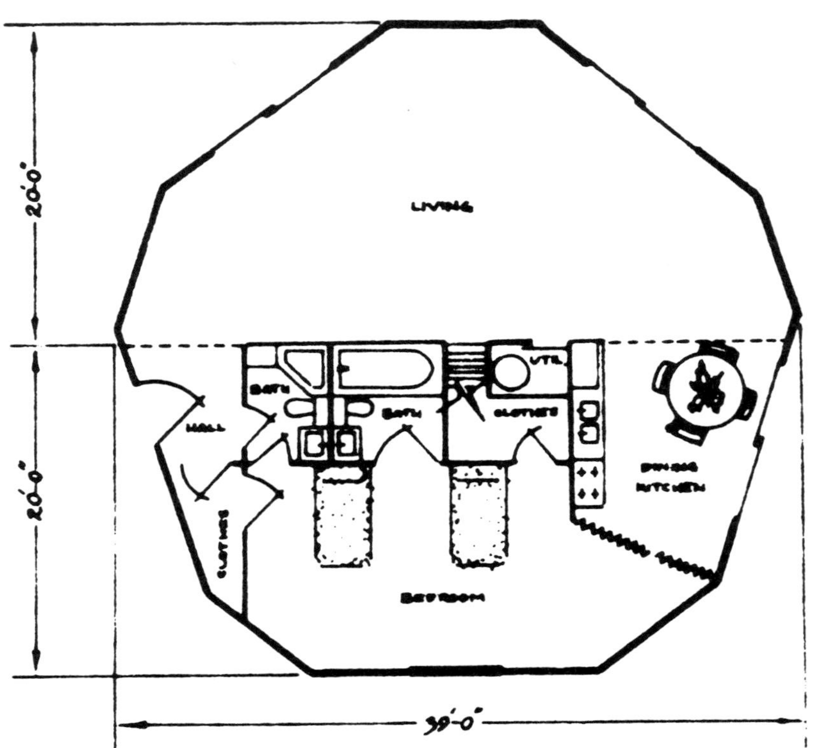

 ·*Home Dome,* floor plan of the ground floor

369 *Home Dome*, view of living room from gallery

370/371: *Home Dome*, library

Roam Home to a Dome

There Once Was a "Square"
With a Romantic Flair
Pure Beaux Arts, McKim, Mead and White
In the Mood that Ensued
He Went "Factory Nude"
Mies, Gropius, "Corbu" and Wright

Roam Home to a Dome
Where Georgian and Gothic Once Stood
Now Chemical Bonds
Alone Guard Our Blondes
And Even the Plumbing
'Looks Good'

Let Architects Sing
Of Aesthetics that Bring
Rich Clients in Hordes
To Their Knees
Just Give me a Home
In a Great Circles Dome
Where the Stresses and Strains
Are at Ease.

Roam Home to a Dome
On the Crest of a Neighboring Hill
Where the Chores are All Done
Before They're Begun
And Eclectic Non-Sense is Nil

RBF

The theme of transporting light constructions by areas continued to occupy Fuller's interest because of the logistical possibilities it represented. The only difference was that in the fifties helicopters replaced the airships that were supposed to transport his 4D house a quarter of a century earlier. The difference between the physical law of an airship, namely, climbing by the principle of "lighter than air," and the downward forced created by a motor seemed to Fuller to be a logical step in the technological development from a passive principle to an active one of "performance." In his eyes, it had always been weapons makers and later the defense industry that was most open and active with respect to new technological developments and that had caused them to accelerate. That, in view of this realization, Fuller did not turn away from the military and find another vehicle for his plans to reform the world, but instead looked to the army – and, characteristically, the navy, which was in his view the creative head on the military body – opened him up to criticism, even from his friends. His answer to this was to say that when the world had been shown its problem and the clearly designed model for its solution, it would immediately step in for the "livingry" and abandon the "killingry." These neologisms are typical for him, and they carry the punch of the apt slogan. This inventiveness with language was his way of showing the industrialized world how to turn swords into ploughshares.

Paper Domes

Last spring we studied and produced a polyester paperboard dome which was assembled without a mast. It was so light that it could be lifted by one person. This dome was made out of flat pieces of corrugated cardboard cut in triangles, coated with polyester and stapled together. This process simplified erection considerably. A second dome was made for the Marine Corps, and although at the beginning they were rather astonished by it, they ended up liking it very much. One of them is now being used in Quantico as a children's playhouse.

On July 1, 1954, the Container Corporation of America told me that they were willing to manufacture two paper domes of my design which I had been invited to do for exhibition at the Triennale of Milan. I developed them so that they could fit into a small packing case. The case had six types of components which coud be folded to form diamond boxes with a six inch interior dead space. The diamonds' edges turbined around each other so that nine of them could make a bigger sized diamond. Triangles printed in the diamonds' faces could be cut open on three radial lines so that they made a triangular set of windows. The outer diamonds fold inwardly, and the inner ones fold outwardly making it possible to have windows wherever one needs them. In the Triennale domes every triangle window was open and translucent bathing cap skins fabricated at our Geodesic shops were very tightly stretched over them. These domes were erected without a mast, starting at their equators and then building them up. When the domes were lighted from the inside they gave a jewel-like effect.

I am quite happy the two domes went out of the country and also that they were produced in relatively few days. The process of manufacturing them included moving from one place to the other, and calculating their exact dimensions on calculation machines. Later it may be possible to produce such domes at the rate of 3000 a day, and with the new developments it will be possible to have a strong, waterproof structure which would be very light and very cheap. The domes produced a critical effect in Italy. At the Triennale there were people from all over the world, including representatives from the Iron Curtain countries.They were aware of the mute message that excited them and caused them to realize that somehow that kind of low cost package which produced a high standard of living and environment control, and living was coming out of American initiative and industry. It also indicated to them a new departure in the overall attitude toward Architecture. The Italians gave it that new name which I mentioned at the beginning: "Architecture out of the Laboratory".

Architecture out of the Laboratory (1955), 30

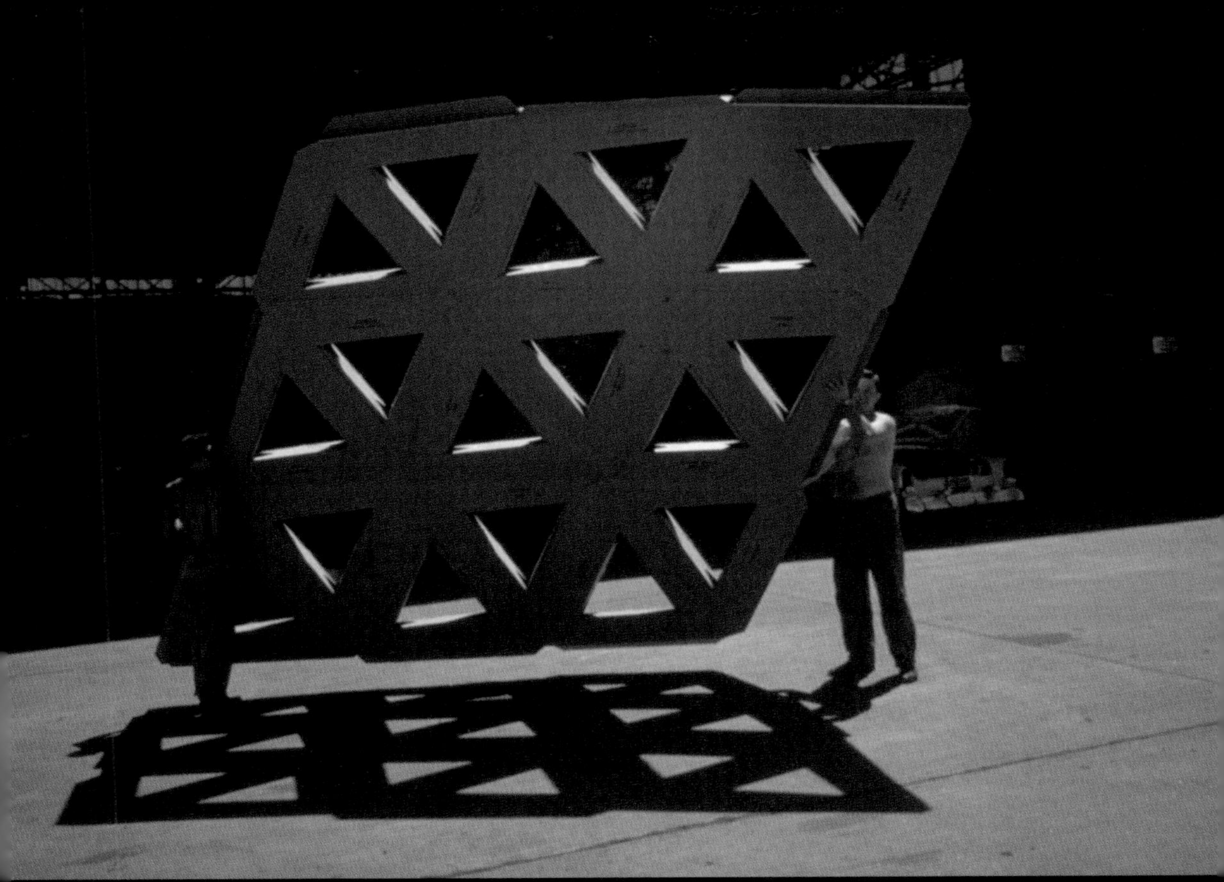

375 ·Two workers carrying a section of the dome's shell across the runway prior to its transport for Europe

·*Triennale di Milano,* 1954, one of the two large domes made of corrugated cardboard, approximate diameter 13 m

377 ·380/381: *Plydome,* experimental construction out of molded plywood, around 1957. The resulting square

·First row: South Pole Dome during construction and after completion, Poly Frame/Temcor (1974): Exhibition Pavilion in Kabul/Afghanistan (1957); Calico Cotton Pavilion in Bombay, exterior and interior (1958).
·Second row: Dome in Aspen, Colorado (1955); Radome, radar dome, part of the DEW-early warning system line at the Polar Circle (1955); Construction of the Kabul dome; Exhibition Pavilion, St. Louis, Missouri, 30 m (1956); Test dome by Jeffrey Lindsay, Canada (1953).
·Third row: Geodesic stable, exterior and interior (1954); Test dome for the Marine Corps (1954); Kaiser Dome, Honolulu, 50 m (1957); Fly's Eye Dome, Monohex Geodesic Structure (1965).
·Fourth row: South Pole Dome, view from interior; Dome from Jeffrey Lindsay at night; Climatron, St. Louis, Missouri (1960).

The industrial magnate Henry J. Kaiser – among other things a producer of aluminum – was one of the many licensees of the geodesic dome structure. In order to demonstrate the superiority of the new principle of construction, he thought up a presentation to generate publicity. In Honolulu, Hawaii, a building was to be constructed in a single day. Together with his assistant Don Richter, Fuller designed a dome just under fifty meters across, the collapsible structure was made of aluminum sheets that were braced by rods on the outside. The individual parts were premanufactured in California and shipped across the Pacific to Hawaii. The assembly proceeded seamlessly; the folding elements were first combined into rings on the ground. These rings were then successively raised on the assembly mast and formed into a dome; the dome was built from the peak down to the ground. In contrast to the thousand-year-old tradition of building in stages from bottom to top, this is probably the most striking articulation of Fuller's conviction that the tensile principle is vastly superior to the principle of pressure. The hoped-for success of this demonstration was modest. After twenty-two hours of construction, the structure was opened with a symphony concert, as a demonstration that they had an ear for a good drum roll in both the literal and figurative senses. Fuller would long continue to praise the excellent acoustics of this space, and its many facets do certainly create an unusually "dry" sound.

383 ·*Kaiser Dome,* Honolulu: 50m-diameter dome, faceted surface support-structure, designed for industrial magnate Henry

·384/385: Grand opening of the *Honolulu Dome* with evening symphony concert, February 1957, the same day the structure had

The Union Tank Car Company was one of the licensees of Fuller's patent. It needed a large space for the overhaul of its container cars. In 1958, the geodesic supporting structure became the largest unsupported covered surface, with a diameter at the base of almost one hundred thirty meters. In contrast to the dome in Honolulu, the one here was an external tractive vault from a supporting structure of tubes arranged in hexagons from which were anchored sheets of metal that were folded inward. Fuller mentioned the remarkable diameter of the supporting tubes, just twenty centimeters, which are not in visible in the aerial photographs. The aerial photographs reveal a proud, rounded form of abundant plasticity in terms of the effects of light and shadow. Fuller was conscious of the morphological proximity to sea creatures like radiolarians. A second, identical dome was built shortly afterward by the Union Tank Car Company in Woodriver, Illinois. Fuller used the dome structure from Baton Rouge in a context that was highly significant. He spoke of a visit he had long been planning to the Andalusian city of Seville, which he finally made in 1962. He had long been aware of the history of the construction of the cathedral, the second largest after St. Peter's: a struggle with the material world that lasted through five generations. Fuller learned that the cathedral of Seville would easily fit within the Union Tank dome of Baton Rouge, even its height. Yet the dome weighed only as much as four of the more than one hundred stone columns that make up the structure in Seville. Fuller tirelessly emphasized such facts, thoughtfully contrasted the one to the other, the past to the present, the old to the new. He himself placed the argument heavily on the technological and on the efficiency of the structure. The interior of the Baton Rouge dome is, in technical terms, a suspended calotte, but it suggests more than that. It suggests the vault of the sky, and it documents one way in which "limitlessness"

·Dome for the *Union Tank Car Company,* Baton Rouge: aerial photograph after the beginning of construction (Fall 1958) and photograph of the finished dome (October 1958)

·388/389: Interior of the Baton Rouge dome, diameter 130 m, height at peak 40 m; in center, the service equipment for overhauling of its container cars can be put to form. This idea is also reminiscent of Schinkel's stage design for Mozart's "The Magic Flute." It is not important here whether Fuller was familiar with the design – probably he was not. What is important is the expression implicit in a building that is so technical. The collection of these raw container cars smelling of grease beneath this overarching screen is more than a chance arrangment. It is astounding as a sign of a kind of thinking that is always discussing how to transcend technology into a new world.

·390/391: *Spruce Goose Hangar,* dome for Howard Hughes' plywood airplane "Spruce Goose," Long Beach, 1976.

seen moored next to the dome.
·(below) Three phases of the construction and interior view

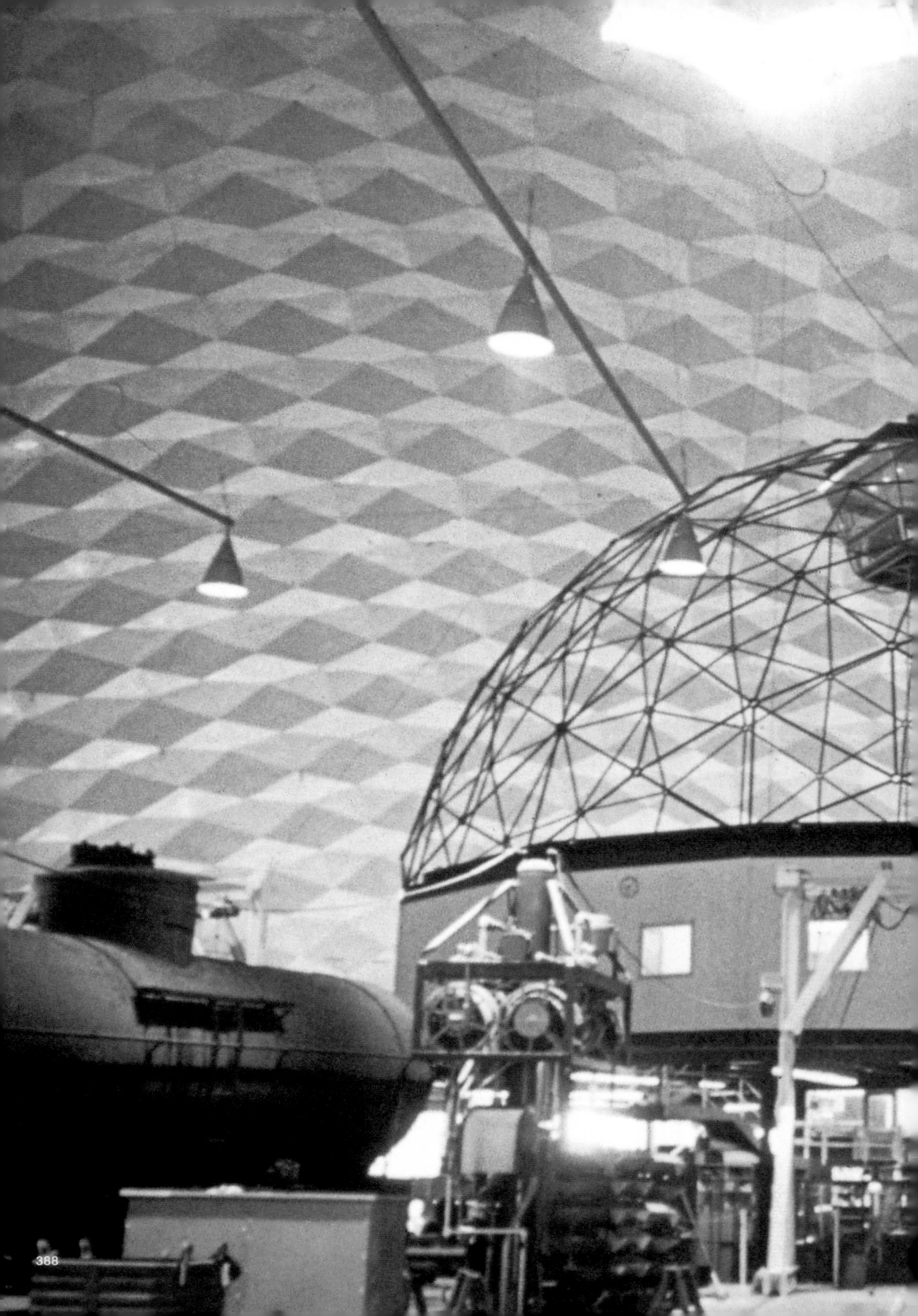

Tensegrity is a concept whose terminology and theory derive from Fuller, but in whose formulation and constructive elaboration many others took part, especially Fuller's students. Tensegrity is an anticlassical approach to structure and construction, and it implies a reversal in perception: what appears to be compactly standing and solid proves to be suspended and ephemeral. Fuller was the first to notice this reversal with respect to the wire wheel, and he recognized it as a revolution in construction. However, for him, tensegrity is also a philosophical model of coherence. By what is something held together, then, if not the compact mass? By increasingly thin tractive members that border on the spiritual.

TENSEGRITY

Wire Wheel

In the high- and low- tide cooperative precessional functionings of tension versus compression I saw that there are times when each is at half tide, or equally prominent in their system relationships. I saw that the exterior of the equatorial compressional island rim atoll of the wire wheel must be cross-sectionally in tension as also must be its hub-island's girth. I also saw that all these tension-vs.-compression patterning relationships are completely reversible and are entirely reversed as when we considered the compressively spoked "artillery wheel" vs. "The tensionally spoked wire wheel." I followed through with the consideration of these (...) functions of structural systems as possibly disclosing the minimum or fundamental set of differentiability of non-redundant, precessionally regenerative structural systems.

As I considered the 12 unique vectors of freedom constantly and non-redundantly operative between the two poles of the wire wheel – its islanded hub and its islanded equatorial rim-atoll, in effect a Milky-Way-like ring of a myriad of star islands encircling the hub in a plane perpendicular to the hub axis – I discerned that this most economic arrangement of forces might also be that minimum possible system of nature capable of displaying a stable constellar compressional discontinuity and tensional continuity. A one-island system of compression would be an inherently continuous compression system, with tension playing only a redundant and secondary part. Therefore, a one-island system may be considered only as an optically illusory "unitary" system, for, of course, at the invisible level of atomic structuring, the coherence of the myriad atomic archipelagos of the "single" pebble's compression-island's mass is sum-totally and only provided by comprehensively continuous tension.

This fact was invisible to, and unthought of by, historical man up to yesterday. Before this mid-twentieth century fact discovery, there was naught to disturb, challenge, or dissolve his "solid-rock," and other "solid things" thinking. "Solid thinking" is even dominant (...) over the everyday logic of many otherwise elegantly self-disciplined nuclear physicists.

As I wondered whether it was now possible for man to inaugurate an era of thinking and conscious designing in terms of comprehensive tension and discontinuous compression, I saw that his structural conceptioning of the wire wheel documented his intellectual designing breakthrough into such thinking and structuring. The compressional hub of the wire wheel clearly islanded or isolated from the compressional "atoll" comprising the rim of the wheel. As these compressional islands were only interpositioned in structural stability by the tensional spokes, I said that this was clearly a tensional integrity where tension was primary and comprehensive and compression secondary and local. This reversed the historical structural strategy of man.

Synergetics (1975), 353

 ·393: RBF with Tensegrity models, Southern Illinois University, 1958

·Wire-spoke wheel with tangentially-braced spokes. Identified by RBF as first Tensegrity structure.

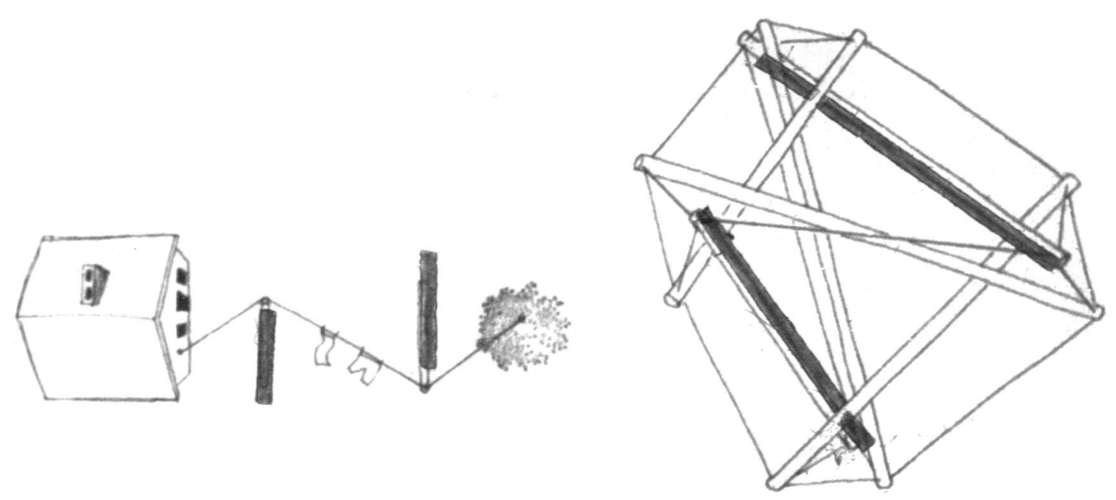

395 ·Placing clothes-line under tension using two poles: the line tends to move away from the end of the pole. ·A spherical Tensegrity: the rods pull the line to form an angle at the rod-ends.
·No rod touches another. Detail from a geodesic Tensegrity Sphere.

Tensegrity

Excerpts from Buckminster Fuller's Discourse on Structural Exploration and Invention

I do have patents covering geodesic structures in America and in many other countries, including England. One cannot patent geometry per se nor any separate differentiated out, pure principle of nature's operative processes. One can patent, however, the surprise complex behaviors of associated principles, where the behavior of the whole is unpredicted by the behavior of the parts, i.e., synergetic phenomena. The latter is what is known as an invention, a complex arrangement, not found in nature, though sometimes superficially similar to nature. Though superficially similar in patternings to Radiolaria and Fly's Eyes, taken out of water. Fly's Eyes will not provide structural precedent or man-occupiable structures.

The processes of engineering, up to the moment of introduction of my invention of geodesic structures, are predicated upon the stress analysis of individual beam and column behaviors, as separate components and thereafter upon comprehensively organized beams, columns, and cantilevers as a solid compressional overall integrity of cohesion, aided here and there by tensionally exaggerated sinews – tension being subordinate and local. Therefore, engineering as academically constituted in 1951 could in no way predict the associated behaviors of geodesics, in which any one, several or many of the components could be removed without, in any way, jeopardizing the structural integrity cohesion of the remaining primary structure.

I have visited approximately all the leading universities and engineering schools in the U.S.A. (only as their spontaneously invited guest, as I allow no promotion or agentry solicitation of engagements) and I have been a guest of major architectural and engineering societies around the world. I am able to state from the direct testimony of its leaders, that world engineering not only

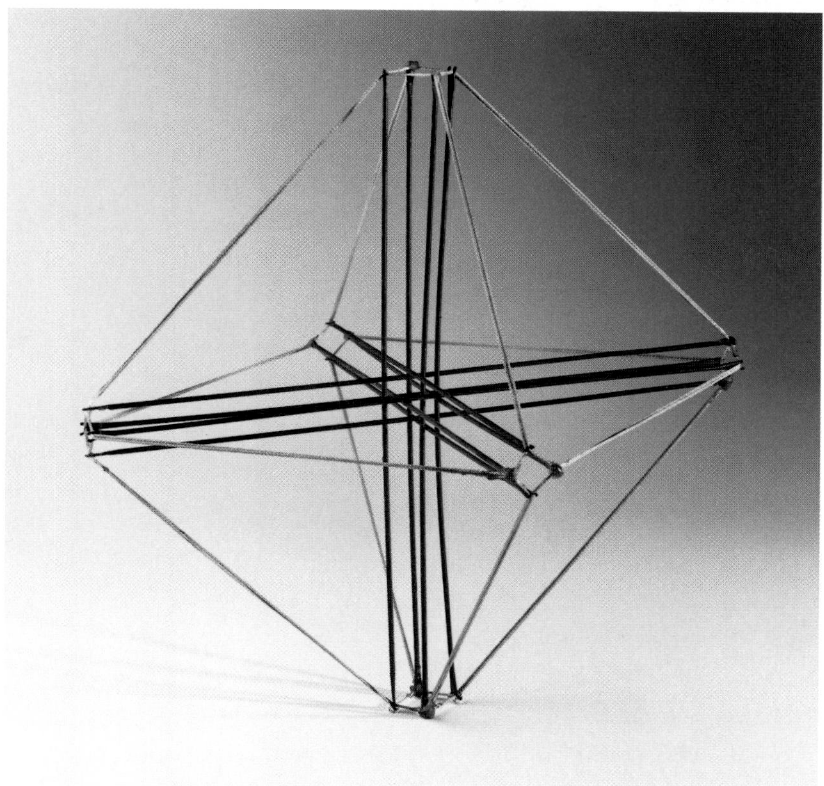

was surprised by the geodesic behavior but clearly stated that it was unable to explain or predict the unprecedented performance per pound efficacy of the geodesic structures by any of the academically known mathematical principles of analysis. Engineers professionally charged with the responsibility of validating my (or my companies') geodesic dome undertakings have only on recourse, in lieu of an extant and proven theory of geodesics and their stress analysis strategies, and that is to direct static load testing accompanied by progressive electric strain gauge readings with loading carried through to "failure." Thereafter formulas are derived, which do not provide any general theory but only special knowledge concerning this particular case.

For these reasons I have had to develop a completely new strategy of educational exposition of synergetic behaviors. It is because I have developed the geodesic theory and not only the invention but this effective strategy of exposition, that I am, and have been, invited to conduct original seminars in the primary educational institutions around the world, as well as before the important annual professional conventions of primary engineering-architectural bodies of these countries.

Whereas, engineers told me before my full-scale demonstrations of geodesic structures, that geodesics would not work, now after one-third of a century of such expositional engagements, discourse and demonstrations, and their consequently mounting journalistically worldwide reportage, it should become increasingly logical that there should now be such an effect upon the general climate of technology that a spontaneous association of the plurality of factors entering into geodesic componentation should be employed by individuals with the now well established foregone conclusion that satisfactory structural stability will thereby result – without the individuals even having to know why. It would result simply because of a new subconsciously established conviction generated by hundreds of experiences of news photographs of succesfully standing structures, ranging from desk models to the largest clear span enclosures of all history, many of

them flown into installation sites at the world's formidable polar, mountain and desert regions. All these have been news simply because these results were synergetic surprises and therefore not obvious. The ability to copy patterns or to make new ventures in higher frequency or more asymmetrical employment of the geodesic structural integrity than has yet been undertaken, is neither invention nor warrant of exemption from the temporary economic, political authority granted to individuals as patents around the world. Patents are the result of all the progressive evidence of history that the majority's commonwealth is benefited by the few in increasing degree in direct proportion to the degree of incorporation of social means of establishing the regenerative initiative of the legally demonstrable inventors. Patenting, though not interoperative, between the countries on either side of the world's Iron Curtain, has survived and is being extended on both sides of the Iron Curtain, with separately total integratability in respect to their two prime domains.

The reports you have received are correct regarding the structures which I have designed for firing to the moon. Your 600 lb. is top for what will be our openable structure of 36' diameter. It may be as low as 300 lbs. There can be larger units. Of primary interest to engineering is the fact that my potential prototypes of satellite and moon structures (on which some patents have been granted and for which others have been applied) are tensional integrity, omnitriangulated, high-tensile-cabled, spherical nets in which local islands of compression act only as local sprit stiffeners. The stiffeners are so oriented that they angle inwardly and outwardly between comprehensively finite, exterior and interior, tensional, spherical nets, thus producing positive and negative waves of action and reaction in inter-stabilized dynamic equilibrium.

Recourse to this discontinuous-compression, continuous-tensioning structure was not obvious to man, therefore the following needs to be observed.

Whereas compression members have an inherent limit ratio of section diameter to length, tension members have no inherent ratio of section diameter to length. The Greeks, who built entirely in compression, discovered that a stone column's slenderness ratio was approximately 188 to 1 of length to diameter. Modern structural steel columns with an integral tensional fibering unpossessed by these stone columns have a limit slenderness ratio of approximately 33 to 1. If we have better metallurgical alloys we can make longer and longer tension members with less and less section – apparently ad infinitum, but not longer compression columns, ad infinitum.

Tensegrity (mimeographed manuscript, 1958), 1–3

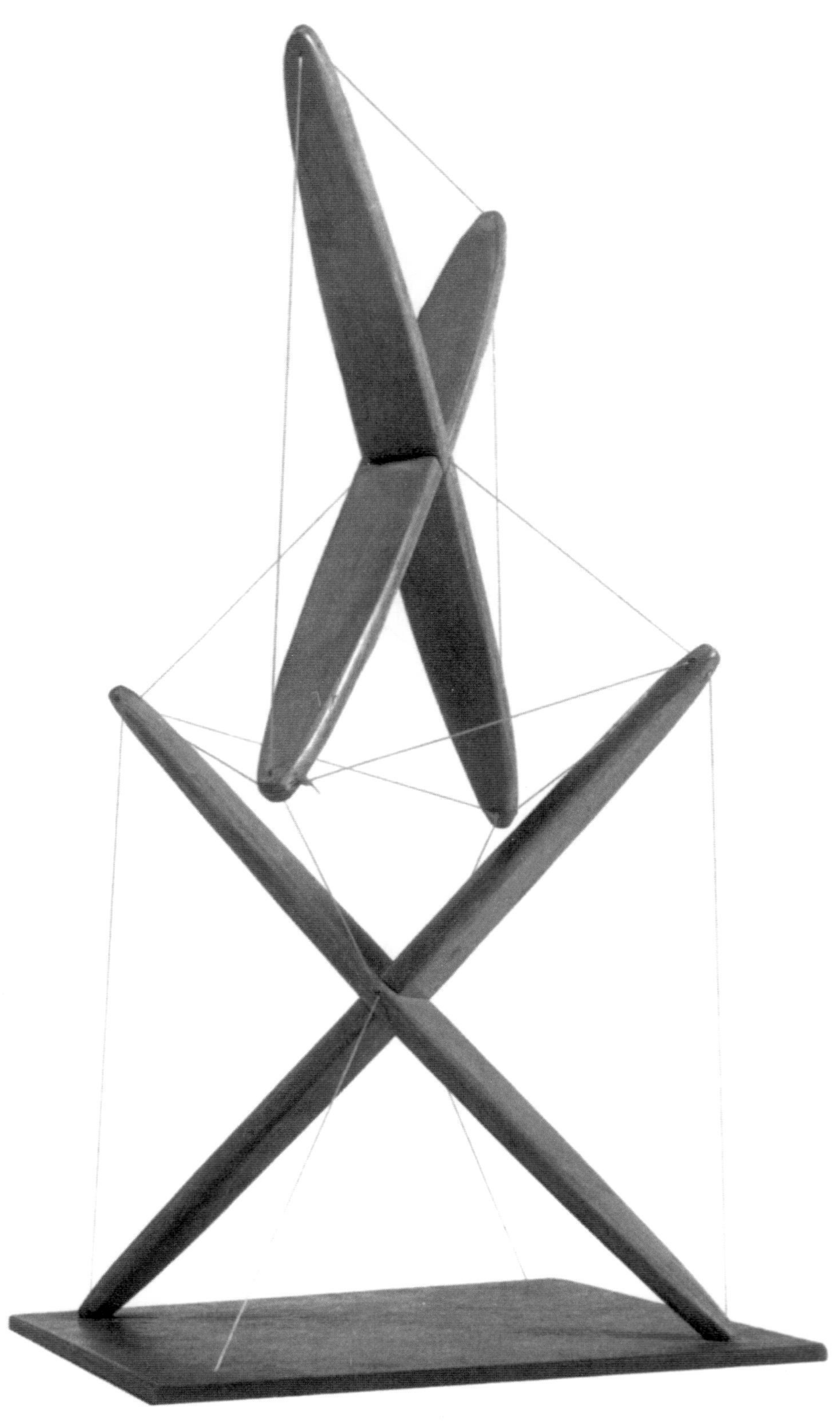

Kenneth Snelson

Fuller recognized the contribution that Kenneth Snelson made, while still a very young student, to the discovery of the tensegrity principle. In Fuller's view, the Dymaxion House was a preliminary step to this principle of tension-integrity, where continuous tensile stress and discontinuous tractive tension interact with a maximum of distribution of labor. Still, the difference between the mast principle of the Dymaxion House and a tensegrity mast is that the former is hierarchical and the latter is not. The anchor mast goes up, and the tension cable goes down from its tip; thus, it is related to a fountain as Fuller himself emphasized in 1928. In a fountain, there is a moment of transition between the deceleration during the ascent and the acceleration of the descent. Fuller called this principle polarized. In the tensegrity principle this hierarchy becomes fragmented; the principle is multipolar. Either the sphere is rounded in the equilibrium of the whole that results from many local forces, or the tensegrity mast ascends from below into a height (or into a distance), push and pull working into a growing dimension by means of antagonistic equilibrium and in regular phases. The image of a rope team of climbers, which Fuller later used for the World Game, is already suggested here – in the truest sense of the word.

·*X-Piece,* Kenneth Snelson's sculpture made of plywood and nylon, December 1948; his pioneering contribution

·*Tensegrity Mast,* Museum of Modern Art, New York, 1959; the construction came out of a 1950 workshop led by RBF

"Don't fig

use them.

ht forces,

”

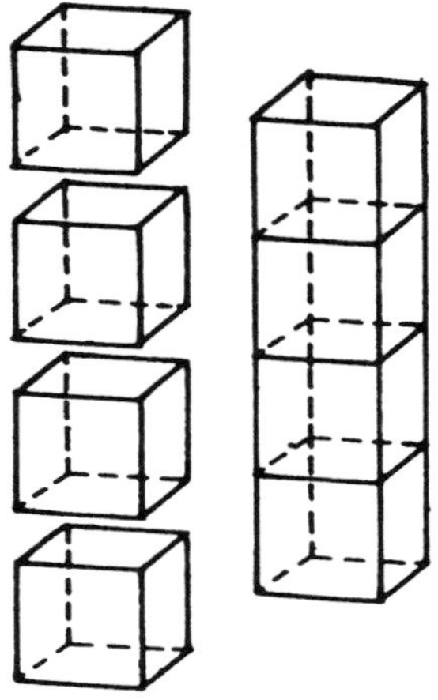

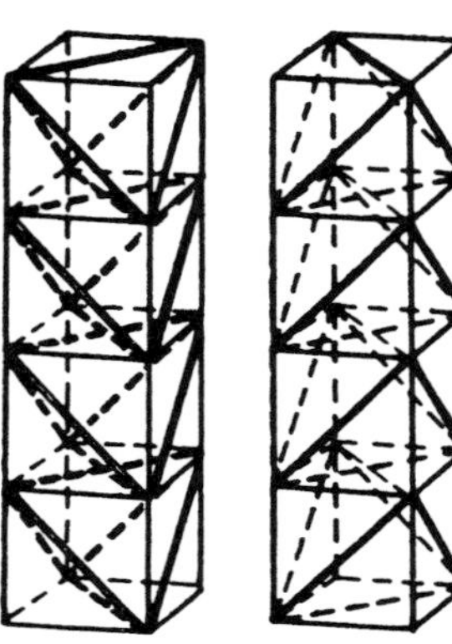

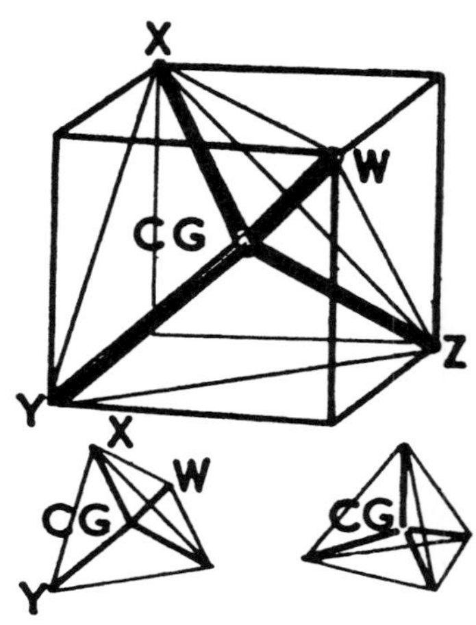

·Schematic construction of the *Tensegrity Mast* out of a “positive” and a “negative” column of cube-inscribed tetrahedrons, which spread linearly and vertically out of two rod-pairs in center, connected by a positionable ball-joint, so that the alternating cells hold together only through the tensionality of the cables

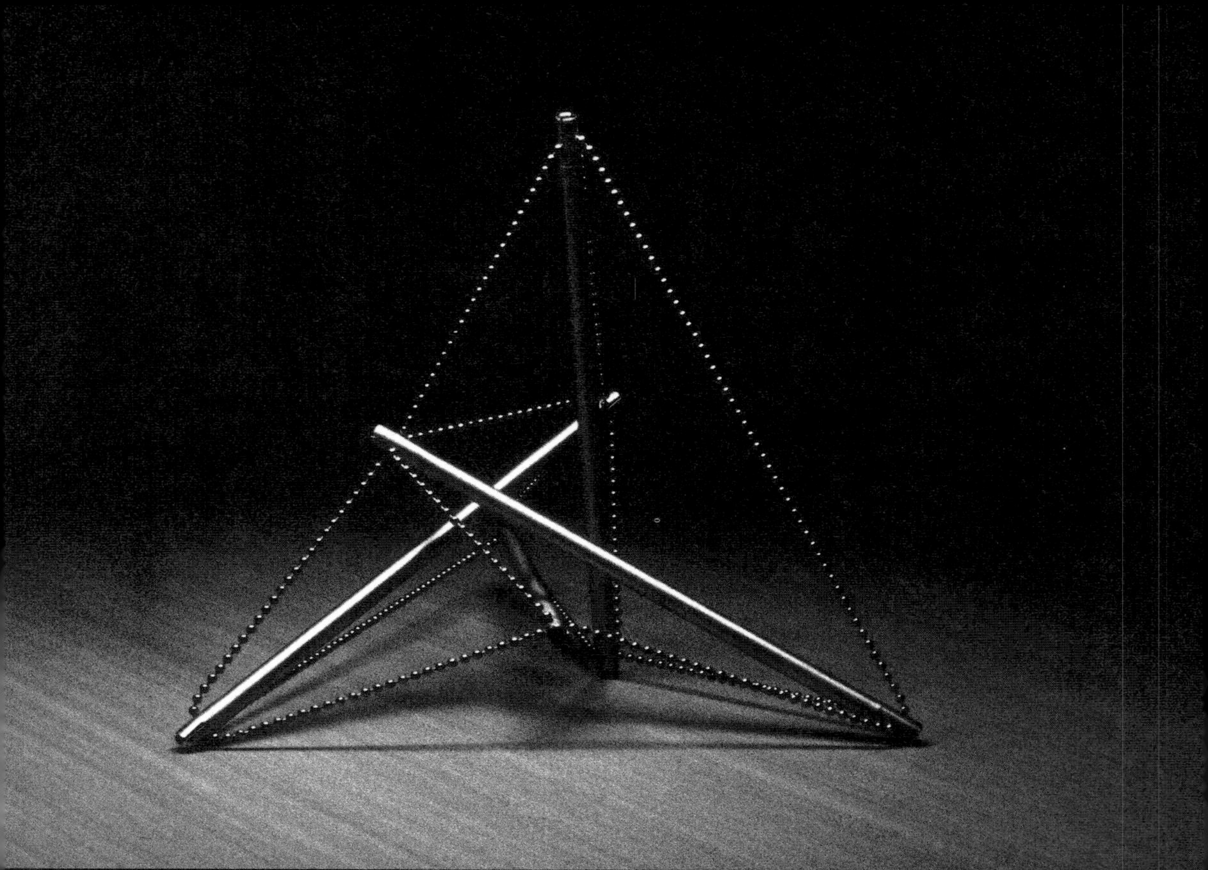

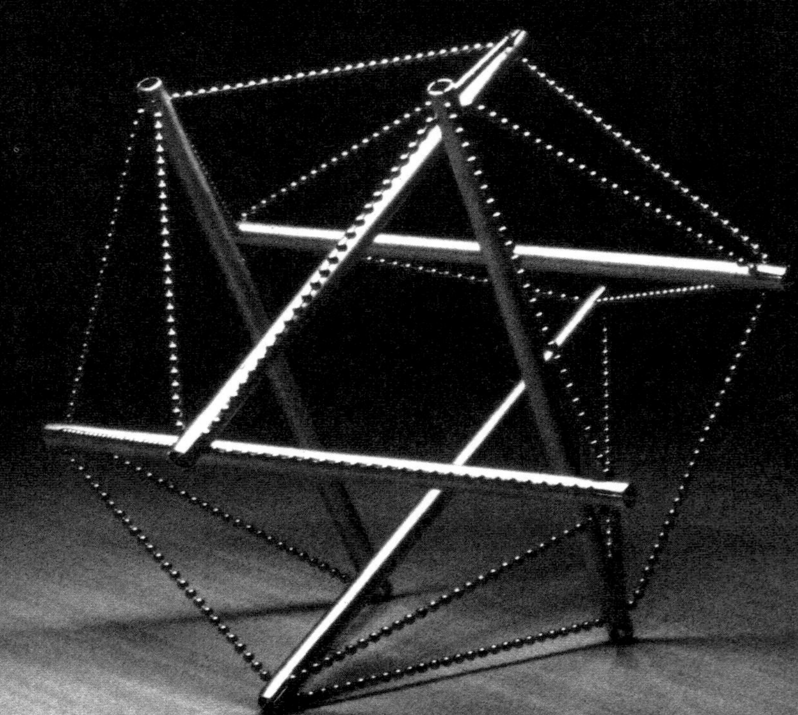

 · All polyhedrons can be modeled using Tensegrity structures. The beaded cords serve as a visualization of the tensionality. Below: Tensegrity-Icosahedron fashioned from three pairs of parallel rods. Model RBF, Black Mountain College, N.C. 1949.

407 · *Tensegrity Hemisphere* using the same components: the multi-rod Tensegrity structures are composed of sub-groups of three rods each. These sub-groupings form triangles, which are sometimes more narrower, sometimes wider. Model ca. 1959.

U.S. Patent – 3,063,521
Application – August 31, 1959
Serial No. – 837,073
Patented – November 13, 1962

Nothing in Universe touches anything else. The Greeks misassumed that there was something called a solid. Democritus thought it could be that there were some smallest things in those solids, to which he gave the name atom. Today we know that the electron is a remote from its nucleus as is the Earth from the Moon in respect to their diameters. We know that macrocosmically none of the celestial bodies touches each other. So, both microcosmically and macrocosmically nothing touches. It was Kepler who discovered the tensional coherence of the solar system despite the millions of miles that intervene between the celestial bodies. Isaac Newton hypothesized the formula for the rate of variance of the interattractiveness of celestial bodies, which proved on comprehensive employment to be correct.

From 1927 on I sought to discover how to produce what I call tensional integrity structures. I have said many times, at some of the top engineering schools in the world, structural analysis is predicated on compressional continuity and nature doesn't use it ever. There is no way for structural analysis to analyze a geodesic dome. This still continues to be true. I think it is reprehensible. The only way to analyze it is with pneumatics and hydraulics. At the molecular level, this is the method of quantum mechanics; it could never be done with crystalline continuity, for such continuity does not exist in Universe.

This brought about my tensional integrity structures, which name I contracted to tensegrity structures. Tensegrity structures are the essence of all geodesic domes. When we increase the

 · *Geodesic Tensegrity* Dome made from polyester fiberglass rods, University of Minnesota 1953. The full sphere would have had such a neglible weight, that it could have been let to float in the air like a balloon.

frequency of modular subdivisions of geodesic domes the edges of the triangles, representing the chords of central angles, get shorter and shorter and the interval between the mid-chord and the mid-arc of the central angles also decrease with the increasing frequency of the modular subdivisions. Because the materials used in the construction of the dome have some substantial dimension, we get to the point where the high-frequency production of the arc-altitude is such that the materials (the individual tensegrity components) touch one another. Every one of these elements is where it wants to be within the structure… there are no tensions anywhere, no slacks, all of the stresses are absolutely even – so we then fasten the two structural components together where they touch and want to be. This takes out the springiness of the geodesic domes and makes them rigidified. Because tension and tensegrity have no limit of clear spanning, tensegrity structures open up completely clear-spanned dome structures of any size.

Inventions. The Patented Work of R. Buckminster Fuller (1983), 179

410 · *Tensegrity Sphere* (Enenticontahedron) composed of 90 rods, Princeton University 1953

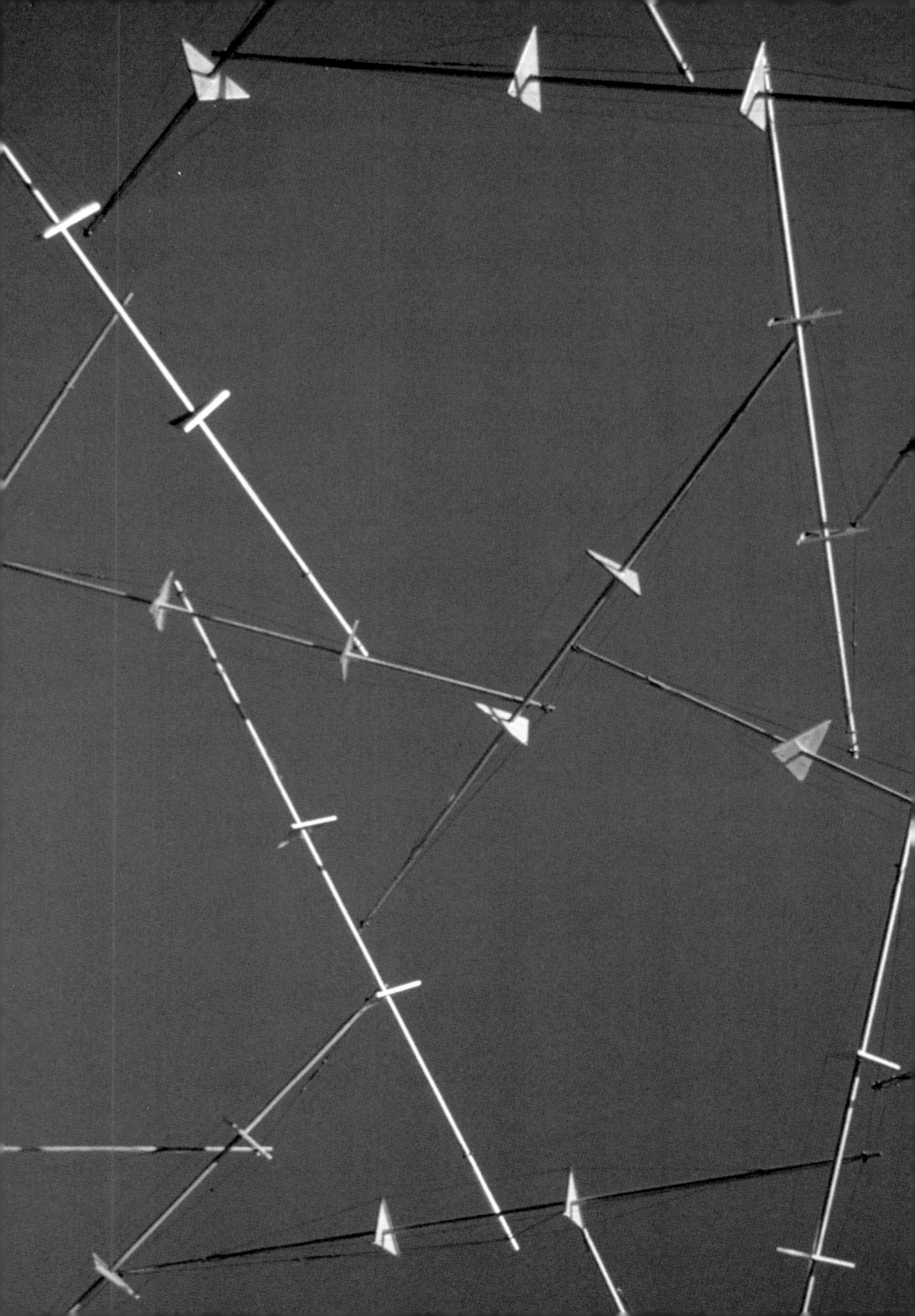

Walter Benjamin spoke of living in a glass house as the revolutionary act par excellence. It also encompasses a terror of total transparency. The Geodesic Restaurant in Wood's Hole, with its completely transparent acrylic glass had to be darkened subsequently, as few guests could tolerate total transparency.
In "Garden of Eden" Fuller pursued the goal of the optimal development of geodesic domes as "environmental controls," as spatial and climatic skins, as regulators and valves of the desired exchange with the environment. With two revolving geodesics placed inside one another, each of which has glass on only one side, it is possible to open and close them as needed. In the intermediate space between the two, more shade could be created in the summer or more light in the winter. The idea was to work together with nature.

GARDEN OF EDEN

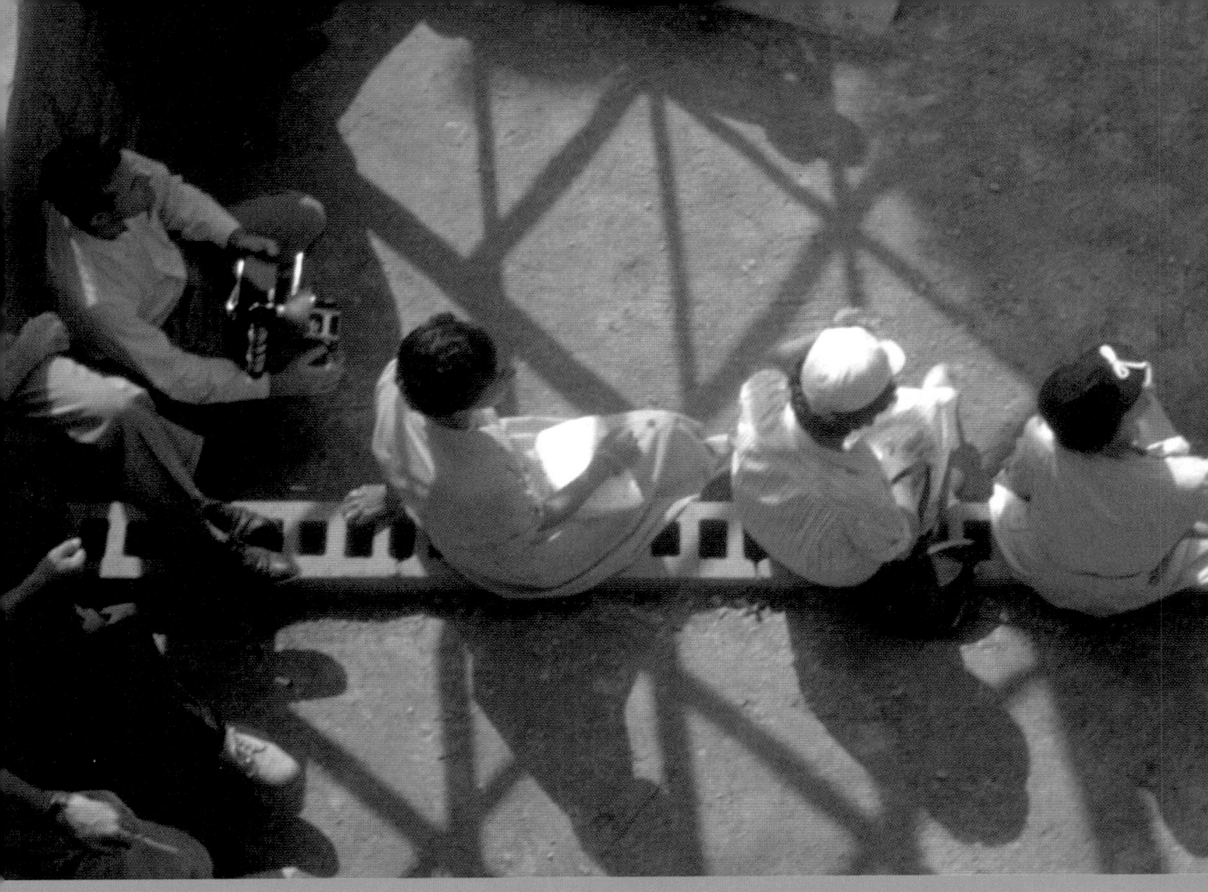

·413: RBF, in the dome of Wood's Hole Restaurant, Cape Cod (Massachusetts), before the final construction

·414/415: *Wood's Hole Restaurant,* night-time photograph after the completion of the dome, before being fitted with furnishings

We spent four weeks designing and fabricating the parts of this dome at Cambridge, two weeks subassembling the diamonds at the Wood's Hole site, and one week fitting them together to form the dome. Most of the time for assembly was spent in shifting the little jack mast that we used to hoist the individual diamonds aloft. We will never do it that way again. We learned from then on to use a boom universally jointed at the center to carry up the members. The members were wood sections, 2″× 3″ and 1″× 8″, all very light and delicate. The structure was skinned in during the following spring and used as a restaurant seating 150 people. The completed dome was 54′ in diameter and weighed 3 1/2 tons.

The night pictures give us a sense of isolation of the light weight structure from the surrounding environment. We tried several skins on this dome, one of them being a three mil Mylar. The structure was painted white so that the whole effect was very unified, ephemeral and pleasing.

When the hurricane came this fall, heterogeneous missiles pierced nine of the forty-five diamonds with three mil Mylar skins. They burst with quite an explosive sound, but only the skin and not the structure was hurt. The balance of the Mylar skin remained undamaged, and the DuPont Company tells us that they are very pleased at the test because the skin, which was simply glued in place, did not tear off entirely. It dilated in and out like a lung, but did not pull away.

RBF, *Architecture out of the Laboratory* (1955), 21

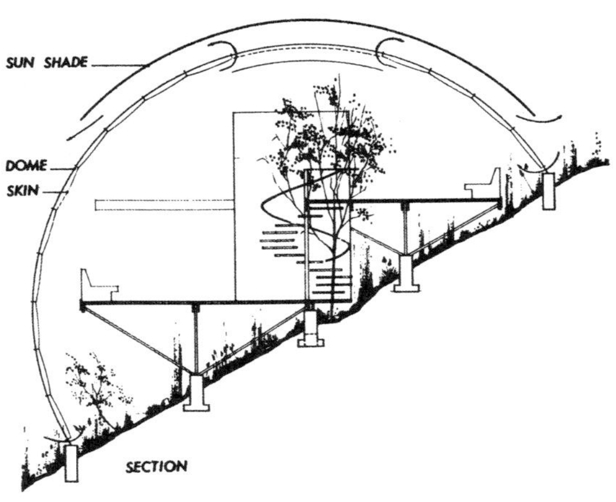
SUN SHADE
DOME
SKIN
SECTION

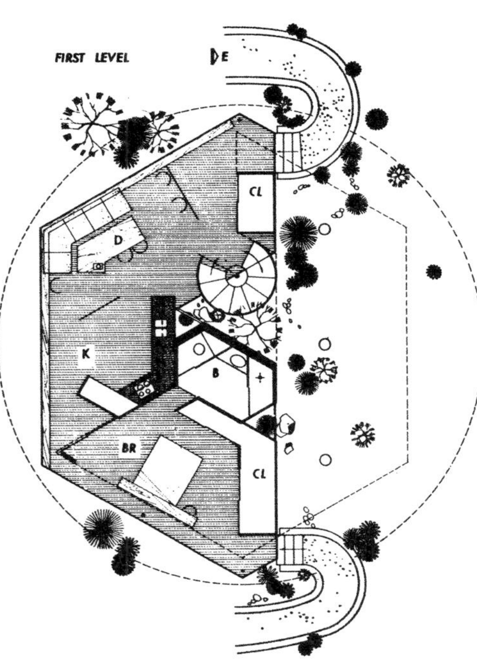

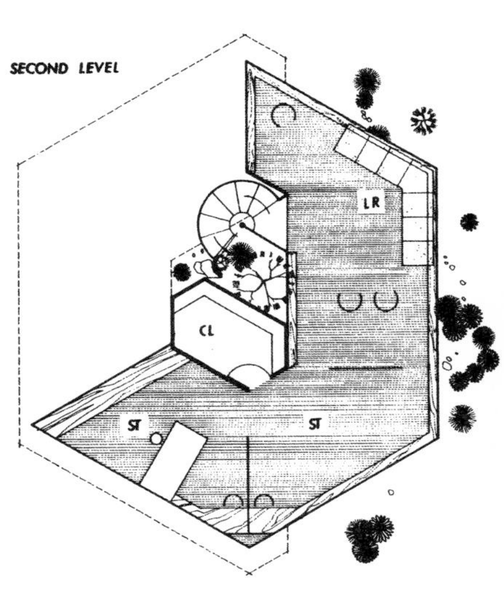

·Floor plan of the upper and lower levels, sectional elevation

A bubble of plastic high in the Hollywood hills

Designs in steel & silver

 ·"Home" section of the *Los Angeles Times,* with a story about the dome

·The dome, seen from Hollywood

In 1965, Fuller was commissioned by the U.S. Information Agency to design the U.S. Pavilion for the World's Fair in Montreal in 1967. After rejecting his proposal to install World Game, Fuller and Sadao, Inc., worked out the design for the Expo Dome: a geodesic three-quarter sphere with a diameter of seventy-six meters and height of sixty-one meters, made from a two-level filigree network of steel rods. In the outer layer, they formed triangular units, and hexagonal units in the inner layer. The dome was sealed with a transparent skin of acrylic glass panels, and the amount of light allowed in could be controlled by moveable, triangular sun shades on the inside, that changed according to the position of the sun, so as to avoid blinding effects. A computer program saw to it that a minimum of facets was closed at any time, and the blocking out of the sun is tracked, so as not to adversely affect the transparent and floating character of the dome. The dynamic adjustment of the interior climate and the unhampered view outward were part of Fuller's program for a structural skin as a valve for the surroundings. The Expo Dome was Fuller's masterpiece, his Taj Mahal, dedicated to his wife, Anne.

423 ·RBF presents President Johnson and Members of Congress with a model for the *'67 Expo Dome,* 1964 ·The Montreal dome shortly before the opening of the Expo. Up to the level of the equator, the construction is a network of horizontally, parallelly arrayed metal rings; above it is a geodesic structure

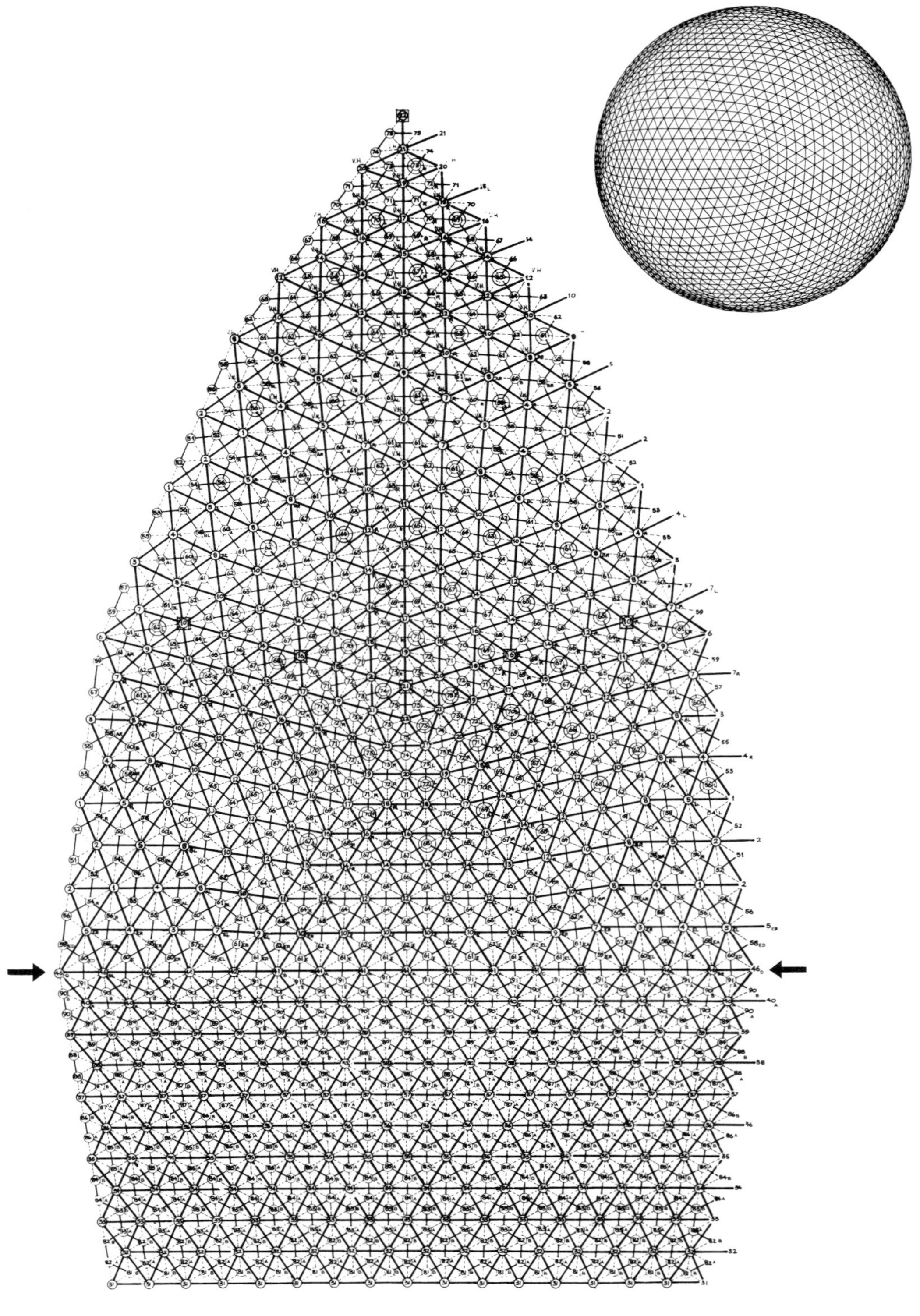

·424/425: The U.S.-Pavilion backlit (from *Paris-Match*,

·"Unpeeling" of a segment of the dome with angle and

427 ·View of the platforms inside the dome, Cambridge

Geodesic Domes: a human being's skin

(…) If industry was to take it on, there are things that we could do in geodesic domes that are spectacular. I haven't let much of it be visible except that anyone looking at the geodesic dome in Montreal saw a very beautiful piece of mechanics. It did all kinds of things to your intuition. You saw there were curtains that could articulate by photosynthesis and so forth, could let light in and out. It is possible, as our own human skin, all of our pores, all of the cells organize, so that some are photo-sensitive and some are sound-sensitive, and they're heat-sensitive, and it would be perfectly possible to create a geodesic of a very high frequency where each of these pores could be circular tangencies of the same size. One could be a screen, others breathing air, others letting light in, and the whole thing could articulate just as sensitively as a human being's skin. And I really think geodesic domes such as that will be developed.

RBF, in *Domebook* 2 (1971)

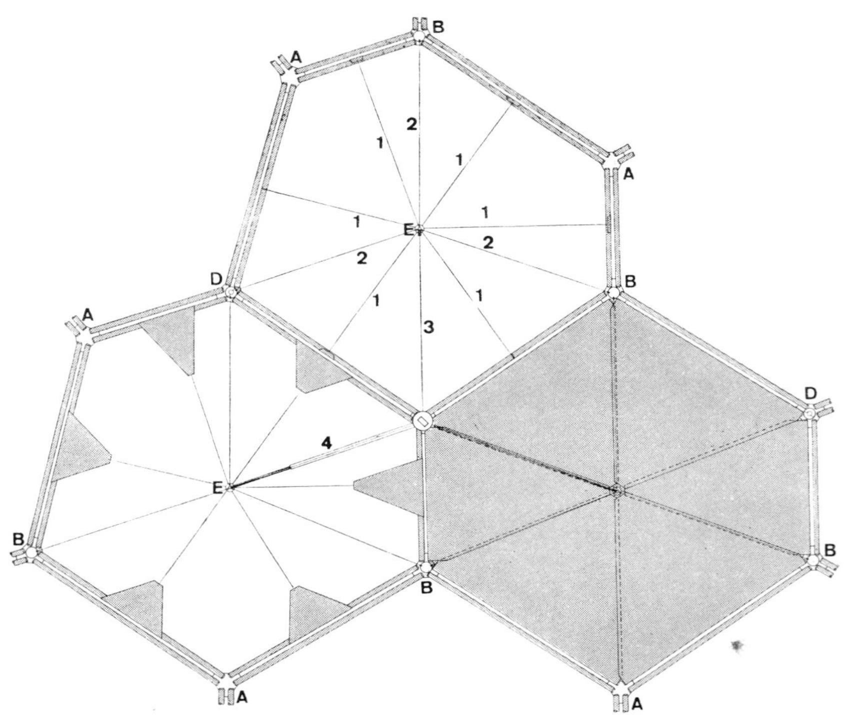

 ·Detail of the dome's outer skin: showing three phases

·Section of the dome, viewed from the interior

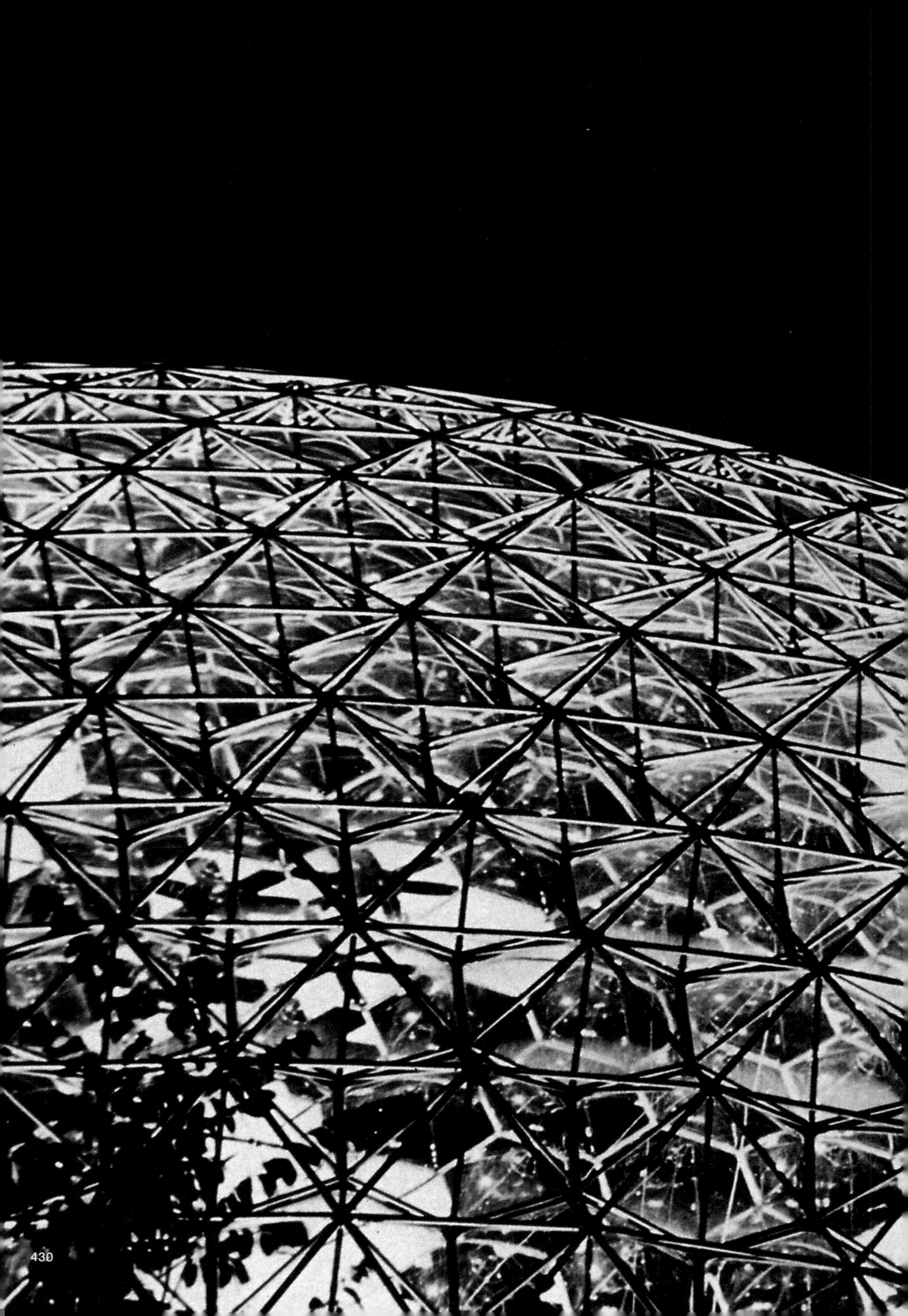

Domed-over Cities

Domed-over cities have extraordinary economic advantage. A two-mile diameter dome has been calculated to cover Mid-Manhattan Island, spanning west to east at 42nd Street from the Hudson River to the East River, and spanning south to north from 22nd Street to 62nd Street.

When we wish to make a good air-cooled engine, we design it with many fins and spicules to carry away the heat by providing the greatest possible external surface area. The dome calculated for mid-Manhattan has a surface which is only 1/85 the total area of the buildings which it would cover. It would reduce the energy losses either in winter heating or summer cooling to 1/85 the present energy cost obviating snow removals. The cost saving in ten years would pay for the dome.

Utopia or Oblivion (1969), 353

"From the inside there will be uninterrupted contact with the exterior world. The sun and moon will shine in the landscape, and the sky will be completely visible, but the unpleasant effects of climate, heat, dust, bugs, glare, etc. will be modulated by the skin to provide Garden of Eden interior."

 RBF, in J. Allwood (1977)

·432/433: Normal Foster and RBF: *Autonomous House,* 1982: Project for a house unreliant on external energy sources, constructed as a double-hull geodesic dome with movable positioning (as a climate-controlling shell) and three freely positioned upper stories. The autonomous house can open and close itself through rotating of the shells, and can thus harvest energy or give it back off to the environment. The house is the culmination in Fuller's work of the theme "house as valving," which was already touched on in the Dymaxion House.

·436/437: Dome over Manhattan, project, 1950 (photographic montage). The dome would have spanned a diameter of two miles and, among other things, would have allowed for the elimination of automobile traffic.

·438/439: *Tetra City:* project for a floating city, planned for a million inhabitants (Fuller and Sadao, 1965).

·440/441: *Cloud Structures:* the greater a geodesic dome, the lighter its structure in comparison to the air it contains. When heated by the sun, some of the expanding air escapes, allowing the 1.5 mile diameter sphere and its thousand inhabitants to float through the air (With Shoji Sadao, 1962).

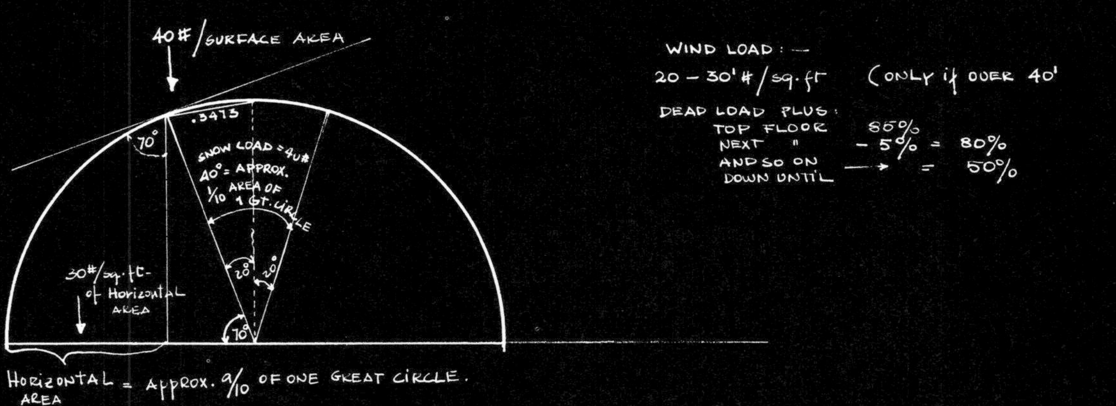

After Fuller had become famous for constructing large geodesic domes, scientists and doctors began to provide examples of geodesic and tensegretic structures in microorganisms. Virus researchers asked for Fuller's advice in explaining the structure of protein shells. Examinations of human tissue showed structures that resembled geodesic and tensegretic basketwork. The interaction of muscles and bones could be understood as a tensegrity structure as could the filaments and needlelike cytoskeletons of cells.

The most spectacular discovery was of the Buckminsterfullerene, a spherical cage molecule of pure carbohydrate. Its discoverer had been inspired by Fuller's Expo Dome to spot the pattern. The "Fullerene" family of molecules is expected to have important consequences on chemistry, electronics, and nanotechnology.

HOW NATURE BUILDS

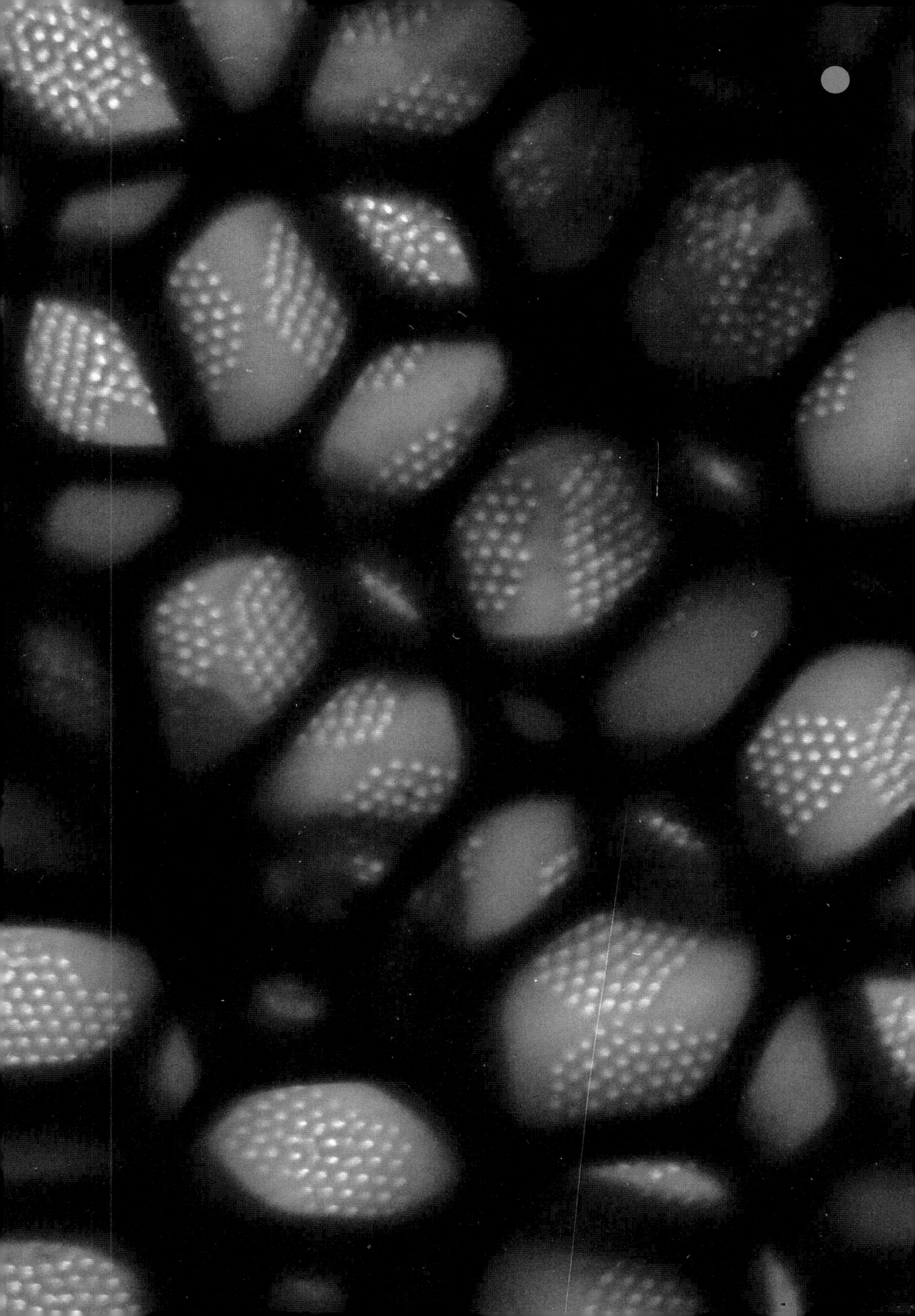

Today I have given you first some fundamental structural principles and subsequently shown you their use by nature. I didn't, however, start by studying these structures of nature seeking to understand their logic. The picture of the radiolarian has been available for 100 years, but I didn't happen to see it until after I had produced the geodesic structures from the mathematical sequence of developments which I reviewed for you earlier. In other words I did not copy nature's structural patterns. I did not make arbitrary arrangements for superficial reasons. What really interests me therefore in all these recent geodesic tensegrity findings in nature is that they apparently confirm that I have found the coordinate mathematical system employed in nature's structuring. I began to explore structure and develop it in pure mathematical principle out of which the patterns emerged in pure principle and developed themselves in pure principle. I then realized those developed structural principles as physical forms, and in due course applied them to practical tasks. The reappearance of these structures as recent scientists' findings at various levels of inquiry are pure coincidence, – but excitingly validating coincidence.

WDSD2: Mexico Lecture (1964), 59

·443: Micro-morphologic exploration with electron microscope yields numerous indications of geodesic structures in nature. Diatoms (algae) with 3500-fold magnification. The recorded image was provided to RBF in 1962 by Dr. Helmcke, TU Berlin.

·445: The German zoologist Ernst Haekel (1834–1919) discovered numerous marine microorganisms and produced drawings to document his examinations under the microscope. His illustrations of Radiolarian became well-known, showing a structure that exhibits the icosahedral symmetry of geodesic domes. Drawing from Haeckel's contribution to an English expedition report, London, 1887.

·446/447: Diatom under the electron microscope, Helmcke 1962. The circular opening at the vertexes of the pentagonal and hexagonal facets was utilized by RBF in his Monohex patent (1965) and the "Fly's Eye" construction, which developed out of Monohex.

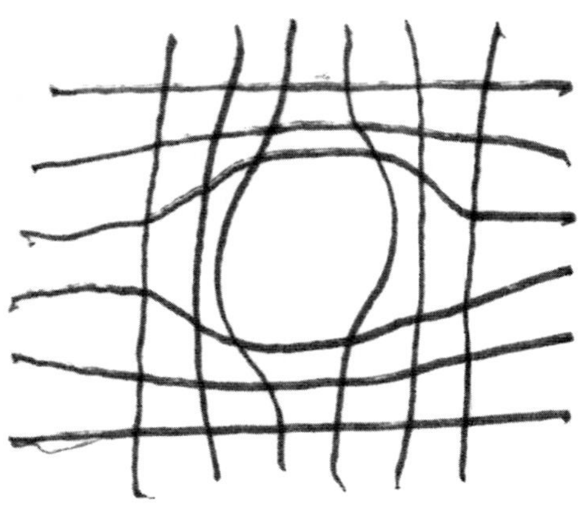

WEAVES
ARE
SPREADABLE
ERGO DISTORTABLE
ERGO UNSTABLE.

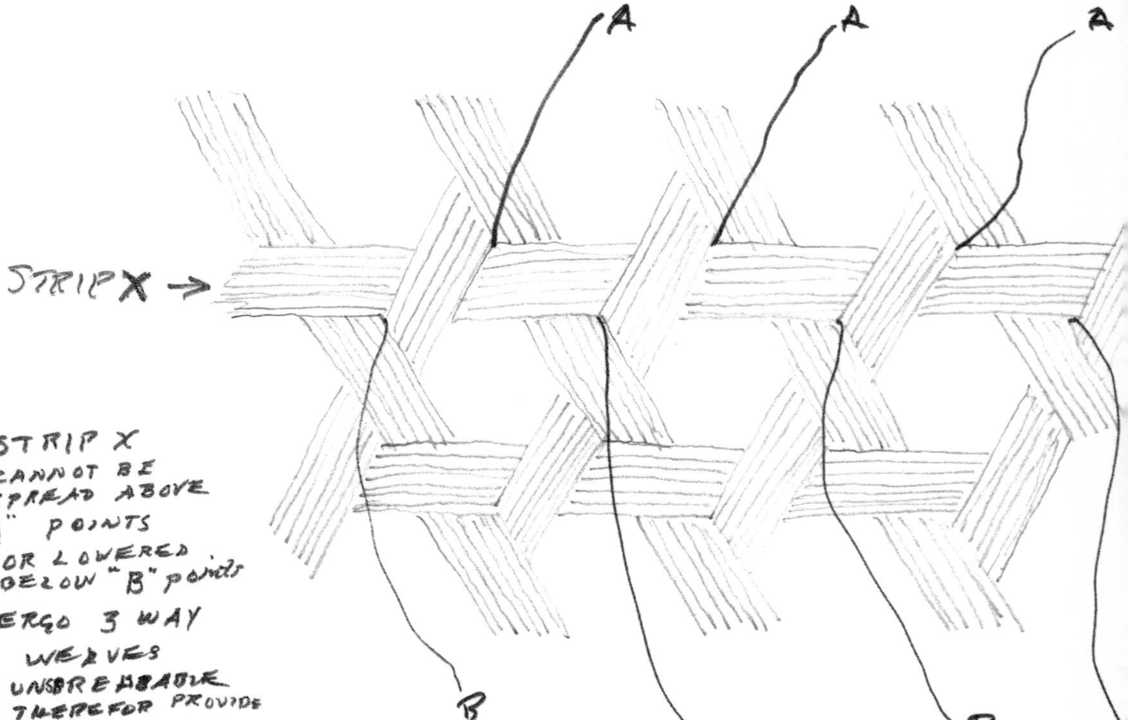

A A A

STRIP X →

STRIP X
CANNOT BE
SPREAD ABOVE
"A" POINTS
NOR LOWERED
BELOW "B" points
ERGO 3 WAY
WEAVES
UNSPREADABLE
THEREFOR PROVIDE
STABLE PATTERN
STABLE STRUCTURE

B B B B

·Synergetic weaves in comparison to euclidic-catechetic weaves. Fuller's arrangement of 60-degree angle modules is found in all weave-types in nature, but also in the wicker-work of primitive cultures. Drawing RBF, undated.

·Weave-analysis by Arthur H.C. von Hochstetter, M.D., who found "principles of a 60-degree lightweight structure" embodied in human connective tissue and, in 1962, entered into a lively exchange with RBF.

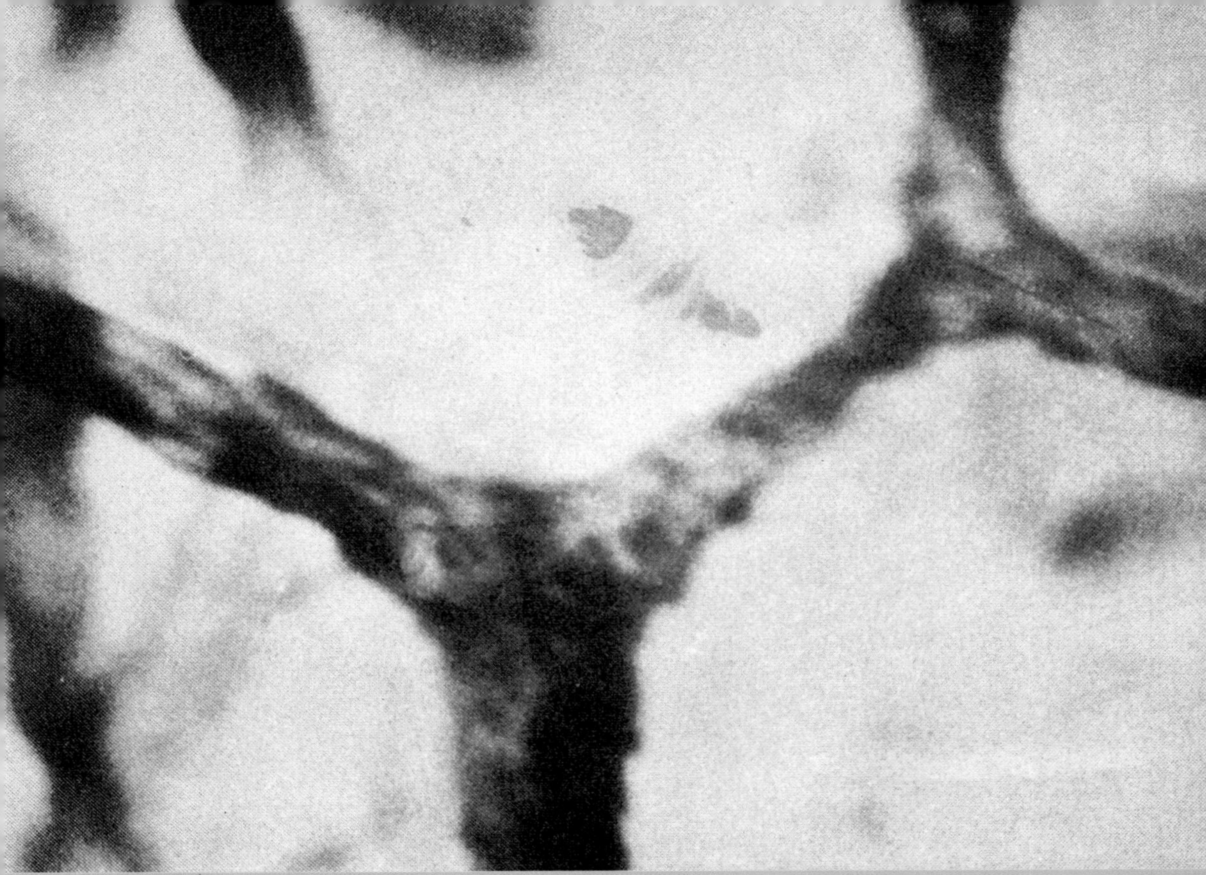

·Geodesic weave-structures: cornea of the human eye (left); fibrous capsule of the human testicle (right).
v. Hochstetter 1962.

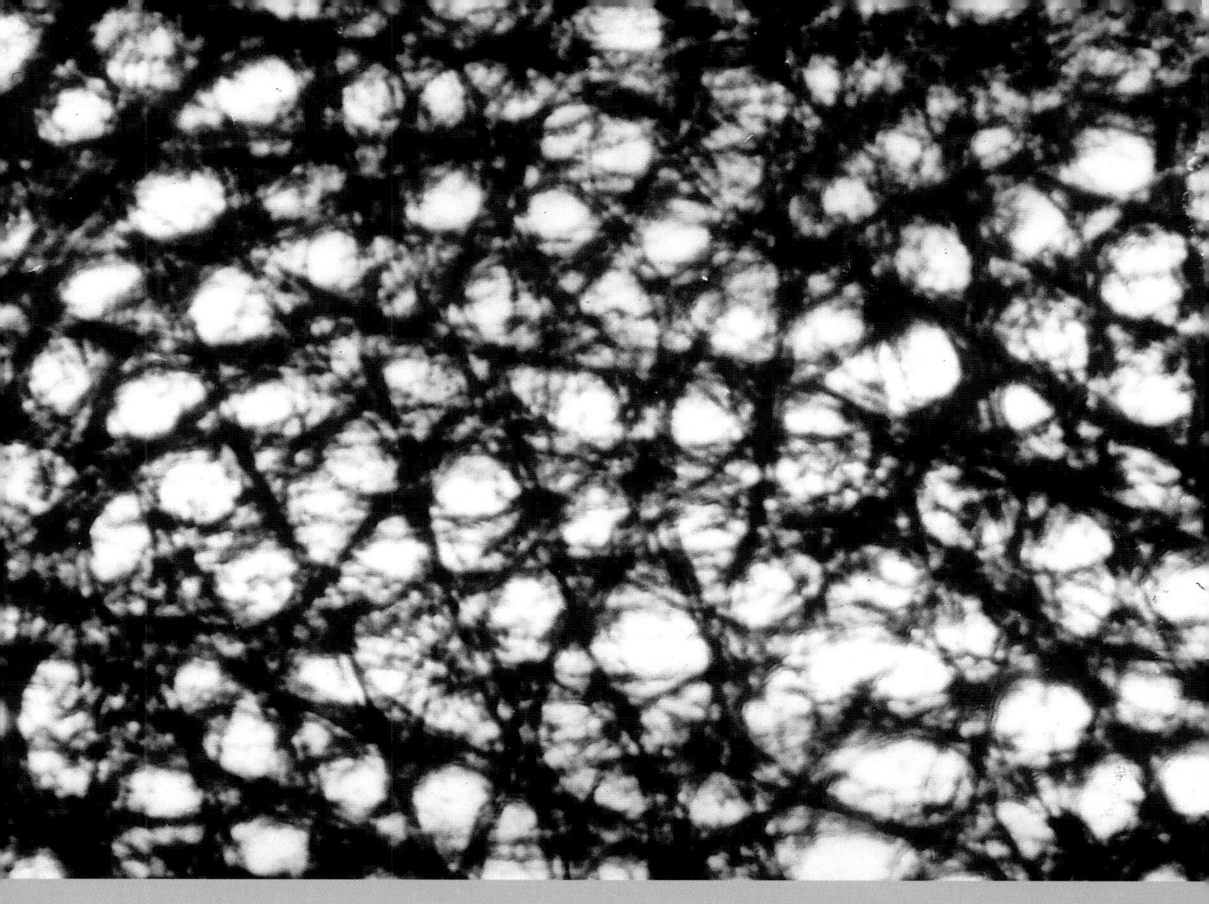

·Construction of the double-shelled Hexagon-Pentagon Dome Construction for the American Society of Metals, Cleveland, Ohio, 1959. Architect John Kelly, construction by Synergetics, Inc.

·The Tensegrity Dome provided for the *Garden of Eden* project, in which Geodesic Domes approximate pneumatic structures and membranes.

·In recent times, there has been an increase in examples of successful modeling with geodesic-tensegretic structures. Donald Ingber's examination of cell growth with Tensegrities, cover page *Scientific American,* January 1998.

·Deformation of the cytoskeleton of cells through application of load: as with a Tensegrity structure, the cell fabric permits a temporary distortion, but goes back into its original form. D. Ingber 1998.

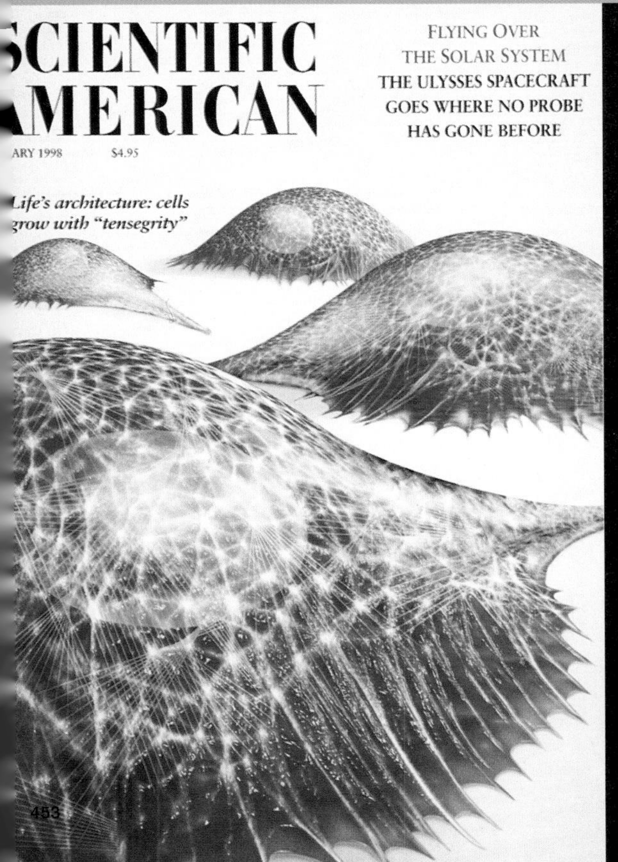

·Fuller's buildings were never to be understood as merely utilitarian but, rather, were meant to instruct. In the case of two scientific discoveries, both of which were honored with the Nobel Prize, Fuller's constructions played the role of the catalyst:

·Discovery of the protein shells of viruses by Donald Caspar and Aaron Klug, 1962, Nobel Prize 1982. Wooden model of the polio virus by D. Caspar, presenting an inverse form of Fuller's *Flying Seedpod.* (Compare to p. 342).

nature

...RNATIONAL WEEKLY JOURNAL OF SCIENCE

...18 No 6042 14-20 November 1985 £1.90

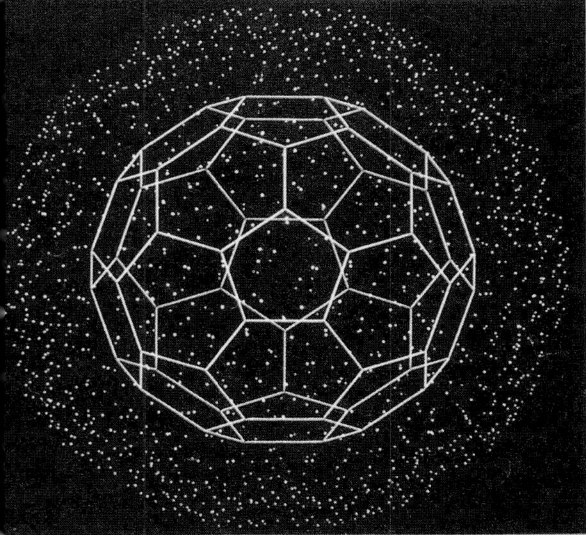

SIXTY-CARBON CLUSTER

AUTUMN BOOKS

Harvey Brooks (transformation of MIT)	Hendrik B. G. Casimir (physics and physicists)
P. N. Johnson-Laird (brain and mind)	Gordon Thompson (dimensions of nuclear proliferation)
Anthony W. Clare	Jacques Ninio

·The spectacular discovery of a new class of carbon molecules, the Fullerene, by Kroto, Smalley and Curl 1985, Nobel Prize 1997. The Expo Dome, which Kroto and Smalley had both seen in Montreal in 1967, led them onto the right track in their search for the unknown molecular structure: to the icosahedral symmetry of the ball-shaped cage molecule.

·Corresponding to their varying number of atoms, there are molecules smaller than the 60-atom *Buckminster-Fullerene,* C-60, as well as much larger cage-like molecules of 70, 240, 540 and 960 atoms. Nevertheless, all feature the 12 pentagons that generate the closed volumetric form; the lattice-work of a uniform hexagonal grid permits no second degree curvature. Fuller's nickname "Bucky" was once again a fitting choice for popular names like "Buckyballs," "Buckybabies," etc. Aside from pure carbon molecules, what is of the highest research interest are the metallic compound, Fullerite, and numerous derivatives. Chemists contend that the discovery of the *Fullerene* has huge consequences for the whole of organic chemistry, speaking of an "end to flat chemistry."

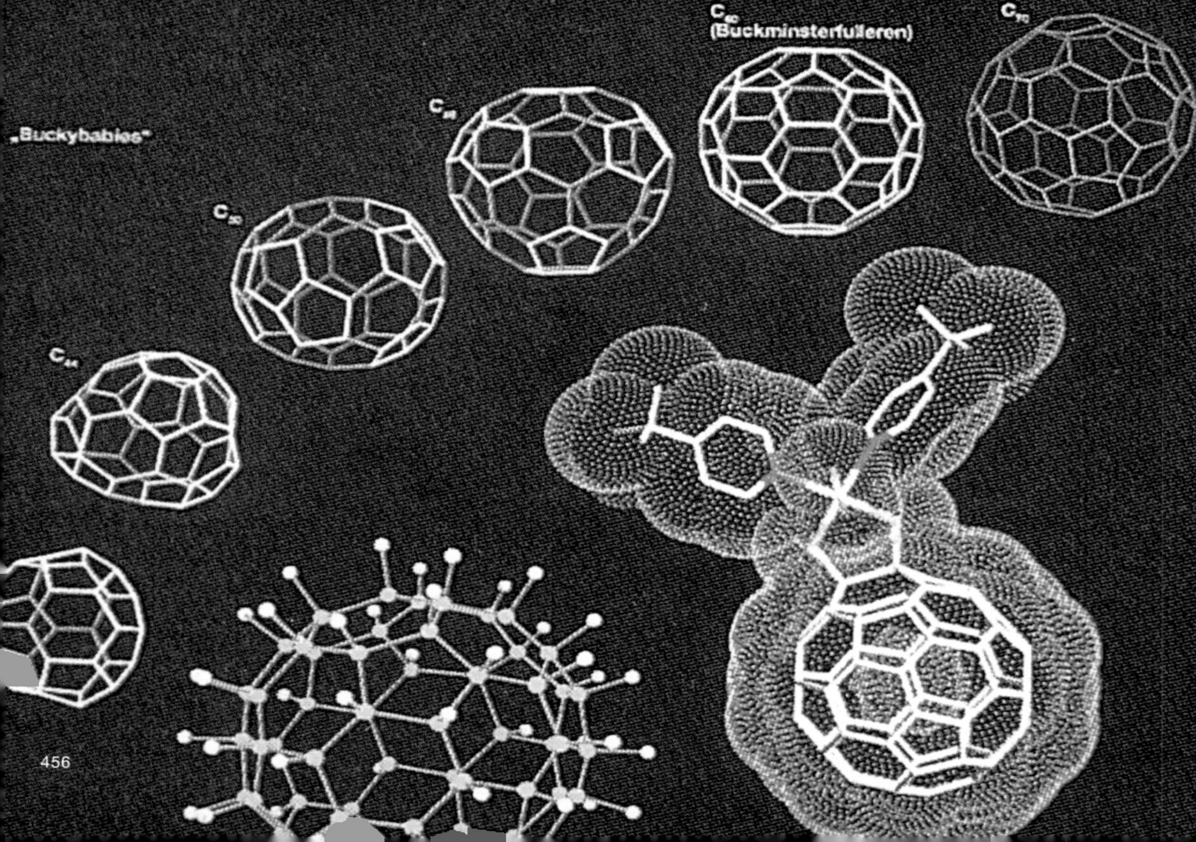

The beauty of bubbles

Looking back at the wake of my ship one day in 1917, I became interested in its beautiful white path. I said to myself, "That path is white because of the different refractions of light by the bubbles of water $-H_2O$ (not $H\pi O$). The bubbles are beautiful little spheres. I wonder how many bubbles I am looking at stretching miles astern?

I began to make calculations of how many bubbles there were per cubic foot of water. I began to find that in calculating the ship's white wake I was dealing in quintillions to the fourth power times quintillions to the fourth power or some such fantastically absurd number of bubbles. And nature was making those bubbles in sublimely swift ease!

Any time one looks carefully at a bubble, one is impressed with the beauty of its structure, its beautiful sphericity glinting with the colors of the spectrum. It is ephemeral – elegantly conceived, beautifully manufactured and readily broken.

Inasmuch as the kind of mathematics I had learned of in school required the use of the XYZ coordinate system and the necessity of employing π in calculating the spheres, I wondered, "to how many decimal places does nature carry out π before she decides that the computation can't be concluded?" Next I wondered, "to how many arbitrary decimal places does nature carry out the transcendental irrational before she decides to say it's a bad job and call it off?" If nature uses π she has to do what we call fudging of her design which means improvising, compromisingly. I thought sympathetically of nature's having to make all those myriad frustrated decisions each time she made a bubble. I didn't see how she managed to formulate the wake of every ship while managing the rest of the universe if she had to make all those decisions. So I said to myself, "I don't think nature uses π. I think she has some other mathematical way of coordinating her undertakings."

RBF: *Conceptuality of fundamental structures.* In: G. Kepes: *Structure in Art and Science* (1965), 66

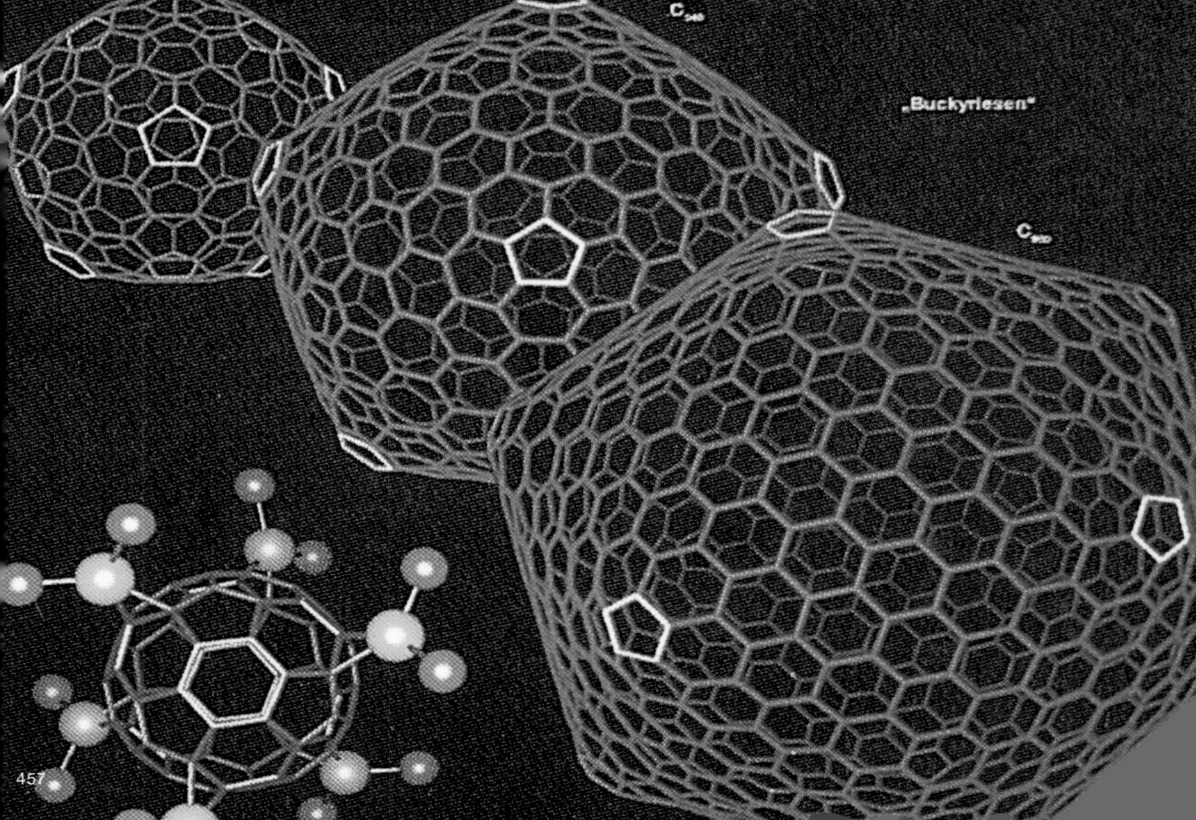

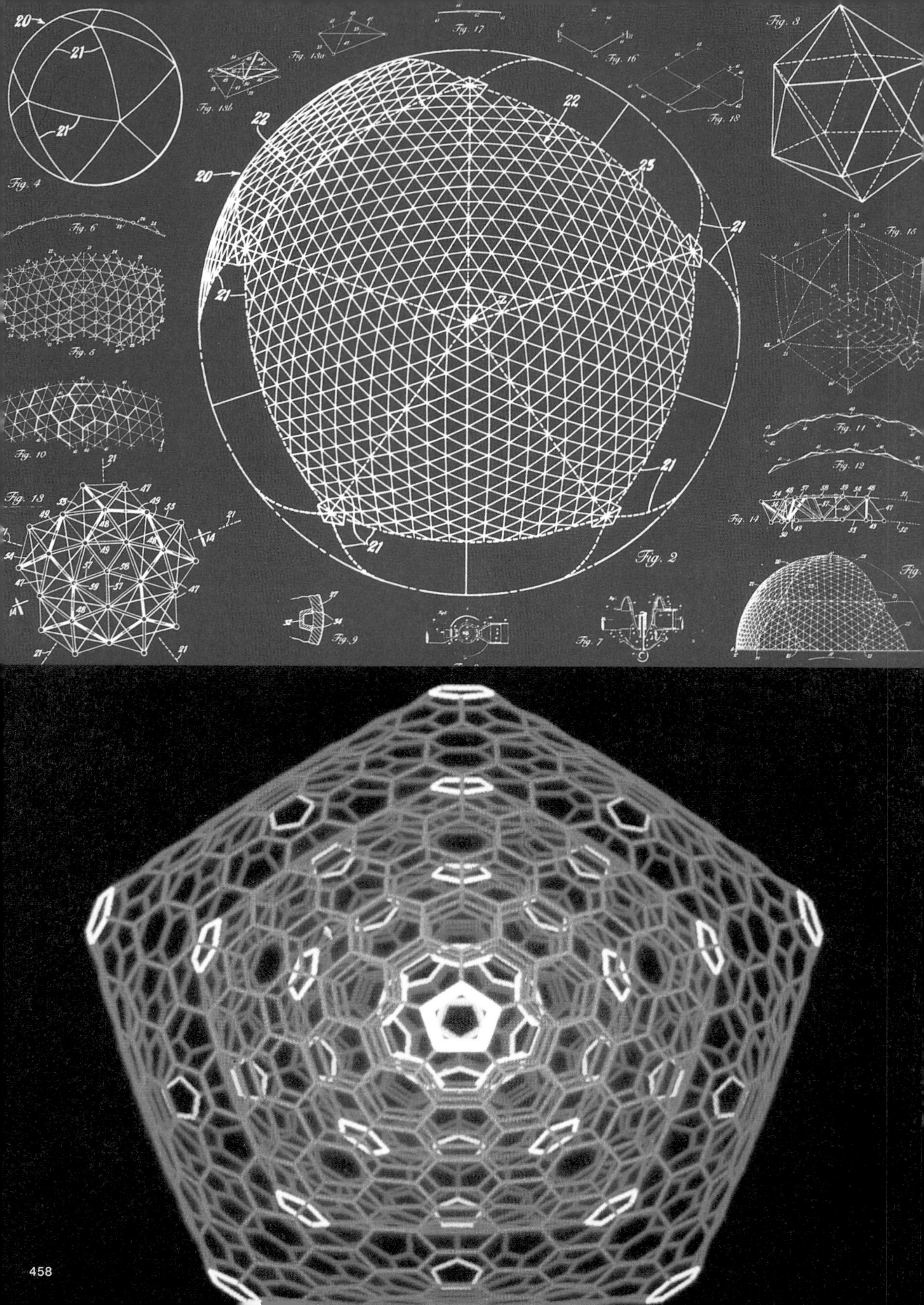
Fig. 2
Fig. 3
Fig. 4
Fig. 5
Fig. 6
Fig. 7
Fig. 9
Fig. 10
Fig. 11
Fig. 12
Fig. 13
Fig. 13a
Fig. 14
Fig. 15
Fig. 16
Fig. 17
Fig. 18
Fig. 18b

·458: Icosahedral Symmetry: drawing from RBF's patent application for "Geodesic Dome," U.S. Patent 2, 682, 235 (1954) – above – and intertwined carbon molecules in a *Hyperfullerene* (taken from F. Cure and R.E. Smalley 1991).

·459: As early as the 1950s, RBF was conducting experimental work with students applying synergetic principles in linear structures such as masts, tubes, etc. Experimental model of a Tensegrity-Tube with spiral torsion. Undated.

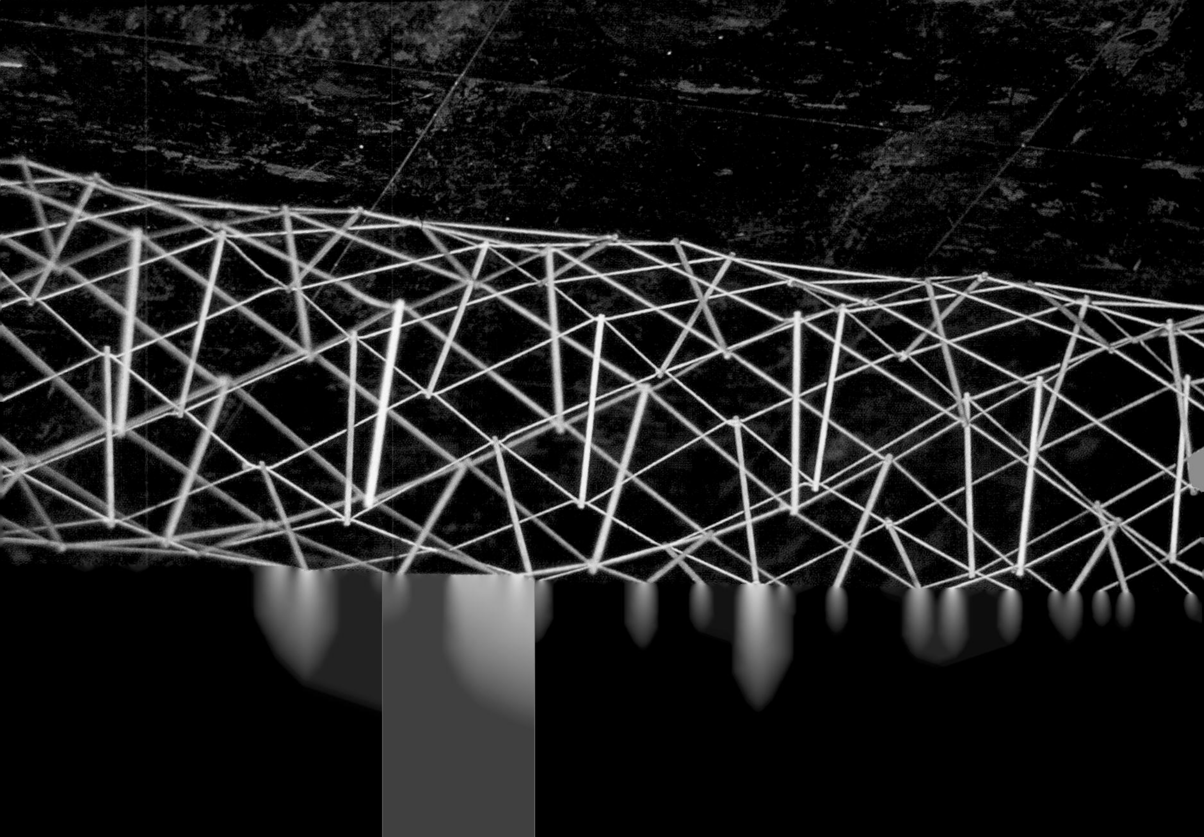

·RBF's modeling of linear structures was made current in 1991 with the discovery by S. Iijima of Nano Tubes made from carbon molecules. With Fullerenes and "Buckytubes," Fuller's geometry suddenly became a central component of the rapidly evolving field of Nano technology. Photograph of model (undated).

·Fuller's work emanated from a study of geometric principles, not from just copying forms from nature. Out of the sequencing of triangles and tetrahedrons, he came upon helix-cycles, each made of ten triple-bound tetrahedral cells. When turned in the same direction, they nest into each other, and tetra-helix cords formed of five-tetrahedron clusters are held together under tension around the transverse axis, but spring back apart from each other as soon as the tension ceases. In 1962, Fuller tried to use this to produce a geometric-theoretical explanation for the replication mechanism of the DNA double-helix.

All of his projects were always developed against the background of a planetary consciousness. Already in 1951, Fuller was using the idea of a spaceship Earth, which has long since become commonplace. We live in a closed, interdependent ecological system with problems we produce ourselves but also with inherent approaches to solutions. One needs, however, an idea of the whole in order to find the right partial solution. World Game was intended to make the conditions of global interaction manifest and to use computer simulation to make the consequences of decisions transparent to all.

WORLD GAME

PROFILE OF THE INDUSTRIAL REVOLUTION
AS EXPOSED BY THE CHRONOLOGICAL RATE
OF ACQUISITION OF THE BASIC INVENTORY OF
COSMIC ABSOLUTES—THE 92 ELEMENTS

The World Game project originated in 1965 as a proposal for the U.S.A. contribution to the 1967 World's Fair in Montreal: an interactive game, supported by a large computer, shown on a large dynamic display of the Dymaxion World Map. It is played by trying to generate the best strategies for improving living conditions for all the earth's inhabitants. The starting point should be the distribution on the globe of natural resources, population, energy, means of transport and communication, etc. Participants – also out of politics – have the chance to play through their own concepts, to then be confronted with a display of the consequences their decisions have on the rest of the world.

For Fuller, World Game is meant to give an answer – or more than one answer – to the mounting social and ecological crisis, which he had been predicting since since around 1950. The crisis resulted from the fact that, on the global scale, only a minority had profited from industrialization. In order to make winners, not only must the social system be changed, but first, once, the way of thinking. Indeed, Fuller planned to adopt the technology of War Games, the mainframe computer, but he aimed to break through the game logic of War Games: in place of the Drop Dead ethic of a merciless society, instituting an ethic of acting together for the advantage of all.

 ·465: RBF teaching in front of his synchronoptic board, "Profile of the Industrial Revolution."

architecture students around the world, *World Design Science Decade* (WDSD). *Geoscope,* built as geodesic sphere with

468 ·NASA control center: example of computerized global data inte[illegible]ation; served as a technical model for Fulle[illegible] co-worker John Meltale for the installation of *World Game.* Photograph ca. 1964.

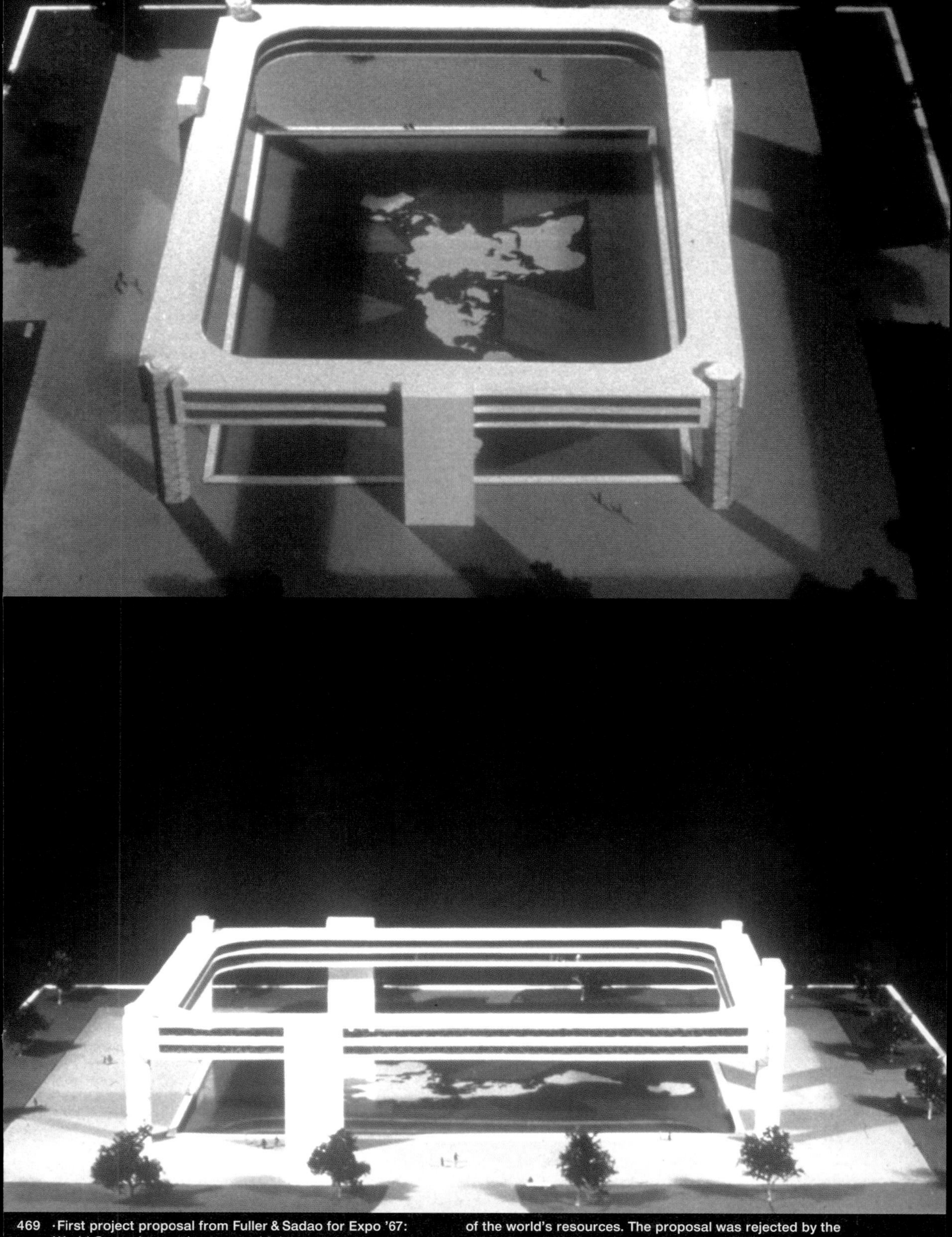

469 · First project proposal from Fuller & Sadao for Expo '67: *World Game*, interactive game with large-scale computer simulation and *Dymaxion* world map as a dynamic display of the world's resources. The proposal was rejected by the U.S. Information Agency. Model 1965.

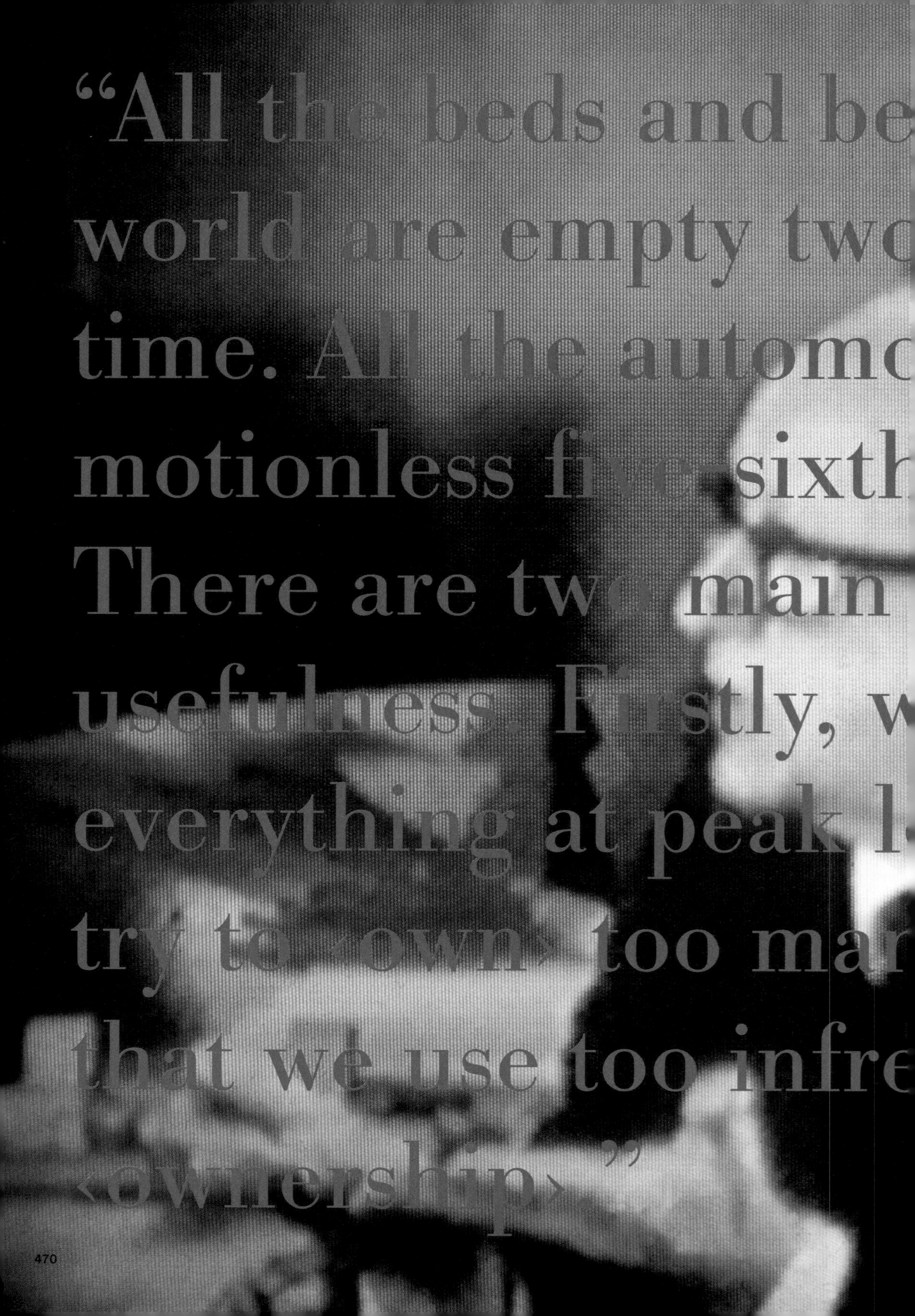

“All the beds and be
world are empty two
time. All the automo
motionless five-sixth
There are two main
usefulness. Firstly, w
everything at peak l
try to ‹own› too man
that we use too infre
‹ownership›.”

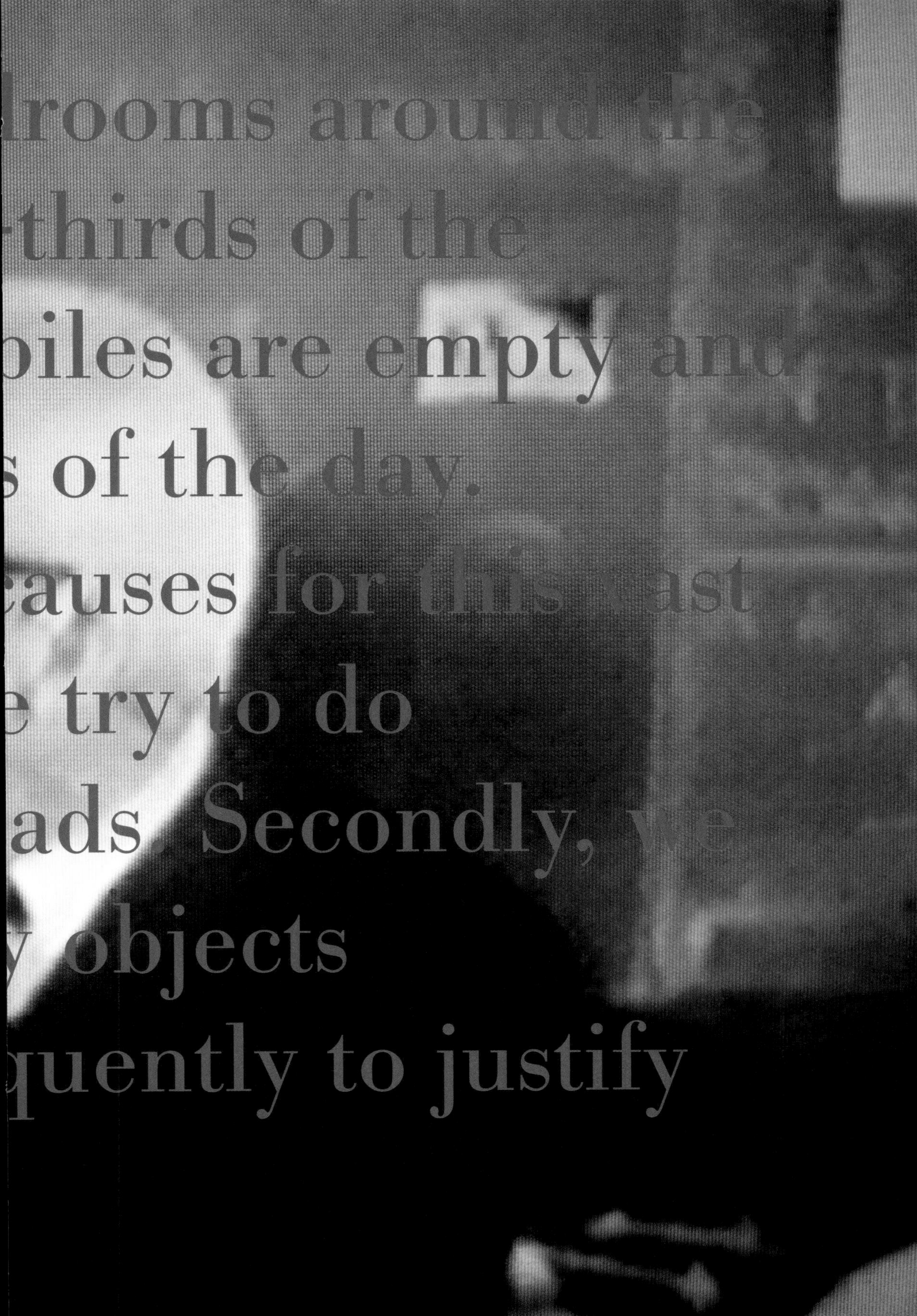

lrooms around the
-thirds of the
biles are empty and
s of the day.
causes for this vast
e try to do
ads. Secondly, we
y objects
quently to justify

How it came about (World Game)

In 1964 the United States Information Agency asked me to consider the design of a building and an exhibition which might be adopted as the United States entry in the Montreal World's Fair of 1967, later known as "Expo'67."

I made a proposal and the exhibition part of it was rejected. I was asked to continue, however, as the architect of the U.S.A. building to house an exhibition designed by others. (...)

Fortunately my U.S.A. geodesic dome for Expo '67 proved a success. Also, but more gradually, however, it is being realized by many that my rejected idea for the American exhibit is looming into ever greater prominence as a highly desirable social facility. I will therefore review the concept and development of my original idea.

I told the U.S.I.A. in 1964 that by 1967 the regard of the rest of the world for the United States would be at its lowest ebb in many decades – if not in the total two centuries of the U.S.A.'s existence. Since each country's World's Fair exhibit would be well published all around Earth, I felt that it would be very important that the United States do something that would tend to regain the spontaneous admiration and confidence of the whole world. (...)

What I proposed was based on my observation that world people had become extraordinarily confident in the fundamental realiability of the computer and its electronically controlled processes. I know that a great many people will contradict me, but I had predicated my conviction – of society's subconsciously established confidence in the computer's reliability – upon vital, therefore undeniable, behavior facts.

I refer to the equanimity with which world-around air jet travelers now commit their lives to the computer's reliability. (...)

 ·First *World Game* Seminar, led by RBF and Edwin

As I now fly around the world in jets in the company of 125 to 135 passengers, we frequently find ourselves coming in for a landing in the night and fog. But everybody is occupied in reading, sleeping, listening to the music or talking. (...)

On the working assumption that humanity now has established implicit confidence in the computers and automated instrumentation, I proposed in 1964 that the United States' Expo '67 exhibition should have a 400 foot diameter 5/8 sphere building similar in shape to the 250 foot diameter building actually built for Expo '67. In the basement of this building would be housed an extraordinary computer facility. On entering the building the visitors would arrive upon a great balcony reaching completely around the building's interior quarter-mile perimeter. The visitors would see an excitingly detailed one hundred foot diameter world globe suspended high within the 400 foot diameter 5/8 sphere main building. Cities such as New York, London, Tokyo, and Los Angeles would appear as flattened out basketball sized blotches with the tallest buildings and radio towers only about one-sixteenth of an inch high.

Periodically the great spherical Earth would be seen to be transforming slowly into an icosahedron – a polyhedron with twenty (equilateral) triangular facets. The visitors would witness that in the processes of these tranformations there are no visible changes in the relative size and shape of any of the land and water masses of the 100 foot diameter miniature Earth. Slowly the 100 foot diameter icosahedronal Earth's surface will be seen to be parting along some of its triangular edges, as the whole surface slowly opens mechanically as an orange's skin might be peeled carefully in one piece. With slits introduced into its perimeter at various places it would be relaxed to subside into a flattened-out pattern. The icosahedronal Earth's shell thus will be seen to gradually flatten out and be lowered to the floor of the building. The visitors would realize that they were now looking at the whole of the Earth's surface simultaneously without any visible distortion of the relative size and shape of the land and sea masses having occurred during the transformation from sphere to the flattened-out condition which we call a map. My cartographic projection of the "Sky Ocean World" functions in just such a manner.

This stretched out football field sized world map would disclose the continents arrayed as one world island in one world ocean with no breaks in the continental contours.

The great map would be wired throughout so that mini-bulbs, installed all over its surface, could be lighted by the computer at appropriate points to show various, accurately positioned, proportional data regarding world conditions, events, and resources. World events would occur and transform on this live world map's ever evoluting face.

I proposed that, on this stretched out reliably accurate, world map of our Spaceship Earth, a great world logistics game be played by introducing into the computers all the known inventory and whereabouts of the various metaphysical and physical resources of the Earth. This inventory, which has taken forty years to develop to high perfection, is now housed at my Southern Illinois University headquarters.

We would then enter into the computer all the inventory of human trends, known needs and fundamental characteristics.

I proposed that individuals and teams would undertake to play the "World Game" with those resources, behaviors, trends, vital needs, developmental desirables, and regenerative inspirations. The players as individuals or teams would each develop their own theory of how to make the total world work successfully for all of humanity. Each individual or team would play his theory through to the end of his predeclared program. It could be played with or without competitors.

The objective of the game would be to explore for ways to make it possible for anybody and everybody in the human family to enjoy the total earth without any human interfering with any other human and without any human gaining advantage at the expense of another.

To accomplish the game's objective, the resources, pathways and dwelling points around the surface of our eight thousand mile diameter, spherical Spaceship Earth must be employed by

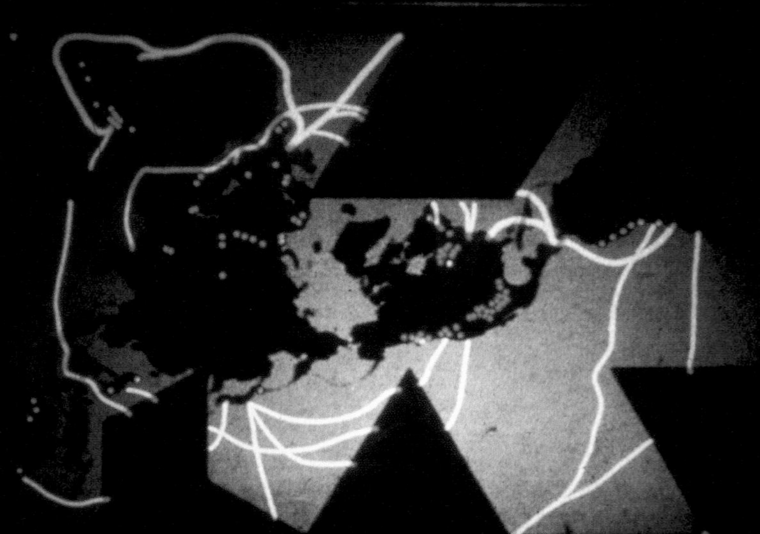

 · *World Game* Seminar 1972. Inventorying of world resources, development trends and world population needs. Visualization on the basis of available data.

·Visualization of global data on Dymaxion Projection.

17 — FOOD MOVEMENT — BREAD GRAINS — = 5×10^5 TON/YR

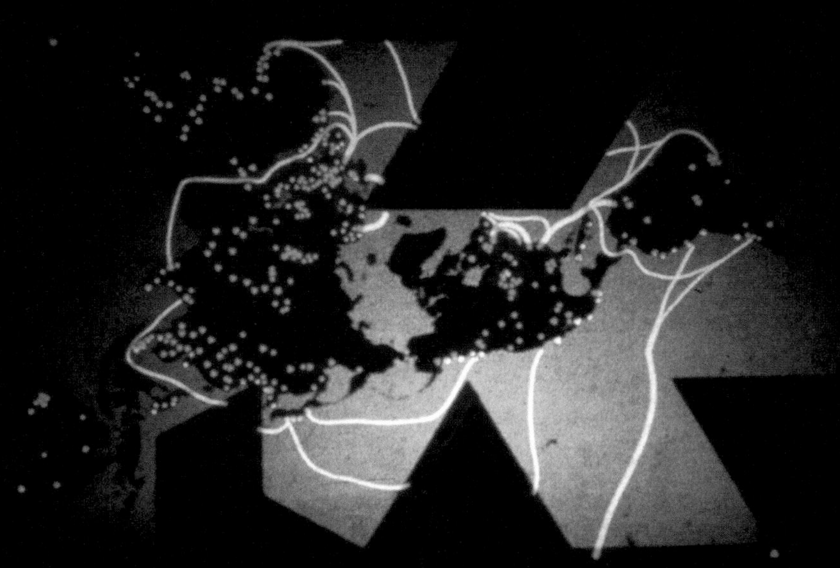

5 — IRON ORE — • SITES ▲ POTENTIAL — DISTRIB

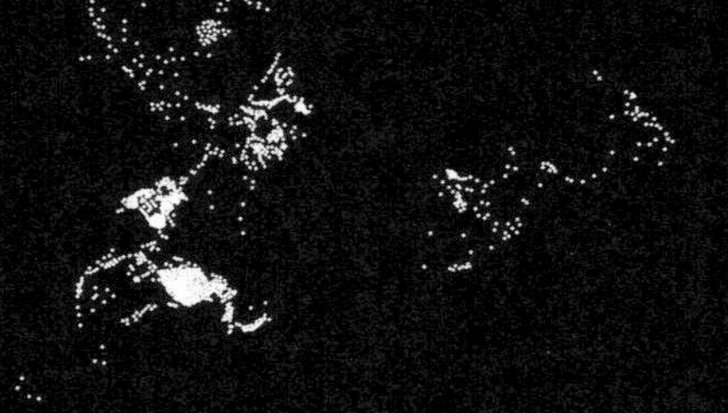

34 – POPULATION – 1970 · = 3·5 × 10⁶

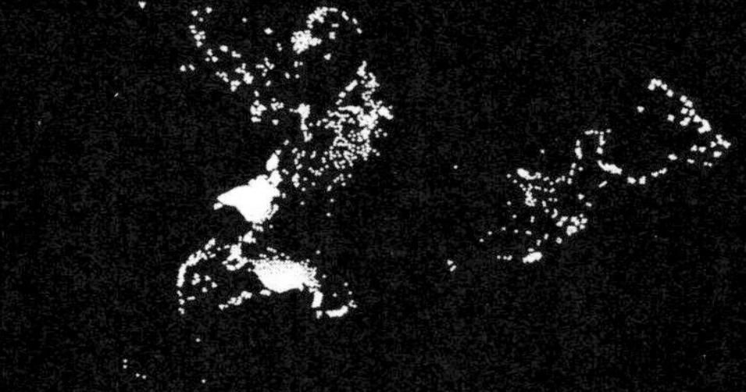

30 – POPULATION – 2000 – PROJECTION · = 3·5 × 10⁶

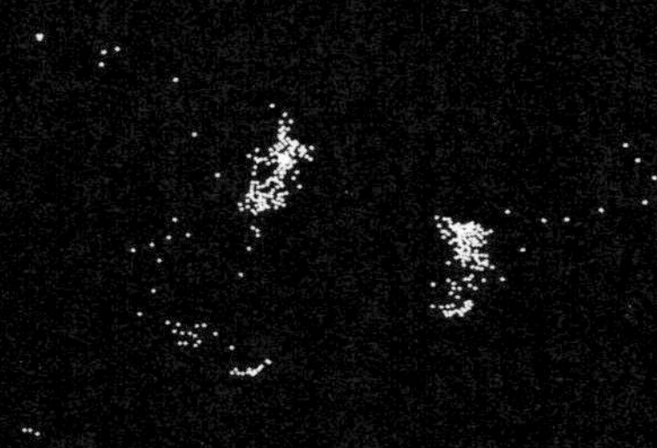

7 – ENERGY SLAVES – 1970

1 – SHIPPING LANES – 1970

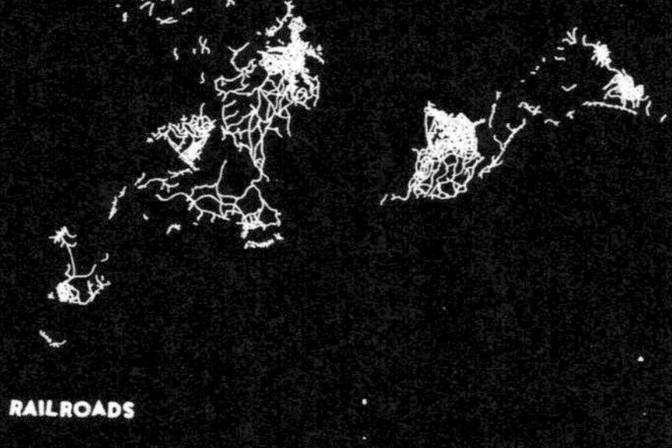

3 – RAILROADS

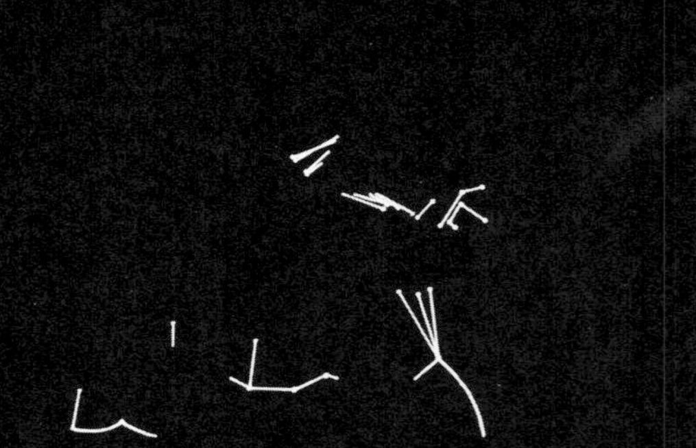

44 – TRANSOCEANIC CABLES – 1964

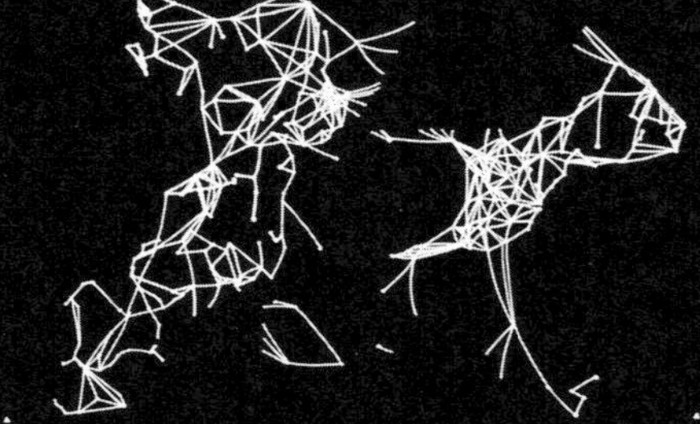

13 – AIRWAYS – 1970

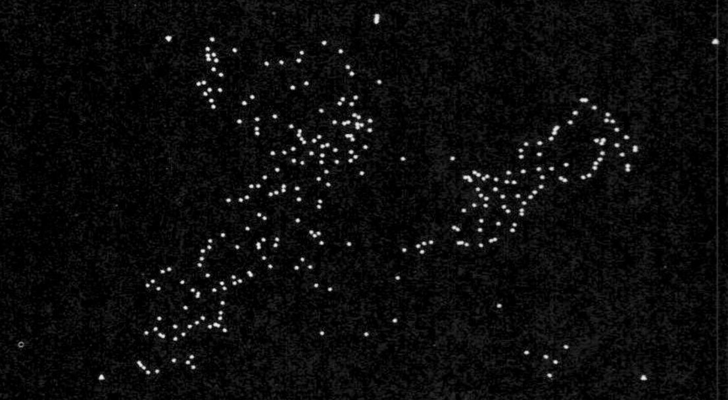

32 – MAJOR INTNT'L AIRPORTS

20 – SATELLITE COMMUNICATION GROUND STATIONS

25 – TELEVISION RECEIVERS – · = $2{\cdot}1 \times 10^5$ | $2{\cdot}1 \times 10^8$ TOTAL |

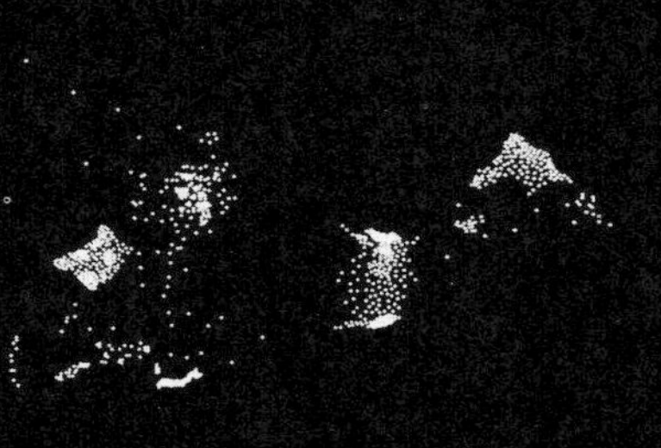

22 – COLLEGES · = 10

27 – % LITERACY ■ 90-100 ▥ 80-90 ▦ 50-80 ▩ 20-50 □ 0-20

14 – FOOD DISTRIBUTION – VEGETABLE PROD · · = 5×10^6 TON/YR

15 – FOOD DISTRIBUTION – ANIMAL PROD · · = 5×10^5 TON/YR

31 – FOOD MOVEMENT – RICE ——— 5×10^5 TON/YR

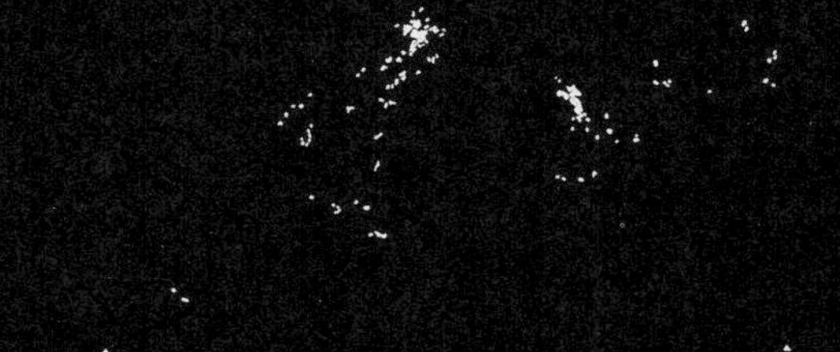

29 – CENTERS OF MFG · AND COMMERCE

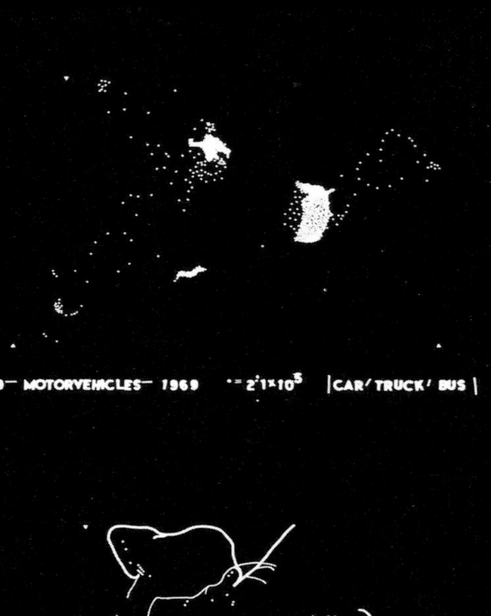

19— MOTORVEHICLES— 1969 · = 2·1×10^5 |CAR/TRUCK/BUS|

28— EARTHQUAKE ZONES AND STABLE LAND MASSES

9— COPPER— 1970 · -SITES — DISTRIB·

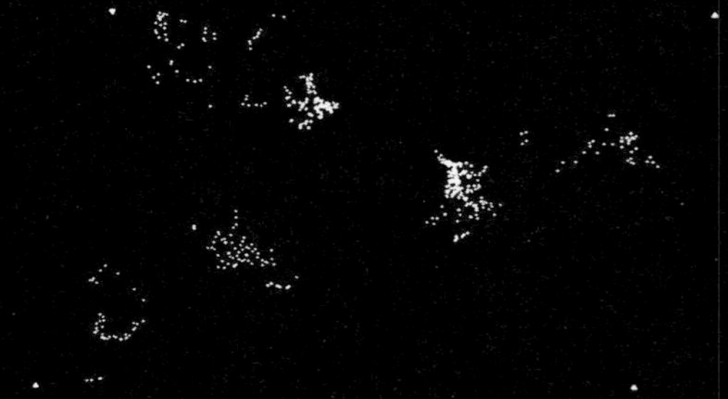

10— THERMAL ELECTRIC PLANTS 1970 · < 10^5 KW • ≥ 10^5 KW

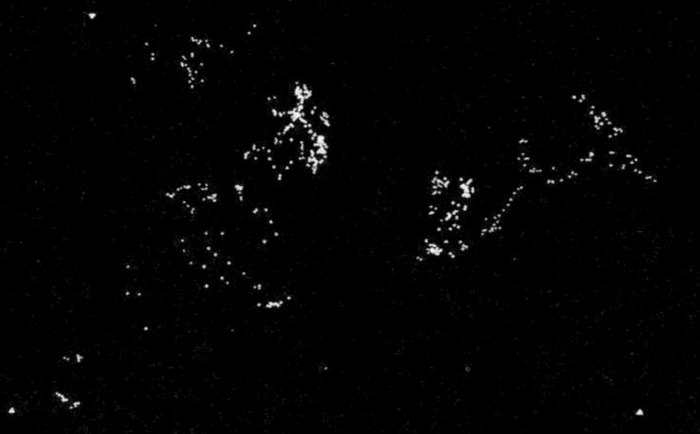

12— HYDROELECTRIC STATIONS · < 10^5 KW ≥ •

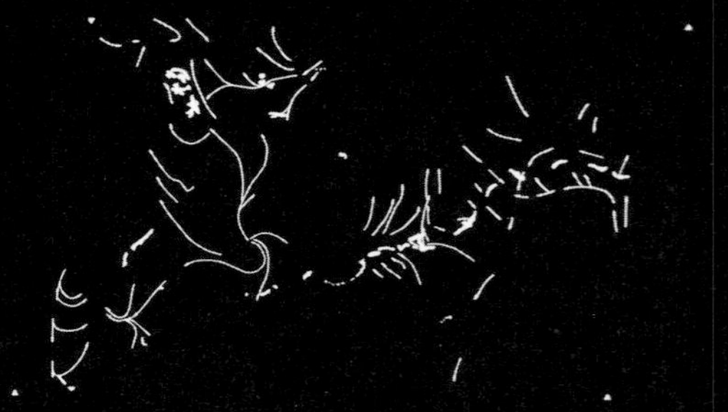

23— POTENTIAL ENERGY SOURCES —WIND ——— — GEOTHERMAL •

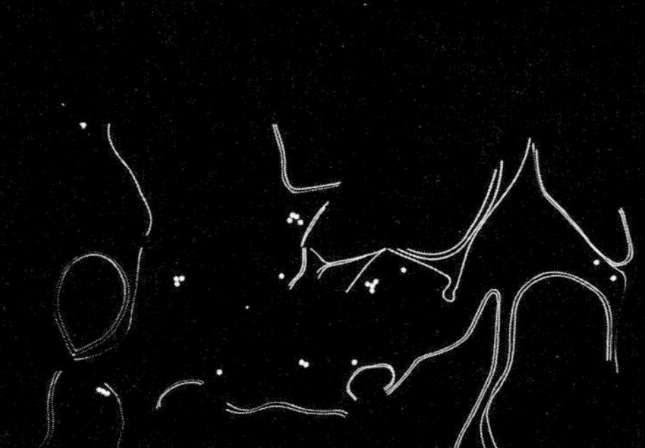

21 POTENTIAL ENERGY SOURCES —TIDAL • —CURRENTS —

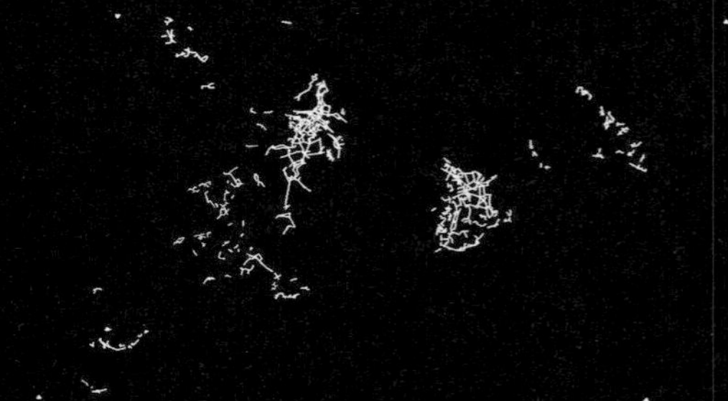

11— INTERCONNECTING ELECTRIC NETS / < 220 KW ≥ /

the players in such a way that the world's individual humans would each be able to exercise complete actional discretion; and would have such freedom of decision regarding the investment of their time in their waking hours that they would be able to travel independently, or in groups, either to-and-fro locally, or continuing intermittently on around the world, dwelling from time to time here or there, finding everywhere facilities to accommodate their needs in an uncompromising manner. The game would seek to use the world's resources, interprocesses and evolutionary developments in such a way that all the foregoing would be possible. It might involve moving whole city-sized groups of buildings from here to there in a few hours.

Let us assimilate World Game playing as I envision it, in order also to envision the kind of information most probably to result from such forward re-integration of our present world inventory of systematically differentiated-out physical and metaphysical variables.

(...) Amongst other grand strategies for making the world work and taking care of everybody is the design science revolution of providing ever more effective tools and services with ever less, real resource investment per each unit of end performance. For instance, a communications satellite, weighing only one-quarter of a ton, is now out-performing the transoceanic communication capabilities of 175 thousand tons of copper cable.

The World Game will explore for, and experiment with, more-for-less uses of our resources so that instead of taking care of only 44% of humanity at a high standard of living, we can take care of all of humanity at an even higher standard of living than man has as yet conceived.

In playing the game I propose that we set up a different system of games from that of Dr. John Von Neuman whose "Theory of Games" was always predicated upon one side losing 100 percent. His game theory is called "Drop Dead." In our World Game we propose to explore and test by assimilated adoption various schemes of "How to Make the World Work." To win the World Game everybody must be made physically successful. Everybody must win.

How it came about, in: *50 Years Design Science Revolution and the Worldgame* (1969), 111

480 *World Game* Seminar 1972: project for a global network of ultra high-voltage highways. Increase in efficiency up to 100 percent through day-night balancing.

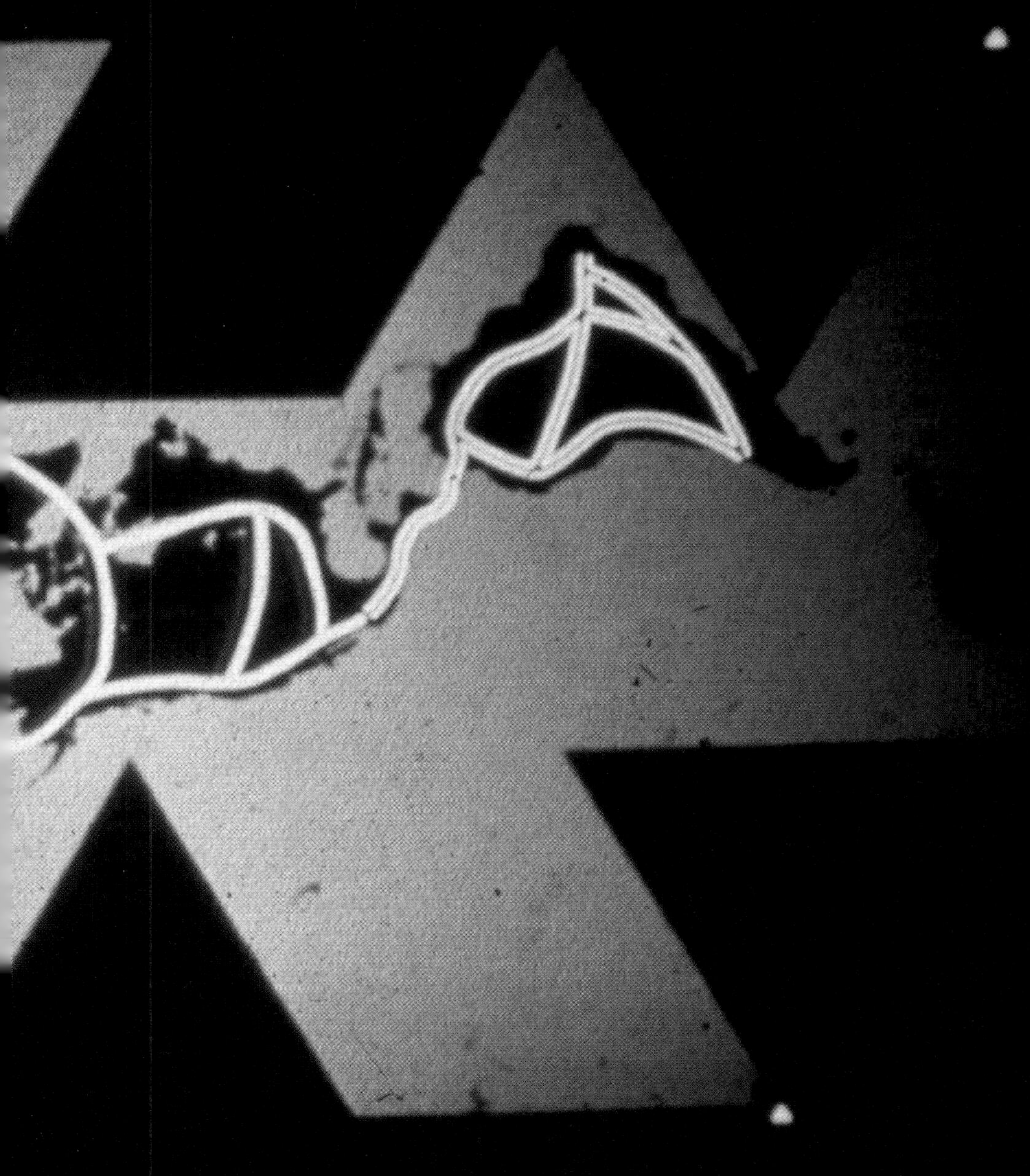

PROJECTED

 ·482/483: Bucky speaking to a group of Hippies, San Francisco 1968

Drop City commune near Trinidad, Colorado, founded 1965. Dome constructions out of pieces of scrapped autos,

"The flow of energy
through a system
acts to organize
that system."

CATALOG procedure

Ordering from the CATALOG

The CATALOG functions primarily as a pointer rather than a seller and prefers to be absent from most of the transactions it encourages.

Address orders to the supplier given with the item unless you know of a better one [if you do, let us know]; if the price is not listed postpaid find out the postage or express cost from the supplier's location to yours [consult post office or express agency for rates]; add state tax if transaction is within your state; and send check or money order with your order.

If the supplier gives you poor service, let us know. That information can be added to his review.

Blank order envelopes are provided at the back of the CATALOG for your convenience and so that suppliers have some idea of the CATALOG's effect on their business – if strong enough it may result in price or service advantages to CATALOG users. Don't use the envelopes if you don't want to.

With some indicated items, books mostly, the CATALOG also will ship. There is no price difference with the service; the CATALOG gets the markup instead of the other guy, is all. For west coast orders it may mean faster service.

Generally, the closer the supplier is to you, the quicker and cheaper the shipping will be. If the item you're getting is at all delicate [Don Buchla tells us], or if you are in a hurry, air express is the best deal. REA mangles..

Anything overseas, do by air.

Subscribing to the CATALOG

$8.00 per year. This includes Fall and Spring issues of the CATALOG and four Supplements. Subscription forms are on page 63. Memorize your zip code.

Suggesting

The validity of the information in the CATALOG is only as good as the transmitted experience of users. For any item, we have to

- *Learn about it.*
- *Get thorough information on it, and*
- *Stay current with its changes and with the improvements of its competitors.*

FROM YOUR EXPERIENCE
HOW WOULD YOU ALTER

per catalog	*THIS CATALOG*
per category	*THESE ITEMS*
per item	*THIS REVIEW*

Reviewing

The CATALOG pays its reviewers $10 an item for: getting familiar with the item, its usefulness, and its competition; evaluating the item; selecting samples of graphics or text (with page references) for the review; and writing a 200 – 300 word review.

Both reviewers and first suggestors of items are credited in the CATALOG.

We invite reviews that improve on present reviews or accompany suggestions for new items. On acceptance for publication reviewers will be paid $10 per accepted review. Unused reviews will not be returned. Polish of submitted material is irrelevant unless it is meant to be camera-ready.

New items that have had some favorable comment, and that we want reviews for, will be listed in the Supplement.

Corresponding

If the content of the WHOLE EARTH CATALOG is mostly products, the content of the Difficult But Possible Supplement is mostly processes.

Commentary from CATALOG users that is of general interest – and not a specific review – will be in the Supplement. Critical comments, new design processes, no-cash techniques, news of specific enterprises, useful fantasies, design student work, time and trouble shortcuts, new uses for common or exotic materials, other realms for the CATALOG to consider, etc., etc., – all welcome. The Supplement could wind up being more useful than the CATALOG.

Advertising

Suppliers, manufacturers, creators of listed items are eligible to advertise in the CATALOG. They may advertise only an item listed or their own catalog. All ads are placed at the back of the book. There are no ads in the Supplement.

Rates:

Selling

The CATALOG and Supplement are available for resale at 50% discount – minimum order 5. Single copy list price of the CATALOG is $5; the Supplement, $2.

Donating

Portola Institute, Inc. is a tax-exempt, non-profit corporation. Donations to Portola or the CATALOG may be deducted.

Retaining subscriptions to the CATALOG are $25 for one year ($17 tax deductible). Sustaining subscriptions are $100+ per year ($92+ tax deductible). Names of retaining and sustaining subscribers will be given in the CATALOG.

CATALOG policy with suppliers and users

The CATALOG is under no obligation to suppliers. Users are under no obligation to the CATALOG.

Suppliers (manufacturers, creators, etc.) may not buy their way into the CATALOG. Free Samples or other blandishments are cheerfully accepted by CATALOG researchers; response not predictable. No payment for listing is asked or accepted. We owe accurate information exchange to suppliers, but not favors.

Our obligation is to CATALOG users and to ourselves to be good tools for one another.

This Issue *of the CATALOG, the first, is heavy on books because they are easy to start with (low cost, simple to get and evaluate). As more CATALOG users report in and we develop better facilities to try stuff out, later issues should contain more information on materials.*

This issue of the CATALOG was prepared by:

Stewart Brand
Sandra Tcherepnin
Joe Bonner

with

Jane Burton	*Steve Baer*
James Fadiman	*Steve Durkee*
Richard Raymond	*Ralph Metzner*
Larry McCombs	*Gurney Norman*
	Robert Albrecht

We are grateful to Ortega Park Teachers Laboratory for temporary use of their mountain idyll for production space.

The WHOLE EARTH CATALOG and Difficult But Possible Supplement to the WHOLE EARTH CATALOG is published at Portola Institute, 558 Santa Cruz, Menlo Park, California 94025 on the following schedule:

CATALOG	*November*
Supplement	*January*
Supplement	*March*
CATALOG	*May*
Supplement	*July*
Supplement	*September*

Preparation of the CATALOG was done on an IBM Selectric Composer and a Polaroid MP-3 camera with instant half-tone system. Nowel Publications, Menlo Park, printed the contents, and East Wind Printers, San Francisco, printed the cover.

Subscription rate: one year – $8. Application to mail at second class postage rates is pending at Menlo Park, California.

WHOLE EARTH CATALOG

access to tools

Spring 1970
$3

487 · *Whole Earth Catalog,* Spring 1970.

Buckminster Fuller

The insights of Buckminster Fuller are what initiated this catalog.

Of the four books reviewed here, Nine Chains to the Moon is his earliest and most openly metaphysical, Ideas and Integrities his most personal, No More Secondhand God the most recent, World Design Science Decade the most programmatic.

People who beef about Fuller mainly complain about his repetition – the same ideas again and again, it's embarassing. It is embarassing, also illuminating, because the same notions take on different uses when re-approached from different angles or with different contexts. Fuller's lectures have a raga quality of rich nonlinear endless improvisation full of convergent surprises.

Some are put off by his language, which makes demands on your head like suddenly discovering an extra engine in your car – if you don't let it drive you faster, it'll drag you. Fuller won't wait. He spent two years silent after illuson· language got him in trouble, and he returned to human communication with a redesigned instrument.

With that, empirical curiosity, and New England perseverance Fuller has forged one of the most original personalities and functional intellects of the age.

I see God in
the instruments and the mechanisms that
work
reliably,
more reliably than the limited sensory departments of
the human mechanism.

And God says
observe the paradox
of man's creative potentials
and his destructive tactics.
He could have his new world
through sufficient love
for "all's fair"
in love as well as in war
which means you can
junk as much rubbish,
skip as many stupid agreements
by love,
spontaneous unselfishness radiant.

The revolution has come–
set on fire from the top.
Let it burn swiftly.
Neither the branches, trunk, nor roots will be endangered.
Only last year's leaves and
the parasite-bearded moss and orchids
will not be there
when the next spring brings fresh growth
and free standing flowers.

Here is God's purpose–
for God, to me, it seems,
is a verb
not a noun,
proper or improper;
is the articulation
not the art, objective or subjective;
is loving,
not the abstraction "love" commanded or entreated;
is knowledge dynamic,
not legislative code,
not proclamation law.
not academic dogma, not ecclesiastic canon.
Yes, God is a verb,
the most active,
connoting the vast harmonic
reordering of the universe
from unleashed chaos of energy.
And there is born unheralded
a great natural peace,
not out of exclusive
pseudo-static security
but out of including, refining, dynamic balancing.
Naught is lost.
Only the false and nonexistent are dispelled.

And I've thought through to tomorrow
which is also today.
The telephone rings
and you say to me
Hello Buckling this is Christopher; or
Daddy it's Allegra; or
Mr. Fuller this is the Telephone Company Business Office;
and I say you are inaccurate.
Because I knew you were going to call
and furthermore I recognize
that it is God who is "speaking."

And you say
aren't you being fantastic?
And knowing you I say no.

All organized religions of the past
were inherently developed
as beliefs and credits
in "second hand" information.

Therefore it will be an entirely new era
when man finds himself confronted
with direct experience
with an obviously a priori
intellectually anticipatory competence
that has interordered
all that he is discovering.

[No More Secondhand God]

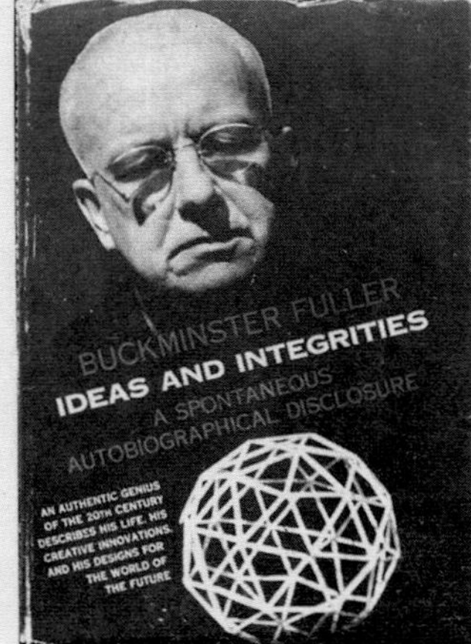

Ideas and Integrities
Buckminster Fuller
1963; 318 pp.

$10.00 postpaid

from:
Prentice-Hall, Inc.
Englewood Cliffs
New Jersey 07631
or
WHOLE EARTH CATALOG

Standing by the lake on a jump-or-think basis, the very first spontaneous question coming to mind was, "If you put aside everything you've ever been asked to believe and have recourse only to your own experiences do you have any conviction arising from those experiences which either discards or must assume an a priori greater intellect than the intellect of man?" The answer was swift and positive. Experience had clearly demonstrated an a priori anticipatory and only intellectually apprehendable orderliness of interactive principles operating in the universe into which we are born. These principles are discovered but are never invented by man. I said to myself, "I have faith in the integrity of the anticipatory intellectual wisdom which we may call 'God.'" My next question was, "Do I know best or does God know best whether I may be of any value to the integrity of universe?" The answer was, "You don't know and no man knows, but the faith you have just established out of experience imposes recognition of the a priori wisdom of the fact of your being." Apparently addressing myself, I said, "You do not have the right to eliminate yourself, you do not belong to you. You belong to the universe. The significance of you will forever remain obscure to you, but you may assume that you are fulfilling your significance if you apply yourself to converting all your experience to highest advantage of others. You and all men are here for the sake of other men."

[Ideas and Integrities]

Nine Chains to the Moon
Buckminster Fuller
1938,1963; 375 pp.

$2.45 postpaid

No More Secondhand God
Buckminster Fuller
1963; 163 pp.

$2.25 postpaid

both from:
Southern Illinois University Press
600 West Grand
Carbondale, Illinois 62903
or
WHOLE EARTH CATALOG

Thinking is a putting-aside, rather than a putting-in discipline, e.g., putting aside the tall grasses in order to isolate the trail into informative viewability. Thinking is FM–frequency modulation–for it results in tuning-out of irrelevancies as a result of definitive resolution of the exclusively tuned-in or accepted feed-back messages' pattern differentiatability.

["Omnidirectional Halo" No More Secondhand God]

Common to all such "human" mechanisms–and without which they are imbecile contraptions–is their guidance by a phantom captain.

This phantom captain has neither weight nor sensorial tangibility, as has often been scientifically proven by careful weighing operations at the moment of abandonment of the ship by the phantom captain, i.e., at the instant of "death." He may be likened to the variant of polarity dominance in our bipolar electric world which, when balanced and unit, vanishes as abstract unity I or O. With the phantom captain's departure, the mechanism becomes inoperative and very quickly disintegrates into basic chemical elements.

This captain has not only an infinite self-identity characteristic but, also, an infinite understanding. He has furthermore, infinite sympathy with all captains of mechanisms similar to his. . . . .

An illuminating rationalization indicated that captains –being phantom, abstract, infinite, and bound to other captains by a bond of understanding as proven by their recognition of eachother's signals and the meaning thereof by reference to a common direction (toward "perfect")–are not only all related, but are one and the same captain. Mathematically, since charactistics of unity exist, they cannot be non-identical.

[Nine Chains to the Moon]

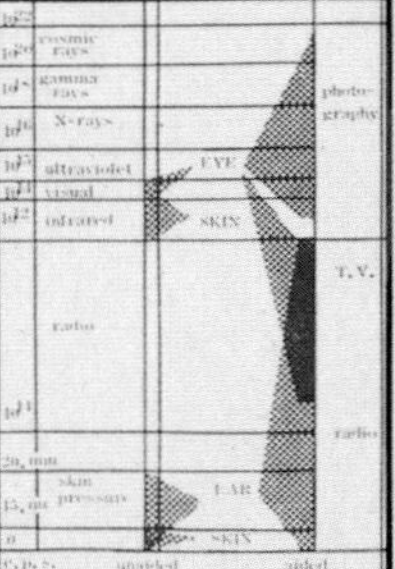

WDSD Document 1

World society has throughout its millions of years on earth made its judgements upon visible, tangible, sensorially demonstrable criteria. We may safely say that the world is keeping its eye on the unimportant visible 1 percent of the historical transformation while missing the significance of the 99 percent of overall, unseen changes. Forms are inherently visible and forms no longer can "follow functions" because the significant functions are invisible. . . . .

There are very few men today who are disciplined to comprehend the totally integrating significance of the 99 percent invisible activity which is coalescing to reshape our future. There are approximately no warnings being given to society regarding the great changes ahead. There is only the ominous general apprehension that man may be about to annihilate himself. To the few who are disciplined to deal with the invisibly integrating trends it is increasingly readable in the trends that man is about to become almost 100 percent successful as an occupant of universe.

Our Air Force Radomes were installed in the arctic mostly by eskimos and others who had never seen them before. The mass production technology made assembly possible at an average rate of 14 hours each. One of these radomes was loaned by the U.S. Air Force to the Museum of Modern Art in New York City for an exhibition of my work in 1959-1960. It took regular building trades skilled labor one month to assemble the dome in New York City.

WDSD Document 2

I define 'synergy' as follows: Synergy is the unique behavior of whole systems, unpredicted by behavior fo their respective subsystems' events.

[Ideas and Integrities]

selfishness (self-preoccupation pursued until self loses its way and self-generates fear and spontaneous random surging, i.e. panic, the plural of which is mob outburst in unpremeditated wave synchronization of the individually random components)

[No More Secondhand God]

 ·Stuart Brand founded the *Whole Earth Catalog* in 1968, inspired by RBF. It was with his writings that the catalog began.

The Last Whole Earth Catalog

access to tools

$5

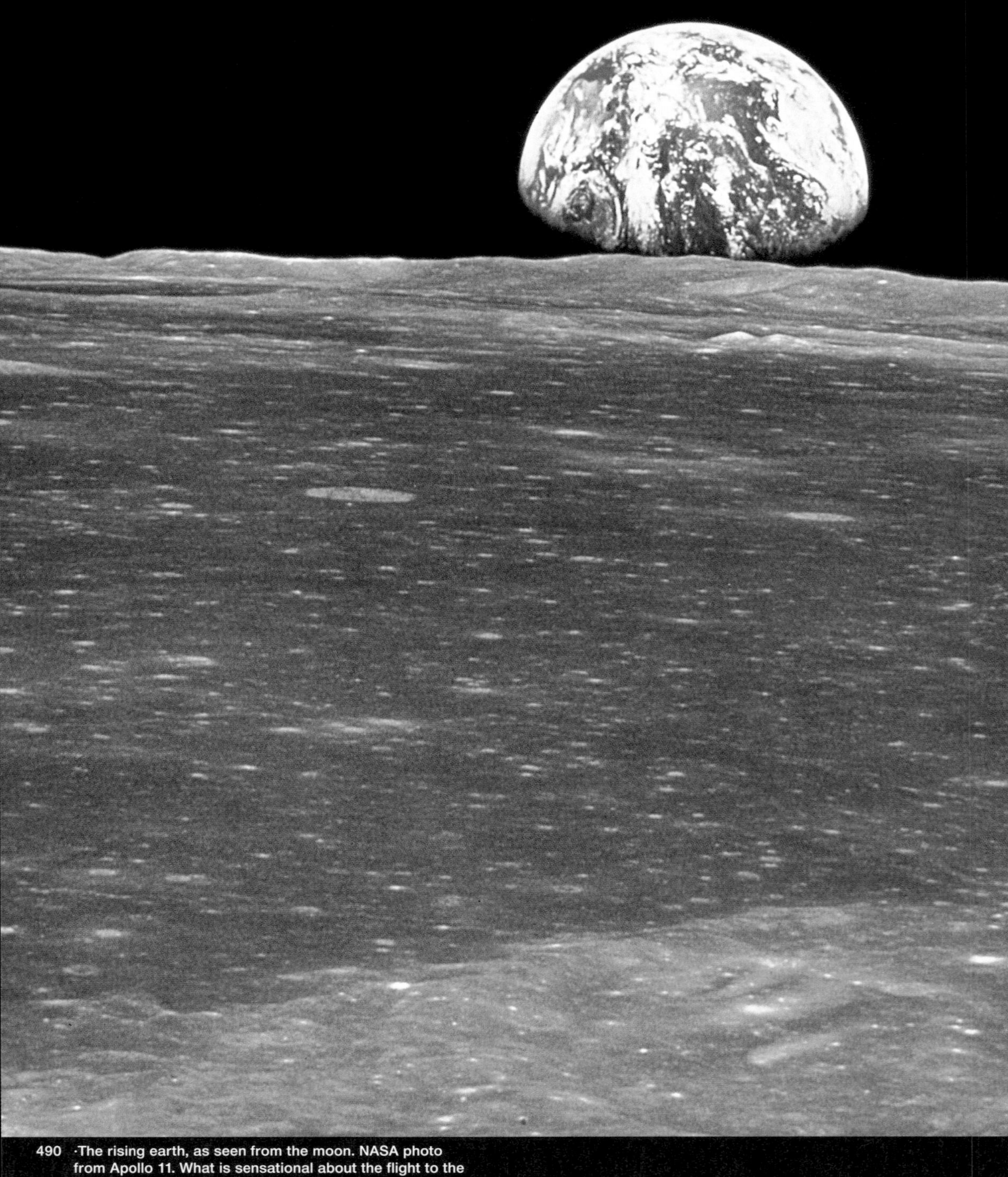

490 ·The rising earth, as seen from the moon. NASA photo from Apollo 11. What is sensational about the flight to the moon is not the moon, but rather the earth.

THE MILLENNIUM

WHOLE EARTH

CATALOG

“Think
act loca

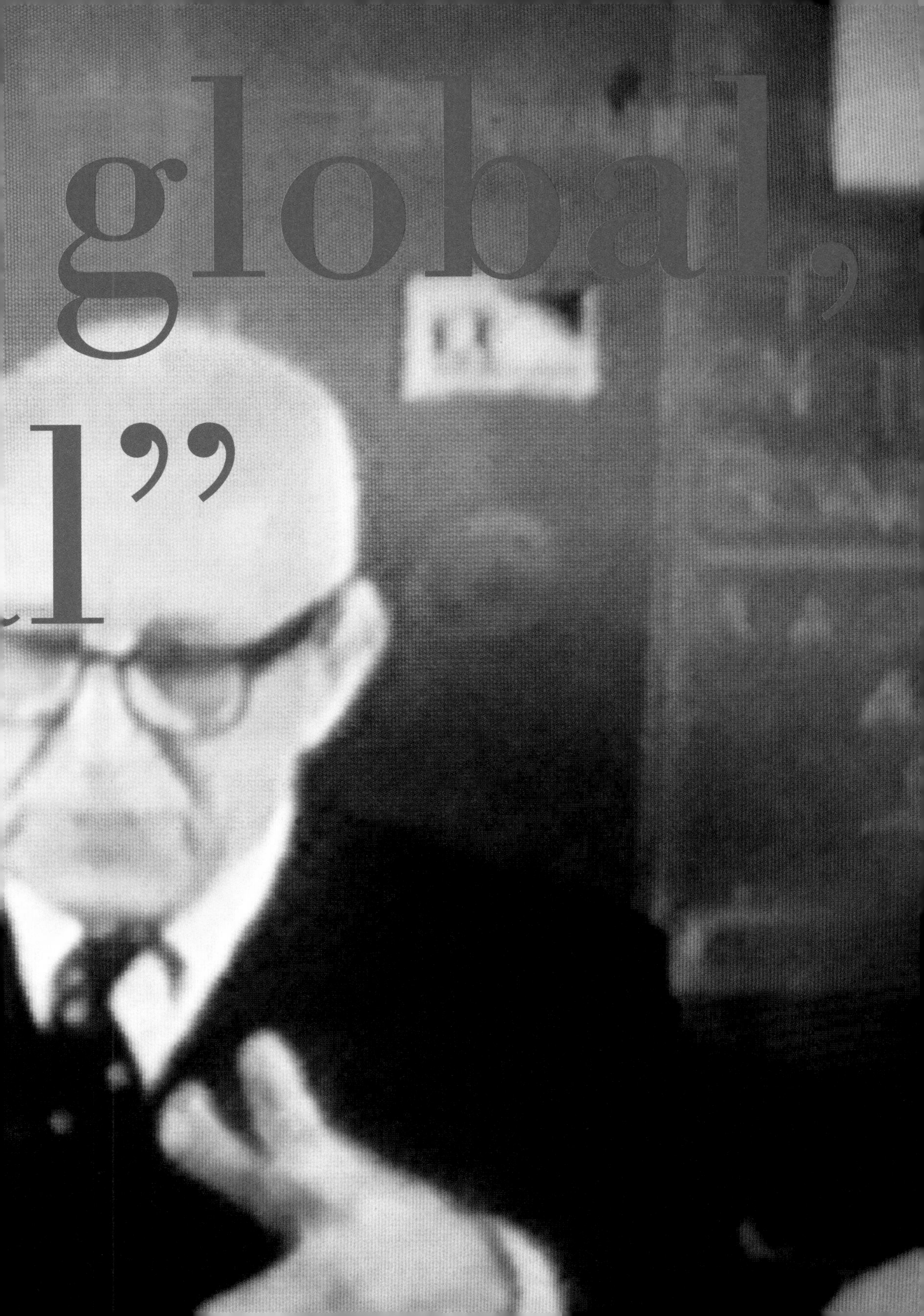

global,

l”

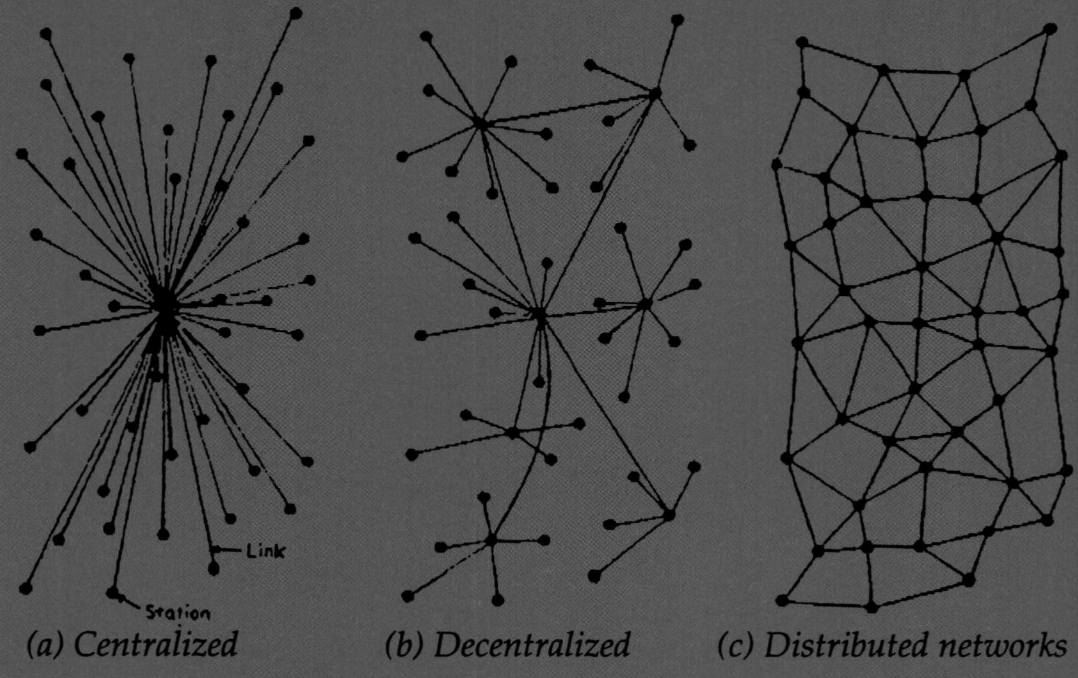

Shrinking Planet Earth

Disparity of the successive present generation's norms with those of previous and yet living generations is swiftly widening the gap between aspirations of the old and newer generations.

To make this disparity and its potential solvability clearer for study, I made some figures that I now find useful in comprehending the enormous velocity of change wrought in our evolving relationship to our respectively altering a priori universes. I started with a sphere twenty feet in diameter as a model, which was meant to represent the slowest relative rate of negotiability of the earth as gauged by the following calculations.

First, I supposed a path to be put about the earth, there being no dry path around the earth. But I wanted to allow a man to walk around the earth at the rate at which the Army says a man can walk daily, and rest and feed. The twenty-foot globe represented the rate at which he would be able to walk around the earth. Then I gave him a horse. The horse also had to sleep and rest and eat – and, using the Army figures again, I found man can negotiate the earth with a horse so fast that the relative size of the earth is reduced to a ball six feet in diameter.

I gave man a fast-sailing clipper ship and the earth came down to the size of a basketball. When I obtained these figures I realized the historical economic advantage that a man with a ship had over a man with a horse and how much greater advantage they both had had over a man on foot, throughout all history. The clipper ship, of course, was a tool; it was the first really large industrial tool that could not be produced by one man. And it did not have to stop to sleep at night like the horse, but kept on going twenty-four hours, day after day.

RIBA-Discourse, in: *Ideas and Integrities* (1963) 80

·Sketches by the computer pioneer Paul Baran from 1969, showing the typology of electronic communications networks. He distinguished between a) centralized b) decentralized and c) distributed networks. The latter are the most suitable for computer networking. They have properties of neuronal

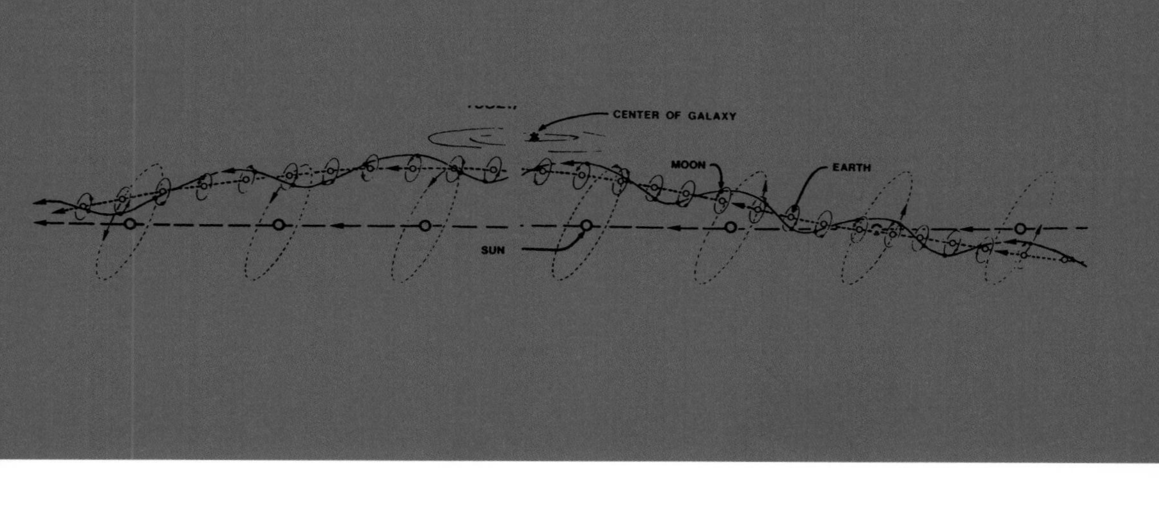
CENTER OF GALAXY
MOON
EARTH
SUN

YEAR	500,000 BC	20,000 BC	300 BC	500 BC
Required time to travel around the globe	A few hundred thousand years	A few thousand years	A few hundred years	A few tens of year
Means of transportation	Human on foot (over, ice bridges)	On foot and by canoe	Canoe with small sail or paddles or relays of runners	Large sail boats wi oars, pack animals, and horse chariots
Distance per day(land)	15 miles	15-20 miles	20 miles	15-25 miles
Distance per day (sea or air)		20 by sea	40 miles by sea	135 miles by sea
Potential state size	None	A small valley in the vicinity of a small lake	Small part of a continent	Large area of a continent with coas colonies

Communica-tions	Word of mouth, drums, smoke, relay runners, and hand printed manuscripts prior to 1441 A.D.	① The Gutenberg 1441 printing press	② The rapid pri 1863 newspaper pr

THE RELATIVE SIZE OF THE W

15,00 AD -1840 AD

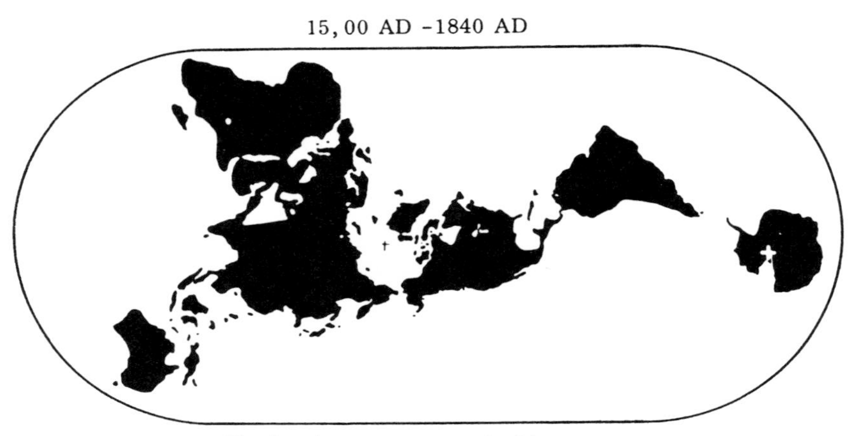

The best average speed of horse drawn coaches on land and sailing ships at sea was approximately 10 m.p.h.

Steam lo
65 m.p.h.
averaged

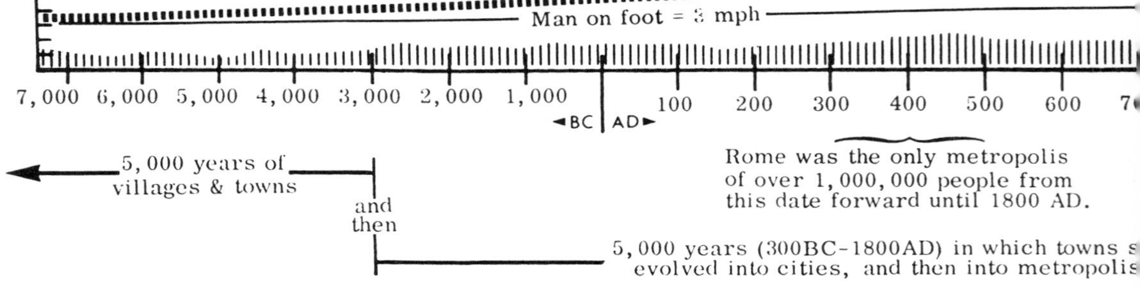

496 The shrinking of planet earth, relative to the means of

498/499: RBF in full cry. Slide presentation, ca. 1969

SHRINKING OF OUR PLANET BY MAN'S INCREASED TRAVEL AND COMMUNICATION SPEEDS AROUND THE GLOBE

1,500 AD	1900 AD	1925	1950	1965
A few years	A few months	A few weeks	A few days	A few hours
Big sailing ships (with compass), horse teams, and coaches	Steam boats and railroads (Suez and Panama Canals)	Steamships, transcontinental railways, autos, and airplanes	Steamships, railways, auto jet and rocket aircraft	Atomic steamship, high speed railway auto, and rocket-jet aircraft
20-25 miles	Rail 300-900 miles	400-900 miles	Rail 500-1,500	Rail 1000-2000
175 miles by sea	250 miles by sea	3,000-6000 air	6000-9500 air	408,000 air
reat parts of a continent with trans-eanic colonies	Large parts of a continent with transoceanic colonies	Full continents & Transocean Commonwealths	The Globe	The globe and more

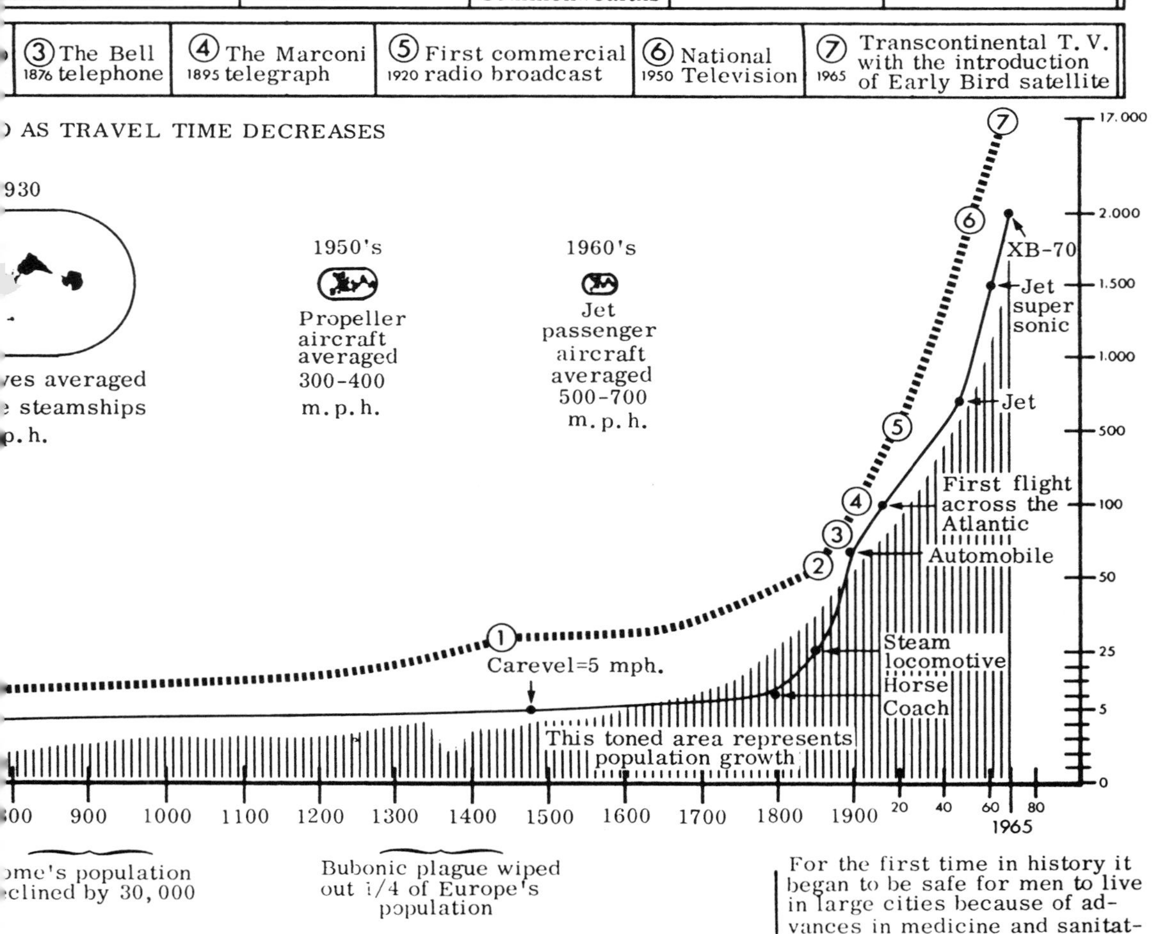

"Children are born true scientists," said Fuller in old age. Nowhere else is the desire to learn so sincere and receptivity so intact. From children we can learn how we learn. As an eighty-year-old, Fuller became a lithograph artist. He sketched for Tatyana Grosman and her United Limited Art Editions (ULAE) a series of sheets called "Tetrascroll" on the story of "Goldilocks and the Three Bears," which he had begun to draw forty-five years earlier as a picture book for children. In twenty-one lithographs and the accompanying commentaries, Fuller summarizes decades of research. The little girl Goldilocks explains to the bears the laws of geometry in nature and the development of human civilization from sea travel.

TETRASCROLL

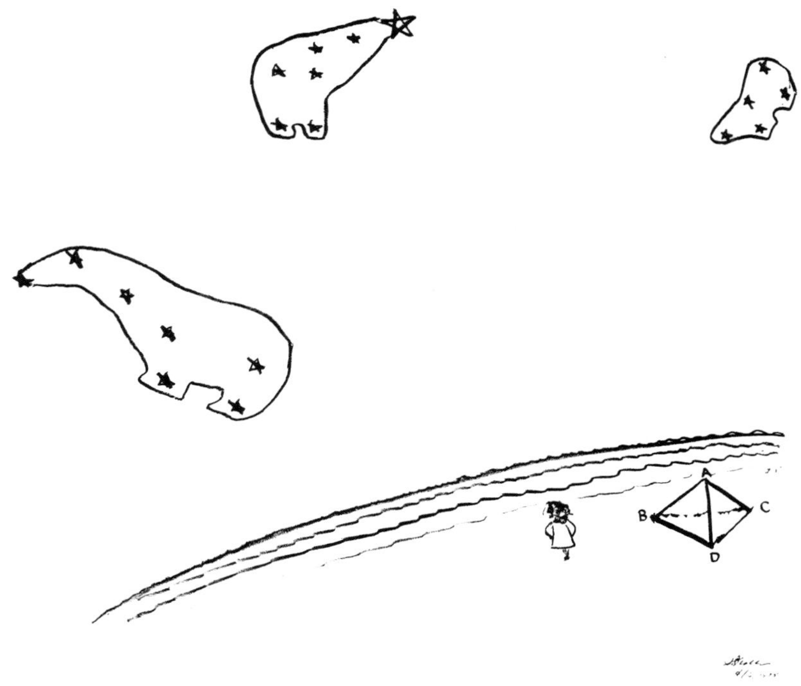

Stone 1

Here is Goldy having a sky party with her three friends, the Polar Bear family. Goldy says the sky party is a "system" because Goldy plus the Three Bears equals four entities (or star events), and it takes four events to produce a system. A system divides all the universe into six parts: all the universe outside the system (the macrocosm), all the universe inside the system (the microcosm), and the four star events A, B, C, D, which do the dividing.

The tetrahedron's four-corner star events do not have to occur at the same time. Goldy found that light traveled six and one-half trillion miles in a year, and was fascinated when an astronomer told her that the star in the nose of the Big Bear is a live show taking place 210 light-years away-and-ago, as the American colonists are first thinking about revolting from English rule: and the pole star at Mommy Bear's nose is a live show taking place 680 light-years away-and-ago, as Dante is writing The Inferno: and the star at Wee Bear's front toe is a live show taking place 43 light-years away-and-ago as Franklin Delano Roosevelt is being elected to the USA presidency for the first time, at the depth of the great 1929–39 Depression: While she, Goldy, is also a live show taking place no time away-and-ago. Altogether Goldy's four live shows constitute a scenario of nonsimultaneous but omni-interrelated events, which can and do define the four corners of a minimum system – the tetrahedron.

She now understands Einstein's concept that Universe is a scenario and not a single simultaneous structure. One picture of a caterpillar does not tell you it is going to transform into a butterfly, and it takes many frames of the cinema to inform you that the butterfly can fly.

Stone 2

Using the vast, water-smoothed surface of the many-miles-long sandy beach, and walking along as she talks, Goldy keeps drawing pictures large enough for the bears to see. Goldy says to the bears, "Let's try an experiment with our tetrahedron."

By pushing successively on the tetrahedron's top vertex, Goldy keeps rolling the tetrahedron ahead of her across the beach. This succession of rollings makes a long, parallel-edged ribbon with a line zigzagging between its edges to produce a succession of adjacent triangles.

Goldy says to the bears, "We have discovered a triangularly subdivided ribbon-printing machine – a wave-printing machine." And Daddy Bear says, "That is also the sand print patterning made by our four (A, B, C, D) bear's feet when we are running. We can start our run with our right hind foot D elevated. We then lunge forwardly over the hinge line running between our two front feet C, D, as foot A goes forwardly and down while foot B is elevated. Because a bear's foot is itself a triangle, Goldy makes a pattern of Big Sky Bear's footprints as he walks or runs eastwardly along the beach. Goldy uses the successive triangles as the frames for the succession of illustrations of her conversation with the bears. She says the ribbon is like a scenario filmstrip with the successive triangular pictures overlapping instead of being vertically separated.

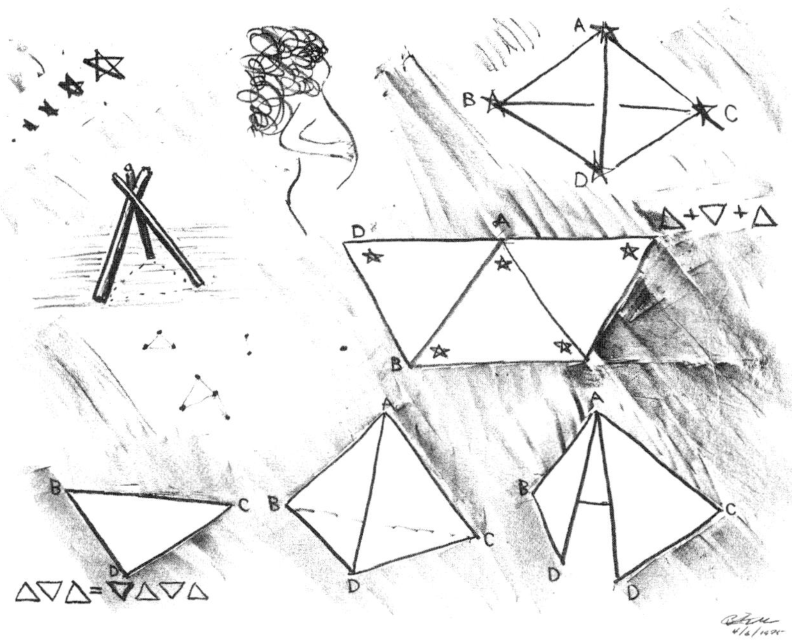

Stone 3

Goldy explains to the Three Bears that synergy means behavior of whole systems unpredicted by the behavior of any of the system's parts when each is considered only by itself. Goldy takes three triangles and brings them together edge-to-edge around a single corner and inadvertently produces the fourth base triangle. Thus she discovers that one-plus-one-plus-one equals four.

Goldy says her mommy is a system and so is her daddy. They both have insides and outsides. Goldy says parents are synergetic: 1 system + 1 system = 3 systems. One outsider plus one insider produces an additional inside-outer – Goldy.

Goldy says to the Three Bears, "If you don't understand any of my words, you can find them in the dictionary." Wee bear replies, "Out here we use cosmic thought communication. We don't have to find words in special-language dictionaries. We use a cosmic thinktionary. All your dictionaries express the universal concepts of our thinktionary but only in special, ethnic-language, sound words. The concepts such as mountain or star or nuance are the same experience engendered concepts in all languages. We understand you perfectly, Goldy."

Concepts are always synergetic systems. Systems are minimum-maximum sets of thinkable, conceptual, omni-interrelevant experience recollections, intertunably differentiated only by time out of nonsimultaneous, unitarily nonconceptual Scenario Universe.

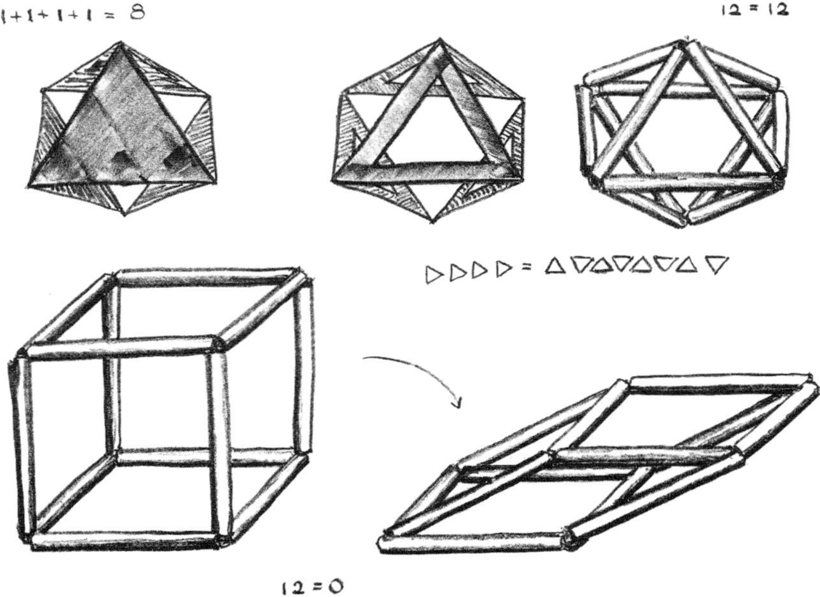

Stone 4

Exploring synergy further, Goldy takes apart the four edge-bonded triangles of the tetrahedron and reassociates them, this time joining them only by their vertexes, which produces the octahedron, or eight-triangular-faceted polyhedron with four inadvertently produced, structurally stable, empty triangular window facets. Here is synergy again with 1 + 1 + 1 + 1 = 8. This clearly seen structural system's symmetrical intertransformation from 4 to 8 – i.e., 1 to 2 – Goldy explains, is the simple principle which, being nongeometrically conceptualized by the physicists, is called by them "a quantum leap." The scientists continue to "fly blindly on instruments." In addition to the eight triangular facets the octahedron has six vertex (star-entity) events and twelve structural edges.

Since half the triangles are empty and only half are filled, Goldy becomes curious and cuts four triangular windows in each of the four filled triangular facets, and she finds the octahedron as yet stable. It seems to Goldy that the twelve edges held together by the six vertexes must be producing the structural stability.

So Goldy next takes two sets, each of twelve equal-length tubes, threads a string through the tubes, and fastens them together in two different ways: (1) with four tubes joined at each of six corners to make the octahedron, which is structurally stable, and (2) with three tubes joined at each of eight corners to make the cube, which proves to be structurally unstable and collapses.

Stone 5

This brings Goldy to her necklace experiment, to discover, if possible, what produces structural stability. When the necklace flexes, the tubular beads do not bend or change their lengths. It is the tension connector angles between the tubes that change and accommodate variable draping of the necklace. One by one Goldy takes the inflexible tubes out of the necklace, which keeps on flexing around her neck and draping over her shoulders until there remain only three push-pull tubes and three tension connector angles. Now for the first time the necklace does not flex or drape around her neck. It is rigid. It is in the form of the triangle, which is the minimum polygon. There is no polygon of two sides and two angles. The necklace triangle has six separate parts: three rigid, push-pull tubular sides and three flexible tension angles, all six of which separate entity-events are interacting to produce a stable pattern. How do they do so?

Any two sides of the triangle constitute a pair of levers fulcrumed tensively together at one end – like a pair of scissors. The longer the two lever arms, the more powerful the shears. So the third side of the triangle is a rigid, push-pull strut taking hold of two adjacent lever arms at their maximum lever-advantage ends, thereby stabilizing the angle opposite with minimum effort. So does each side of the triangle most effortlessly stabilize its opposite angle. Since a structure is a pattern-stabilizing complex of events, a triangle is structure. Structure is triangle. There are no other such minimum-effort, six-foldedly combined, minimum-limit-of-a-series, cosmic cases such as this one.

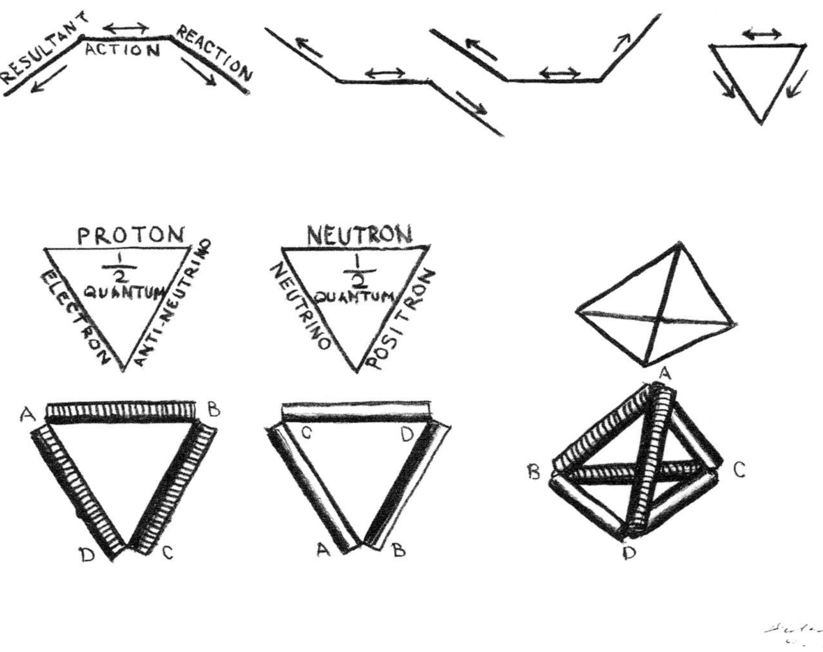

Stone 6

In nonsimultaneous Scenario Universe every event action has both reaction and resultant "side effects" (precessions), which may be graphically represented by three angularly associated vectors that can take either open or closed forms.

The atomic proton and neutron are unique, always and only co-occurring, prime energy events of Universe. They have different mass but may be co-intertransformed by means of their two different actions and the latter's, two each, different energy "side effects."

The proton and the neutron are both actions, and each has its respectively different reaction and resultant "side effects." Each of these two, three-vector teams equals one-half of an energy quantum (or one half-spin).

Because it consists of two such half-quanta of energy, the vector-edged tetrahedron exactly equals one quantum of energy and is also one minimal structural system of Universe. The six-vector tetrahedron is also synergetic in that two vector triangles combine to make the four triangles of the tetrahedron. This is not magic. The two additional invisible triangles that become visible by associating the visible pair are always secreted in the invisible complementarity reserves of the 99.9 percent invisible Universe of utterly abstract weightless principles. Employing vectors, the two convergent sides of any given angle can only be considered as potentially realizable by a third and invisible vector which invisibly holds apart the outer ends of the angle describing convergent vectors.

Stone 7

Next Goldy says to the Three Bears, "Unlike you constellations of stars who intercoordinate with combined gravity and precession and know the most about both, few people gravitationally cohered around precessionally steered planet earth comprehend either gravity or precession."

Goldy takes a closed cylinder made of flexible rubber which is filled with water. She presses its ends toward one another, and the cylinder bulges outward radially at its mid-girth, tensively stretching the rubber outward radially in a plane perpendicular to the axis of Goldy's compressing. Next she pulls the two ends of the cylinder away from one another, and the cylinder contracts radially at its mid-girth, becoming compressed in a plane perpendicular to the axis of Goldy's pulling.

These right-angle results of applied effort are called precession. Precession is the integrated effect of bodies in motion on other bodies in motion. The gravitational interattraction of the sun and earth results in the earth traveling around the sun in an orbit perpendicular to the line of the sun's and the earth's gravitational interattraction.

Precession is regenerative. Goldy drops a stone into the water. A circular wave is generated, growing outwardly in a plane perpendicular to the axis of the stone's line of fall. Then the outward motion of the wave precesses the surrounding water, causing the water to rise as a complete circular ridge in an upward direction perpendicular to the plane of the wave's horizontally outward growth, and the rising water in turn immediately reprecesses the surrounding water into further outwardly radiating horizontal growth.

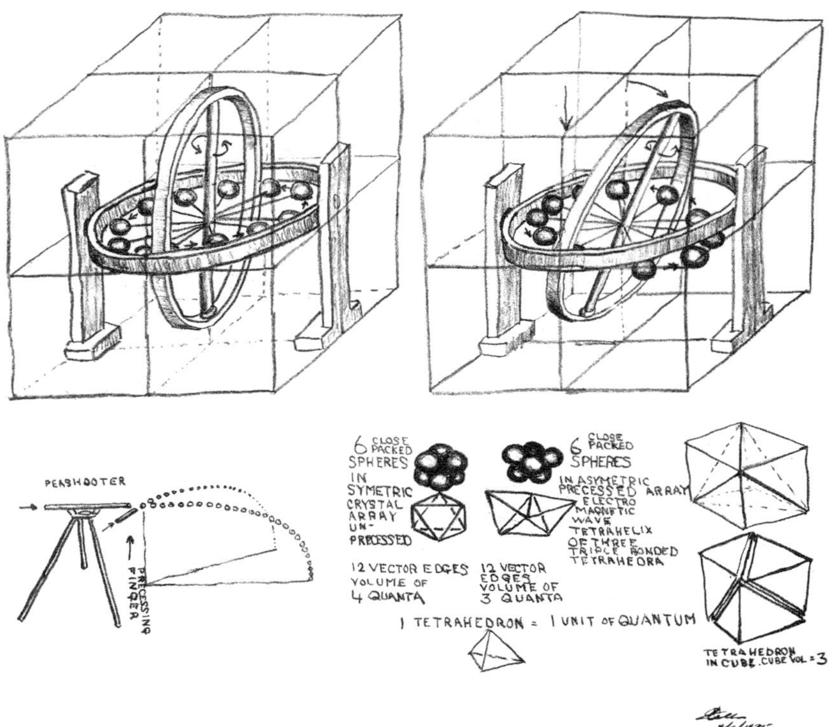

Stone 8

Goldy starts the air gun's impelment of the rotatable shaft's turbo-blades, and the "grass-skirt" ring of balls rises to a vigorously maintained horizontality of rotation. Goldy lowers a mechanically guided steel finger to touch downwardly, around axis Y, once only on the top of each of the balls as each passes horizontally and tangentially by the point in the plane of rotation nearest to Goldy. Each separate ball is deflected downward a discrete angular amount, which results in the plane of the ball's rotation tilting downward to the right around axis X as the powerful angular momentum of the ball's pulling outwardly on the rotating shaft causes the shaft to rotate rightwardly from its previous verticality in plane C, having rotated the same discrete angular amount as that which the plane of the rotating balls has deflected downwardly and to the right around axis X to accommodate maintenance of its tensionally enforced perpendicularity to the plane of the ball's rotation. Thus in turn, the inner rotatable annular ring in which the shaft is mounted is forced to accommodate the shaft's reorientation, and it rotates the same exact angular amount around axis X in plane C as the angular change of the plane of the rotating balls and the shaft. Goldy stops the rotation and returns the shaft and inner annular ring to their vertical starting point, with the same "grass skirt" of attached balls hanging down limply around the shaft. Once again Goldy starts the air gun's blowing of the turbo-blades. She finds that as she pulls the top of the annular ring rotationally toward herself around axis Y, it refuses to yield toward her and yields instead only leftward, as its rotating shaft and the plane of the rotating balls all tilt rightward and downward around axis X, X' as before in plane C in a direction at right angles to the plane of rotation of her pulling. With the machine in static (motionless condition) Goldy pulls the top of the vertical shaft toward her, and both it and the inner and outer rings immediately rotate toward her around axis Y between the two main stanchions.

Stone 9

Goldy next shows the bears how the three-face-bonded tetrahedra-arc in its initial, neutral, nontransmitting state becomes spirally extended positively or negatively to attain its information-transmitting state, only with the addition of one more face-bonded tetrahedron. She then shows that with every twenty tetrahedra the tetrahelix completes approximately one 360 degree helical revolution (352° 40′ exactly), which tetrahelix is the mathematical model employed by the DNA-RNA helix, discovered by virological scientists (Watson-Crick-Wilkerson) to be always transmitting the specific information controlling the design of all biological species, with that 7° 20′ of angle (less than 360°) being twist-sprung to introduce the unzipping force necessary to offspring (or give birth to) any given species' off-molded offspring from the parent.

Goldy shows how the extended tetrahelix's skin can be stripped off and laid out flat as a three-row, omni-triangulated, wavilinear ribbon. Goldy then identifies the positively or negatively asymmetrical tetrahelix patterning with lightning. Goldy then introduces Naga – the sea serpent – god of the oceanic world of the ancients. Naga is the wave. Naga is a live tetrahelix. At sea the wavilinear profile of Naga's back always rims the horizon. The ancient Hebraic language of the earliest biblical scripture came to contain the word "nachash", which means serpent, or "whisper," or "divine" (the ch being a guttural or g sound) – i.e., Naga and Nachashol (or Nagashol) means "the sea," and the root verb nacha (= Naga) means to "lead," "conduct," "guide," and Nacha – pronounced Naga – is also the name of the ancient seafarer N(O)(A)(C)H – Noah.

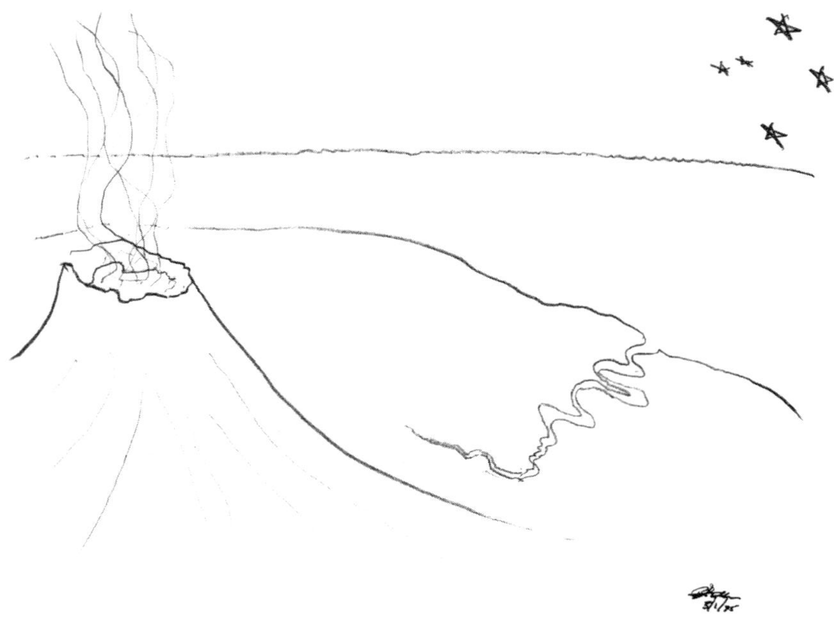

Stone 10

When the god of the sea Naga tidally pentrates the river mouths of the land, as seen from the high mountain, his snakes shape is clearly revealed by the river's shape. The Japanese word for river, Nagala, indicates that the ancient water people looked upon the riverbanks and beds as constituting the female organ of the land being sexually intruded by Naga, god of the sea, as the oceanic tides pulsed inwardly and outwardly for great distances at the lower extremities of the rivers. The early humans sensed and revered the greater pattern events of Universe as manifes-ting an ever and everywhere presence of a knowing and life-giving, supporting, and terminating competence vastly greater than that of humans. They saw themselves and all that they could see including the sun, moon, and stars as having only minuscule local parts in an organic whole whose shape and size transcended both the ranges of their vision and the scope of their imagining.

Stone 11

Goldy reasons that because all human beings are always born naked, ignorant and helpless yet chromosomally programmed to experience periodic hunger, thirst, curiosity, and the procreative urge – and are thereby provoked to search and learn only by trial and error, to discover for themselves how best to cope in general – their nakedness and organic structuring primarily 60-percent water, which requires frequent replenishment, combines with the fact of water's small temperature range between freezing and boiling to make it immediately clear that naked humans could not be born nakedly and survive north of the freezing line or in arid desert or where any large carnivorous animals could swallow their for-months-helpless babies. By far the most favorable localities and conditions for successful inauguration of human life, to be found anywhere around planet earth, would be those environmental conditions unique to the barrier-reef-encircled and mountain-island-centered coral atolls formed atop extinct volcanoes in the southern seas of the Pacific and Indian oceans. Here we have the beautiful, clear, shoal water lying protected within the circular Great Barrier reef of coral against which great ocean waves thunder and spend their massive power. Lagoons, temperatured ideally for prolonged human immersion, lie inside the reef and abound with fish and all other most ideal primitive environmental conditions for nurturing naked humans. Beautiful shoal beaches make easy walking into and out of the water while high palm-tree coconuts, full of milk, fall thumping to the ground, and humans may find a host of fruits and no wild animals. Here humans quickly learned that wood floats and stones sink.

Stone 12

Goldy and the bears now discuss the fact that when the atoll-incubated humans first began to penetrate the mainland, they dug vast systems of canals into the shorelands to extend and complement their normal sea life support with additional life support grown in the freshwater-fed fields and paddies lying adjacent to their saltwater canals, which latter served to float their life-support boat cargoes to their floating water markets. Their half-water, half-land life gave these oceanic people the opportunity of capturing, taming, and domesticating both animals and vegetation by inbreeding them for emphasis of desired life-support characteristics. Thus developed the earliest, primarily waterborne, canal states which were ruled by kings who were, by demonstrated facts, the most physically capable of protecting their kingdoms against invaders. Because no one knew why the physically big and strong ones happened to be born big and powerful, the legend readily developed that the kings were divinely selected – ergo, were living demigods.

Goldy now draws a picture of what is often mistakenly spoken of by twentieth-century European and American tourists as "temple." Goldy says, however, that these local demi-universe models were only to serve the monarch at the time of his death – as a great "step" ladder into heaven their cosmological model started with a large watery lake surrounding the miniature world. Centrally within this lake the god of the sea, Naga, sculptured in stone, completely surrounded the perimeter of their square, islanded world. Angkor Wat in Cambodia is the best remaining example of these cosmological models. It is completely surrounded by a square body of water which long ago led off by canals to the sea.

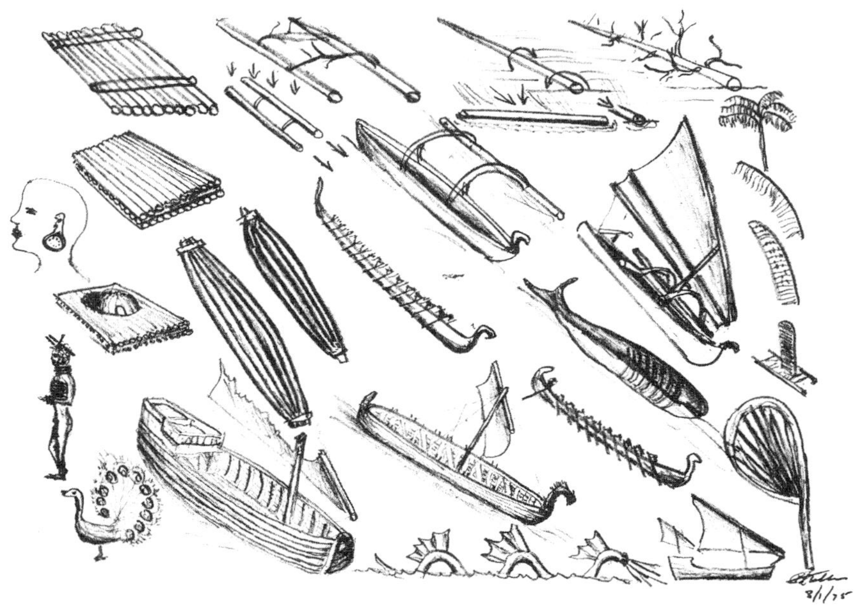

Stone 13

Atoll humans quickly learned that wood floats and stones sink, that a single float-ing log rolls and that two floating logs held cross-connected by their branches do not roll, but provide a structurally stable floating device from which they could fish or dive for edible mollusks. The atoll dwellers soon learned to build both multilogged rafts and dugout log boats as well as outrigger-stabilized dugout canoes. Then they learned how to stitch together the palm tree's fronds to make combined masts and sails and found that their outrigger canoes could sail zigzaggingly to windward while rafts could only ride with the ocean currents or drift to leeward with the wind. The outrigger canoes could be worked (or could "beat") to windward by a succession of individual "tackings," ... at a firmly sail-filling angle, which is about thirty degrees one side or the other of the direction from which the wind is coming. This produced a low atmospheric pressure on the forward and leeward side of their sails, which, as with twentieth-century airplane wingfoils, pulled their boats forward in the direction of least resistance, which direction could be modifyingly controlled by steering oars or paddles. The atoll sailors, ... in their much-earlier-developed, large, dugout, outrigger canoes, learned to paddle them into the wind to become, undoubtedly, the first successful westward sea travelers against the prevailing winds. But this windward travel against the waves took both musclepower and much food to support it, which curtailed the length of voyaging. Nonetheless, it did make possible for the first time in history the predetermined (and celestial-navigation-maintained) direction of travel independent of the direction of winds and ocean currents. Team paddling powered their voyages of discovery around islands, along seacoasts, and inland to explore the rivers. It established the principal mode of travel of the canaled, water-state kings of those four-cornered cosmological models (of kingly ascent into the realm of the gods).

Stone 14

Thus the human sailormen worked ever westward and ever farther off shore until in the lower latitudes of the Indian Ocean their westwardly headed voyaging was reversed by the powerful, eastwardly flowing Antarctic's world whirlaround of wind and water. Six thousand miles south of, unbeknownst to any other, terrestrial humans this eastward world whirlaround distributed those voyagers into the Pacific, Atlantic, and Indian oceans.

Many of those sailors not washed overboard from their craft and drowned were wrecked and marooned on the myriad of lonely, barren islands scattered widely apart in the world's southern seas. Thus we find waterborne humanity first paddling, then rowing, then sailing determinedly westward into the wind to follow the life-giving sun's sky course, gradually working farther off shore and at their southern extremes of voyaging being unexpectedly caught in and swept along by the world-around eastward forces to bring about human occupation of many of the very remote islands of our planet. Those voyagers marooned on lonely islands who found the food, water, and sheltering means to survive became scavengers of subsequent shipwrecks on the treacherous rocks of those islands. Hundreds and even thousands of generations of remote inbreeding developed highly differentiated physical and cultural features.

The historically unprecedented swiftness of 20-century development of the world around integrating transportation and communication means is now trending to rapidly integrate and cross-breed back to an average world-human all these isolated increments of humanity of the islands as well as of those inbred among isolated inland tribes. In this way ever-evolutionary Universe has contrived first to conserve by isolation all the lessons learned regarding humanity's artifact-inventing ingenuity in coping with the most extreme conditions and secondly to synergetically integrate all knowledge for the mutual advantage of all humanity.

Stone 15

Traveling ever farther westward against the prevailing winds seemed to say that merchants and those who patronized them were deliberately contradicting the winds of God's will. This sailor-popularized apprehension greatly frustrated realization of the vast wealth potential in the east-west high-seas trade. For the miracle answer the kings, nobles, and merchants of Mesopotamia turned to the navigator-astronomer-priests of Babylon. To convince everyone in the new western world that accounts of any earlier religions or people elsewhere on earth were false, the priests said that the beginnings of humans in Universe had occurred nearby to Babylon in a garden called Eden. Their story from then on is well known. What seems pure nonsense in the Garden of Eden story of the creation of a woman from a man's rib is explained as follows. Vessels of the sea are always female because they contain their crews in their interior wombs. The female "Eve" was the high-seas, world-around-sailable vessel; her great strength developed when navigators discovered the backbone-mounted rib cage employed by Nature in the design of whales, porpoises, seals and other sea creatures. So man built his high-seas vessel, "Eve," with strong wooden ribs rising sidewise from her keel, planked "her" in, then leather-thong-fastened the planks' edges together, tied them tightly into the ribs, and pitched her seams. Thus "Eve" the ship, built from Adam's rib cage design, was temptingly "led on" by Naga the serpent, god of the sea, around the world, with Adam aboard. Thus Naga showed Adam, by means of Eve, that the earth is as round as the apple.

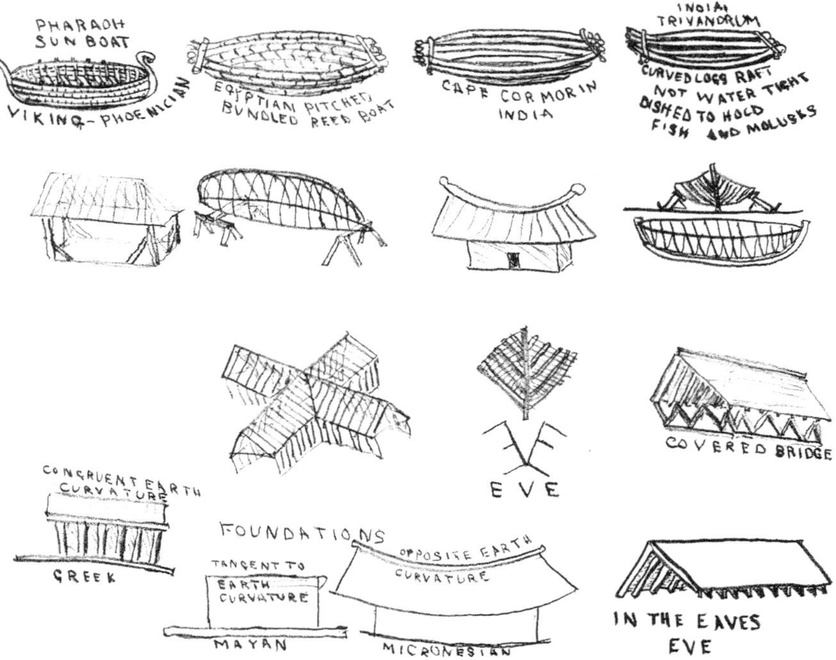

Stone 16

When the blond-haired Vikings reached Scandinavia, they beached and overturned their large, powerfully keeled, transversely ribbed, thong-tied and pitch-seamed, longitudinally lap-straked boats to provide a watertight roof for their snow-covered winter homes. Mounting the boat ends on wooden "horses," the Vikings often clustered four boats end to end in the form of a cross and draped skirts of skins from the gunwales to the ground. This became the cruciform prototype of subsequent cathedrals, whose stoneribbed ceilings' eaves (Eves) became known as the "nave" – the boat. Most of Austronesia was tropical. There were no snow loads to be supported. The beached boats were not overturned. The Japanese Austronesians also had reached far enough northward to have to cope with winter snows in the same manner as had the Vikings. The Japanese name for the house "roof" is the same as that meaning "bottom of the boat." During their long years of voyaging, having no written means of amassing records, the water people composed many verses chronicling their experiences and those learned relayingly from previous generations. The Naga chants became the ragas of the Japanese and Balinese, meaning "the tales of the old people" or the sagas of the Vikings. Rehearsed daily, millennium after millennium, successive verses were identified with successive ancestors and logs in their rafts or ribs in their boats. Later turning their boats upside down on the land to protect against the storms, they developed therefrom ribbed, roofed, and columned buldings for non-seagoing communal purposes, with the ancestors' features carved on their respective columns.

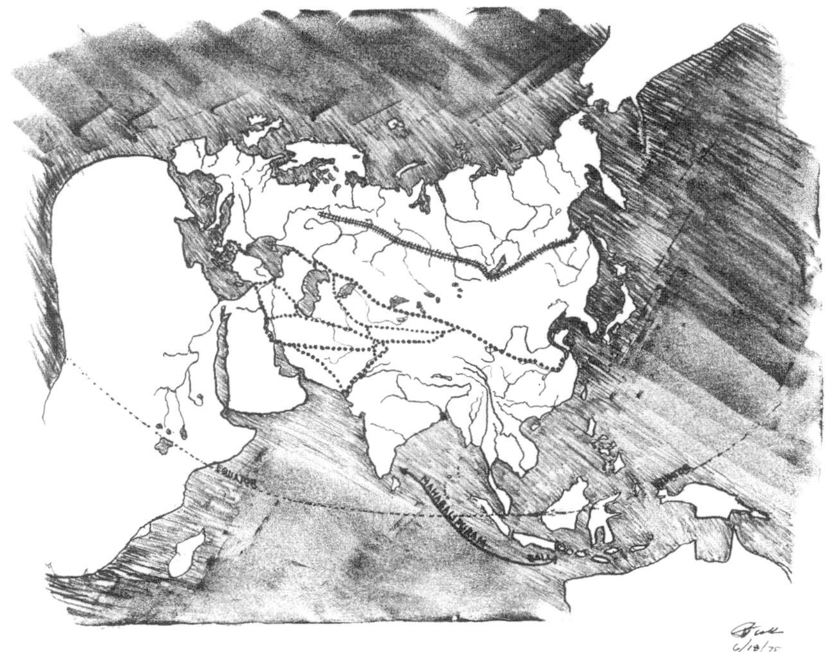

Stone 17

The Austronesians caught and domesticated mainland animals millions of years ago. Their horsemen hunters penetrated inland earliest and fastest, reaching Europe, dwelling and painting in caves wherein they bleached out during the last Ice Age. In the subsequent Stone Age these horsemen built strategic citadels, exploited the herders and the fruit, nut, leaves, and root gatherers. Goldy now draws a map for the Three Bears showing the horsemen's multimillennia-earlier five east-to-west caravaning routes and how the evolving Indian Ocean ships could more swiftly carry much larger cargoes than caravans and how, despite the religiously biased way in which the priests and scribes chose to write Eurasian history, all of its military struggles were waged between empires secretly puppeted by the grand masters of the alternate east-west overland and water routes. Thus was Emperor Constantine puppeted into shifting Roman Empire headquarters to command the Bosphorus and reestablish the overland route. Seeking to reestablish the water route the "Holy Grail" drive of the "Crusaders" sought to wrest overland access to the Indian Ocean from the Moslem Saracens. After 4,000 years of struggling all Near-East-transiting trade was mothballed when Henry "The Navigator" of Portugal inaugurated sailing from Europe to the Orient via the Atlantic around Africa. Thereafter, a succession of world-ocean masteries culminated in the East India company's "British Empire"-puppeted supremacy over all the world's oceanic trade routes. This oceanic mastery suddenly terminated in World War II, when the line-of-supply control shifted into the air and is now shifting again into the airless ocean of outer nothingness, intervening relativity's cosmic events.

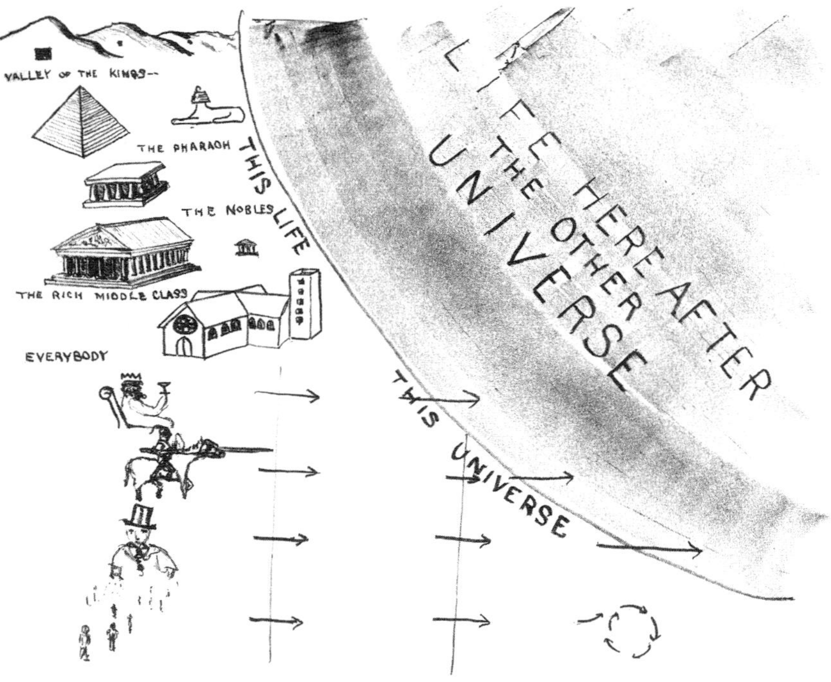

Stone 18

The Austronesians assumed that the gods were not concerned with the people but only with posthumous deifying of the king. With a few millennia of development and proliferation of technological know-how, tools, physical materials, and building capability, civilization realized that in addition to equipping the god-monarch for safe passage into the afterlife, it would be feasible also to assist the nobles to reach the next world. This was done by preparation of elaborate tombs in which were secreted all the riches, food and tools that the king (and later, nobles) would need in the next life. As millennia passed, so much know-how was accumulated in this life that during the last millennium B.C. it became technologically feasible to prepare elaborate mausoleums to enable the rich middle class to migrate safely into the next world. Finally it became technically feasible to prepare all of humanity for entering into the afterlife. This occurred at year zero. The concept of a cosmic intellectual integrity governing all Universe that is utterly concerned only with all humans which cosmic integrity, if properly worshipped and served, could arrange for the happy afterlife of every devout believer, inspired inception of a plurality of new people's religions and two millennia of church, cathedral, temple and mosque building. In addition to getting all "worthy" people into the next world, ever-multiplying technical know-how made it successively feasible to take care of the present life, first of kings (divine right of kings), then nobles (magna carta), then of the well-to-do middle class (Victorian period) and in the twentieth century for all humanity. This eliminated the necessity for two worlds – one universe now embraced all.

Stone 19

Goldy now elucidates some post-Eden history for the bears. Mycenae was able to control the floatable line of supply to besiege the static walls of Troy. Homer's epic probably represents the first introduction to public knowledge of the changeover in the grand strategy of the world's power structure from supreme dependence on the power of the almost invincibly massive high walls of the great city-states to supreme dependence on the naval architects' engineeringly superior control of the lines of supply by deep sea vessels – the waters covering three quarters of planet earth reached all the river mouths of all continents and islands, outperforming the separate land-terminalled overland routes. For long the local, fertile lands commanding, city-states masters had starved the famished outsiders to death. Now the more-with-lessing Mycenean sea masters starved the Trojan city-state insiders. Because the Trojans were the progeny of the overland horsemen, the Mycenean sailormen produced the famous Trojan Horse within which symbol of seeming acknowledgement of Trojan superiority they hid some fighting men while deceptively withdrawing their maritime fleet.

Goldy and the bears soliloquize on the fact that this 2000 B.C. historically lethal moment of conversion of supremacy from absolute dominance by massiveness to a doing-more-with-less mastery marked the beginning of a four-millennia development that would culminate at the end of the twentieth century A.D. in doing so much with so little as finally to be able to support all humans at an economically sustainable higher standard of living than any have ever experienced, thus to eliminate altogether the fundamental scarcity syndrome and all lethal interstruggling of humanity, allowing humanity to become preoccupied with greater problems of Universe, with which ultimately to cope, humans had been given their minds.

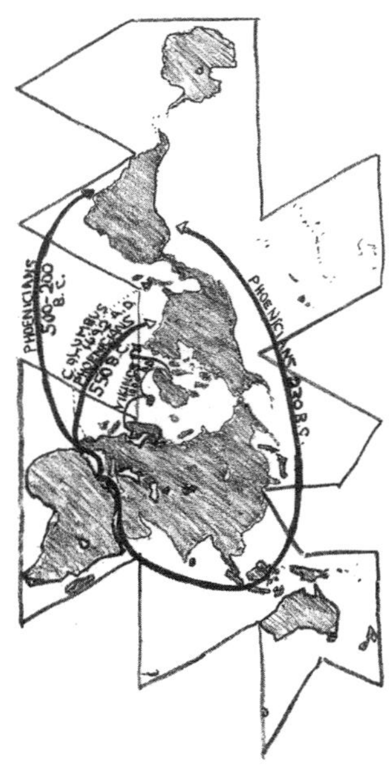

Stone 20

Goldy says that with the first coinages of humanity now being brought out from Mediterranean ocean bottoms, one of the most frequent is that of the Phoenician sailor's symbol – the double-headed, two-way-facing Janus whose overall contour is that of the Cretan king's symbol, the vertical-axis hexagon. Dry-land anthropologists and archaeologists have misinterpreted the Janus symbol as meaning only that the sailor was a notorious liar, professing to be going one way but actually going the other. To the maritime archaeologists and anthropologists the Janus symbol clearly reveals that the sailors knew that they could face in any direction and, traveling constantly in that great-circle direction, come back to where they started because the earth is a sphere. The sailor's symbol was the six-triangle hexagon, for only triangles are structurally stable. With their keeled, triangularly trussed framing and triangularly rigged ships, navigators have encircled the spherical earth for aeons. Land people have been historically local and ignorantly subservient to the earth's sphericity, commencing their wall buildings with stone, the landholding and -guarding people build compressively, squarely, redundantly and heavily, commencing their shipbulding with wood, water people build tensively, curvilinearly, triangularly, and with utmost efficiency of realized function per each ounce of weight, minute of time, and erg of energy designedly invested in their "environment controlling" vessels. They must do ever more with ever less in order to float controllably in any manner of storms while doing so much with so relatively little.

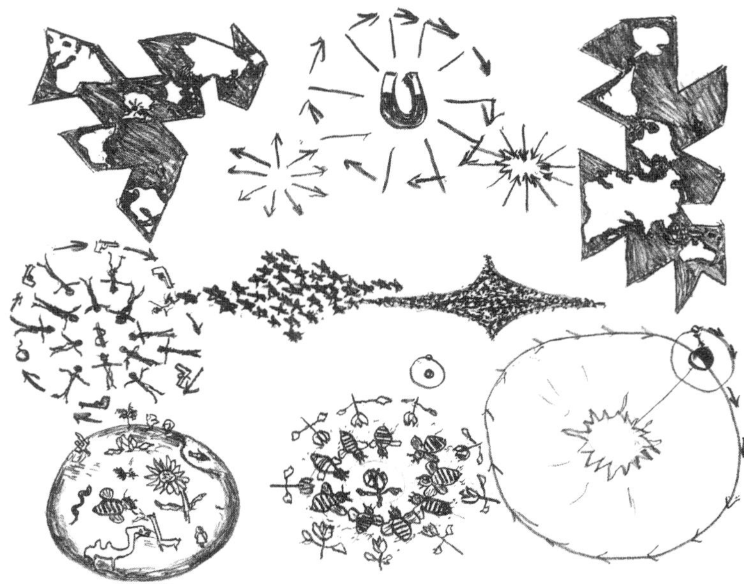

Stone 21

Goldy and the bears agree that the Austronesian water people constitute the prime ongoing organism of human evolution designed and conditioned by multimillion years of experience to withstand the mania of rationalized selfishness and eventually capable through ever-more-with-ever-less artifacts of rendering altogether obsolete opportunistic myopia in general. The water people cannot pretend to themselves, as do the businessmen, that a rectangle is an inherently stable structure, that the wind, sun, or any other operating facilities of Universe belong to anyone, or that there exists a seemingly forever lethal inadequacy of human life support purportedly inherent in the metabolics of planet earth's biospheric system.

As long as humanity has been convinced of the existence on our planet of an unalterable, lethally human-life-support inadequacy, just so long is selfishness to be rationalized as being not only "realistic" but essential in securing the vital needs of those dependent upon the family "bread-winner" or "bean-guardian," those who eschew selfishness and commit themselves to the golden rule are considered to be either ignorant or foolish. Since, however, it is now looming into ever more widely held knowledge that there is and can continue to be ample life support for all, both today and tomorrow, for the first time in three million years of known human presence on this planet, selfishness can no longer be rationalized as inherently valid. This is one of the most dramatic changes in all history.

Selected bibliography

Abbreviations

BFA	Buckminster Fuller Archive
DCH	Dymaxion Chronofile
DI	Dymaxion Index
HE	Hamilton Extracts
MF	Manuscript Files
RBF	Richard Buckminster Fuller
SP	Synergetics Papers: BT (Blue Trunk), GT (Green Trunk)
WDSD	World Design Science Decade

Books by RBF

·*Nine Chains* to the Moon. Philadelphia et al. 1938 New edition: Carbondale, Ill., 1963
·*Untitled Epic Poem on the History of Industrialization.* New York 1962
·*Education Automation.* Carbondale, Ill., 1962
·*Ideas and Integrities. A Spontaneous Autobiographical Disclosure.* Ed. Robert Marks, Englewood Cliffs, NJ, 1963
·*No More Secondhand God and other Writings.* Carbondale, Ill., 1963
·*Operating Manual for Spaceship Earth.* Carbondale, Ill.: Southern Illinois University Press 1969
·*Utopia or Oblivion. The Prospects for Humanity.* New York et al. 1959
·*The Buckminster Fuller Reader.* Ed. James Meller, London 1970
·*I Seem to be a Verb.* New York 1970
·*4D Timelock.* (1928) Albuquerque, NM, 1970, 1972
·*Buckminster Fuller to Children of Earth.* Garden City, NY, 1972
·*Intuition.* New York 1972; revised edition: Garden City, NY, 1973
·*Earth Inc.* Garden City, NY, 1973
·*Synergetics. Explorations in the Geometry of Thinking.* In Collaboration with E.J. Applewhite. New York, London 1975
·*Tetrascroll. Goldlocks and the Three Bears. A Cosmic Fairy Tale* (1975). New York 1982
·*And it Came to Pass – Not to Stay.* New York, London 1976
·*Synergetics 2. Explorations in the Geometry of Thinking.* In Collaboration with E.J. Applewhite. New York, London 1979
·*Buckminster Fuller Sketchbook.* Philadelphia, PA, 1981
·*Critical Path.* New York 1981; London 1983
·*Inventions. The Patented Work of Buckminster Fuller.* New York 1983
·with Anwar Dil: *Humans in Universe.* New York 1983
·*Grunch of Giants.* New York 1983
·*Cosmography. A Posthumous Scenario for the Future of Humanity.* New York et al. 1992

Reference books regarding RBF

·*Dymaxion Index 1927–1953,* Bibliography and Published Items Regarding Dymaxion and Richard Buckminster Fuller. Fuller Research Foundation, New York (1953) (printed as manuscript)
·*Basic Biography: Richard Buckminster Fuller.* Private edition, Philadelphia, PA, 1983
·R. Buckminster Fuller, John McHale: *World Design Science Decade* (WDSD) *1965–1975.* Document: 1-6. 6 vols. World Resources Inventory, Southern Illinois University. Carbondale, Ill. 1963–1967
vol. 1: *Inventory of World Resources. Human Trends and Needs.* Phase I, Document 1 (1963), vol. 2: *The Design Initiative.* Phase I, Document 2 (1964), vol. 3: *Comprehensive Thinking.* Phase I, Document 3 (1965), vol. 4: *The Ten Year Program.* Phase I, Document 4 (1965) vol. 5: *Comprehensive Design Strategy.* Phase II, Document 5 (1967), vol. 6: *The Ecological Context: Energy and Materials.* Phase II, Document 6 (1967)
·R. Buckminster Fuller: *50 Years of Design Science Revolution and the World Game.* World Resources Inventory, Philadelphia, PA, 1969
·R. Buckminster Fuller: *The World Game: Integrative Resource Utilization Planning Tool.* World Resources Inventory, Southern Illinois University, Carbondale, Ill. 1971).
·*The Artifacts of R. Buckminster Fuller. A Comprehensive Collection of His Designs and Drawing* in Four Volumes. Edited by James Ward. New York, London 1985
·*Synergetics Dictionary. The Mind of Buckminster Fuller.* Complied and edited by E.J Applewhite. Vol. 1–4 New York, London 1985

Private editions RBF

·*4D Timelock.* (Private print, mimeograph) Chicago 1928 MF 28.01.01 (4D Files, reboxed), BFA
·*Dymaxion Comprehensive System. Introducing Energetic Geometry, Washington,* DC, March 14, 1944 (Private print, mimeograph) MF 44.03.01, BFA
·*Fluid Geography.* (Reprint from an article in the American Reptune, April 1944). Appendix: *The Dymaxion Map,* patented January 29, 1946. Wichita, Kansas (1946)
·*Designing a New Industry.* A Composite of a Series of Talk by R. Buckminster Fuller, 1945–1946. Fuller Research Institute, Wichita, Kansas, 1946
·*Earth, Inc.* Fuller Research Foundation. New York (1947); Appendix I: *Energetic Geometry.* p. 17–18

Articles about and interviews with RBF

·*Universal Architecture.* In: T-Square, vol. 2, no. 2, February 1932, p. 22–25, 34–41
·*Universal Architecture,* Essay 3. In: Shelter, vol. 2, no. 4, May 1932, p. 33–42; essays 1 and 2: p. 59–86
·*Shelter's Editorial and Economic Policy.* In: Shelter, vol. 2, no. 4, May 1932, p. 1–2
·*Correlation.* In: Shelter, vol. 2, no. 4, May 1932, p. 3
·*Dymaxion House.* In: The Architectural Forum, March 1932, p. 285–288
·*Putting the House in Order.* In: Shelter, vol. 2, no. 5, November 1932, p. 2–8
·*Conning Tower – «Hoop Skirt» Room.* In: Shelter, vol. 2, no. 5, November 1932, p. 64–65
·*Streamlining.* In: Shelter, vol. 2, no. 5, November 1932, p. 71–78
·*Teleology.* In: Shelter, vol. 2, no. 5, November 1932, p. 105–110
·*Scientific Shelter for USSR.* In: Shelter, vol. 2, no. 5, November 1932, p. 94–95
·*Dymaxion Houses.* In: Architectural Record, January 1934, p. 9–11
·(RBF and Fortune staff:) *U.S. Industrialization.* In: Fortune, vol. 21, no. 2, 1940, p. 50–57
·*World Energy. A Map by R. Buckminster Fuller.* In: Fortune, vol. 21, no. 2, February 1940, p. 57
·*Theory and Practice of the Spinning Wheel.* In: Fortune, vol 21, no. 5, May 1940
·*The Mechanical Wing.* In: The Architectural Forum, October 1940, p. 92
·*Fluid Geography.* In: American Neptune, vol. 4, no. 2, April 1944, p. 119–136
·*Comprehensive Designing.* In: trans/formation, vol. 1, no. 4, 1950, p. 18–23
·*Man on Earth.* In: Ladislav Sutnar: *Transport – next half Century 1951–2000.* Prepared and visually presented by Ladislav Sutnar. Foreword by Buckminster Fuller. New York 1950, o.p.
·*The 90 Percent Automatic Factory.* In: Student Publications of the School of Design, North Carolina State College, Raleigh, NC, vol. 2, no. 1, Fall 1951, p. 29–33
·*Architecture out of the Laboratory.* In: Student Publication, University of Michigan, Ann Arbor College of Architecture and Design, vol. 1, no. 1, Spring 1955, p. 9–34
·*(Letter to John McHale)* New York, January 7, 1955 (*«Influences on my work»*) In: Architectural Design, July 1961, p. 290–296
·*Tensegrity.* Introduction by John McHale. In: Portfolio & Art News Annual, no. 4, 1961, p. 114–148

· *The Prospect for Humanity.* In: Saturday Review, 28.8./19.9./ 3.10.1964
· *Conceptuality of Fundamental Structures.* In: Gyorgy Kepes (ed.): Structure in Art and in Science. New York 1965, p. 66–88
· *The Dynamo and the Computer.* In: Appalachian Review, vol. 2, no. 1, Fall 1967, p. 6–10
· (RBF et all.) *Der US-Pavillion auf der Weltausstellung in Montreal.* In: Bauen + Wohnen, Nr. 10, October 1967, p. 339–406
· *City of the Future.* In: Playboy, January 1968, p. 166–168, 228–230
· *World Game. How it came about.* In: *50 Years of the Design Science Revolution and The World Game.* Philadelphia, PA, 1969, p. 111–118

Archival materials and manuscripts RBF

· *Lightful Houses.* Manuscript, undated (1928), 73 p. MF 28.01.01 (4D Files reboxed), Folder IX, BFA
· *Cosmopolitan Home Corporation. Lightful Products. Cosmopolitan Home Corporation. Products «Lightful» Trade Mark.* Set of two manuscripts, 5 and 4 p., undated (1928) MF 28.01.01 (4D Files reboxed), Folder IX, BFA
· *Dymaxion Car Cover Piece.* Mansucript, 1 p., undated MF 34.01.01, BFA
· *Ballistics of Civilization.* Set of manuscripts, undated (1939) MF 39.03.01, BFA
· *Dymaxion Comprehensive System. Introducing Energetic Geometry.* Prepared especially for the Northeastern Engineer, January 1944. Mimeographed manuscript, 15 p., SP, BT Folder 16 (2), BFA
· *«Wichita»* Draft. Set of 11 sheets of notes and sketches (construction of Wichita House: ventilation system, support mast) HE, Box Nr. 3, Vol. 18, BFA
· *Fuller Research Foundation – Its Philosphy and Objectives.* Manuscript, undated (ca. 1946–50). MF 46.08.01, BFA
· *Letter to Einstein,* Jan. 17, 1948 (draft). Handwritten draft for letter on note paper, *«Dear Mr. Einstein...»*, 9 p., sign. RBF SP, BT, Folder 28 (2), BFA
· *Geodesic Structures – «Your Private Sky» – Jeosahedronal Data.* Manuscript with sketch, 4 p. June 15, 1948. SP, BT, Folder 15 (3), BFA
· *Eureka – Eureka – This is what Archimedes sought and Pythagoreans and Kepler and Newton – And Eureka again!!!* Set of four sheets of paper with handwritten notes, calculations and sketches. Signed and dated, April 25, 1948. SP, BT, Folder 14 (7)
· *The Autonomous Dwelling Facility.* Manuscript, 6 p. undated. MF 49.04.02, BFA
· *The Comprehensive Designer.* Manuscript, 7 p., June 1, 1949. MF 49.06.01, BFA
· *BF's Noah's Ark no. 2.* Written January 1950 to July 1950. Handwritten manuscript with sketches, 12 p. in envelope. MF 51.02.01 (Geodesic Grid Data), Folder III, BFA
· *Project Noah's Ark no. 2.* To Richard Hamilton in confidence and faith. Signed and dated on cover: BF August, 22, 1950. Manuscript, 22 p. text and 24 p. illustrations, bound as brochure. MF 51.02.01, BFA

Books about RBF

· Aaseng, Nathan: *More with Less. The Future World of Buckminster Fuller.* Mineapolis 1986
· Applewithe, E.J.: *Cosmic Fishing. An account of writing Synergetics with Buckminster Fuller.* New York 1977
· Baldwin, Jay: *Bucky Works. Buckminster Fuller's Ideas for Today.* New York 1994
· Edmondson, Amy C.: *A Fuller Explanation. The Synergetic Geometry of R. Buckminster Fuller.* New York 1992
· Grimaldi, Roberto: *R. Buckminster Fuller 1895–1983.* Roma 1990
· Hatch, Alden: *R. Buckminster Fuller at Home in Universe.* New York 1974
· Kahn, Ed. Robert; Wagschal, Peter: *Buckminster Fuller on Education.* Amherst, MA, 1979
· Kenner, Hugh: Bucky. *A Guided Tour of Buckminster Fuller.* New York 1973
· Kenner, Hugh: *Geodesic Math: And How to Use it.* Berkeley, CA, 1976
· Krausse, Joachim: *Design-Strategie am Werk. Eine Einführung in die Planungs- und Entwurfsarbeit von Buckminster Fuller.* In: R. Buckminster Fuller: *Bedienungsanleitung für das Raumschiff Erde.* Reinbek 1973, p. 127–178
· Lord, Athena: *Pilot for Spaceship Earth.* New York, London 1976
· Marks, Robert W.: *The Dymaxion World of Buckminster Fuller.* New York 1960
· McHale, John: *R. Buckminster Fuller.* New York 1962;
· Pawley, Martin: *Buckminster Fuller.* London 1990
· Robertson, Donald W.: *Mind's Eye of Richard Buckminster Fuller.* New York 1974
· Rosen, Sidney: *Wizard of the Dome. R. Buckminster Fuller. Designer for the Future.* Boston, Toronto 1989
· Sleden, Lloyd Steven: *Buckminster Fuller's Universe: An Appreciation.* New York 1989
· Snyder, Robert: *Buckminster Fuller. An Autobiographical Monoloque/Scenario.* New York 1980

Other literature

· Ashton, Dore: *Noguchi East and West.* New York 1992
· Banham Reyner, *The Dymaxlorat.* The Arts Magazine, no. 38, Gerber 1963. p. 66–69
· Bayer, Herbert: *World Geo-Graphic Atlas.* Container Corporation of America, Chicago, Ill., 1953
· Cage, John: *A Year from Monday. New Lectures and Writings.* London 1968
· Einstein, Albert: *Wissenschaft und Religion.* In: Berliner Tablatt, November 11, 1930; reprinted in: ders., *Mein Weltbild.* Hgg. von Carl Seelig, Berlin 1988, p. 15–18 (engl.: *Religion and Science.* In: New York Times Magazine, November 9, 1930, sect. V, p. 1)
· Farrell, Barry: *Interview* (with RBF), in: Playboy, February 1972
· Fuller Snyder, Allegra: *DOME Talks with Allegra Fuller Snyder.* In: Dome, vol. 8, no. 2, Winter 1995–96, p. 20–41
· Hafner, Katie und Matthew Lyon: *Where Wizzards Stay up Late. The Origins of the Internet.* New York 1996
· Harris, Mary Emma: *The Arts at Black Mountain College.* Cambridge, MA, London 1987
· Ingber, Donald E.: *The Architecture of Life.* In: Scientific American, January 1988, p. 30–39
· Kahn, Lloyd (ed.): *Domebook 2.* Pacific Domes, Bolinas, CA, 1971
· Krausse, Joachim: *Gebaute Weltbilder von Boullée bis Buckminster Fuller.* In: ARCH+ 116, March 1993, p. 20–82
· Murphy, Charles J.V.: *The Polar Watch.* In: Fortune, December 1957, p. 118–120, 245–255
· Rheingold, Howard: *The Virtual Community. Finding Connections in a Computerized World.* London 1993
· Sadao, Shoji: *Geodesic Domes.* In: Joseph A. Wilkes (ed.): *Encyclopedia of Architecture. Design. Engineering. Construction.* Vol. 2, New York 1982, p. 577–585
· Tomkins, Calvin: *In the Outlaw Area. A Profile of R. Buckminster Fuller.* In: The New Yorker, January 8, 1966, p. 35ff.; reprinted in: *The Artifacts of R. Buckminster Fuller.* Ed. James Ward, Vol. 1, New York, London, 1985, p. XI–XXI
· Wachsmann, Konrad: *Wendepunkt im Bauen.* Wiesbaden 1959; engl.: *The Turning Point of Building.* 1961
· *Whole Earth Catalog.* (Ed. Stuart Brand) Menlo Park, CA, 1968 ff.

Photo Credits

Editor's remark: The majority of the photographs used in this book and provided from the *Buckminster Fuller Archive* are nto signed. The identification of the photographer is in many cases impossible. Numerous photographs are slides, namely after 1949, many of which probably have been taken by RBF himself. RBF used 135 and 6x6 film material.

Beechcraft Photo: 235 – 7, 247
Butler Manufacturing Corp.: 213–7
Fortune Magazine: 138, 221, 223–6, 229–231, 246, 248–9
Norman Foster and Partners: 432
Bernard Hoffmann: 203, 207, 209
Clemens Kalischer: 321, 322, 323, 325$
Kaufmann+Fabry: 201
Hazel Larsen Archer: 317
Archie Lieberman: 8
Life Magazine: 34, 251, 254, 256–63
F.S. Lincoln: 30, 123, 145, 169, 179, 186–188, 191–193
Barbara Morgan: 153
Museum für Gestaltung Zürich (Betty Fleck): 155, 253, 290, 396–9
NASA Photo: 485
Koni Nordmann: 215–5
John Philips: 213, 218
Shoji Sadao: 423
Julius Shulman: 418–20
Stockade Building System: 27-8, 76-7
Whole Earth Catalog: 483-4, 486-9

Index of the Texts and Excerpts from Texts by RBF

My First Invention 44, 63
Letter to Vincent Astor 52
Letter from R.B.F. to his Sister, Rosamund Fuller 69
Dirigible 102
Comics (funnies-Philosophy) 103
Buckminster Fuller's 1928 Prophecy 106
On Mimeographing 118
The Dymaxion House 135
Greenwich Village Avantgarde 148
Romany Marie 154
Doing the Most with the Least 164
Mobile Shelter 164
Conning Tower 176
Man on earth 182
House = Porch Dwelling Machine 200
(Bathroom) 210
Metals 222
Wichita House 238
Dymaxion Projection 252
Construction of Dymaxion World Map 270
Closet Packing of Spehres 278
Geodesic structures 334
90% Automatic Cotton Mill 336
Flying Seedpod 342
Geoscope: point of view 346
Ford Rotunda Dome 357
Roam Home to a Dome 371
Triennale di Milano Paper Domes 374
Wire Wheels 394
Tensegrity 396
Nothing Touches Each Other 408
Woods Hole Restaurant 417
Geodesic Domes: a human being's skin 428
Domed over cities 434
How Nature Builds 444
The beauty of bubbles 457
How it came about 472

Transporting heavy loads is a challenge in every respect. The increasing expectations of the market in terms of feasibility, speed, reliability, environmental compatibility and global coverage show with ever greater clarity that the existing transport system has reached its limits.
CargoLifter is the solution to these urgent transport problems. The idea is as obvious as it is fascinating: a flying crane that can lift large and bulky goods and take them non-stop to their destination without having to use roads, bridges, transshipment cranes, port facilities or airports.
A new world-wide logistics system for heavy and special loads will be constructed as of the turn of the millennium. Thus, on the one hand, account will be taken of the interlinking of the continents in the heavy-transport sector. On the other hand, CargoLifter will also shift the focus to sustained economic management: in comparison to conventional transport methods, CargoLifter can move heavy and bulky loads with scant regard for the infrastructure and fewer transshipment processes at the same time as being kinder to the environment. CargoLifter will use existing markets and, by being on the cutting edge of technology, create new markets.

CargoLifter – a new dimension in transportation.

SKY CRANE

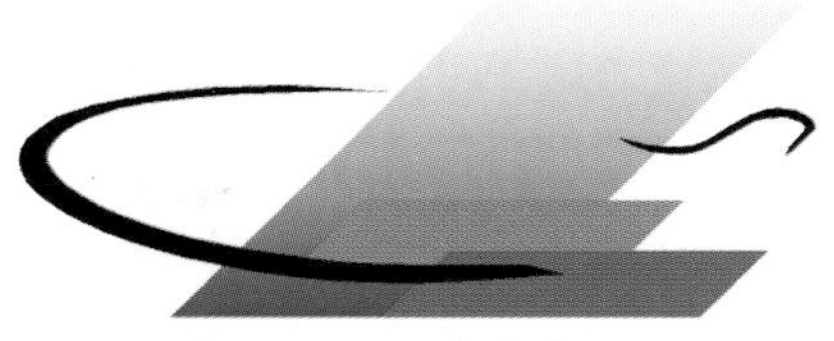

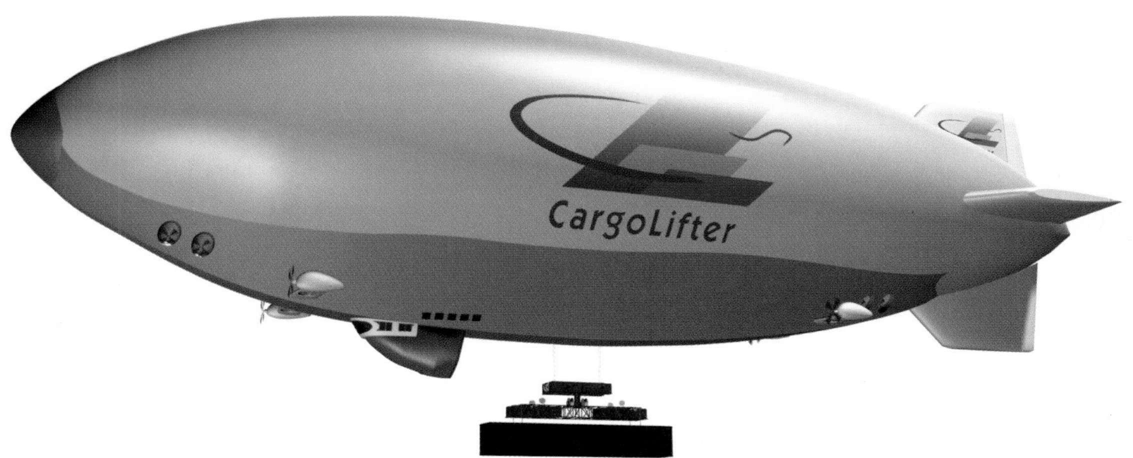

CargoLifter AG
Grüneburgweg 102
D-60323 Frankfurt/Main
Germany
Telefon: +49-(0)69-150570
Fax: +49-(0)69-1505818100
www.cargolifter.de

CargoLifter Inc.
4600 Marriot Drive, Suite 225
Raleigh, North Carolina 27612
USA
Phone: +1-919-7880613
Fax: +1-919-7880614
www.cargolifter.com

Your Private Sky

R. Buckminster Fuller
The Art of Design Science

A collaborative project of:
Museum für Gestaltung Zürich, Design Museum London, Kunsthalle Tirol, Hall, Stiftung Bauhaus Dessau Zeppelin-Museum Friedrichshafen, Lars Müller Publishers in cooperation with the Buckminster Fuller Institute/ Buckminster Fuller Archive, Santa Barbara, Ca.

A cultural engagement by CargoLifter AG, Frankfurt/Main

Supported by Franke Schweiz AG, Aarburg CH, and Jakob AG, Drahtseilfabrik, Trubschachen CH

Publication
Edited by Joachim Krausse and Claude Lichtenstein
Concept: Joachim Krausse, Claude Lichtenstein, Lars Müller
Design: Integral Lars Müller/Lars Müller, Christian Guler, Annette Schmidt
Translation: Steven Lindberg, Julia Thorson
Setting: Heidy Schuppisser, Baden
Lithography: Ast und Jakob AG, Bern
Printing: EBS Editoriale Bortolazzi-Stei, Verona (I)
Binding: Buchbinderei Burkhardt AG, Mönchaltorf

Printed in Italy

Exhibition
Concept: Joachim Krausse, Claude Lichtenstein
Design: Alexandra Gübeli/Zurich Museum of Design
Assistance: Karin Frei, Christina Hemauer
Secretary: Doris Brem
Graphics: Tobias Strebel, Martin Woodtli
A/V Technology: Jörg Schellenberg
Slide show: Koni Nordmann, Martin Peer (Kontrast)
Construction: Assemby crew, MfGZ, led by Jürg Abegg
Rudolf Fluri (corrugated cardboard dome)
Model building:
Architecture: Heinz Caflisch, Franz Engler, Remo Darungs, Carmen Gasser
Sectioning of sphere: Students of HGKZ, Instructor U.B. Roth: Roman Aebersold, Olivier Blaser, Tonja Gnannt, Jonathan Hotz, Veronika Hunziker, Roland Jaggi, Thomas Stäubli, Joachim Vogt
Project direction: Claude Lichtenstein

Lars Müller Publishers
5401 Baden/Switzerland
mail: books@lars-muller.ch

ISBN 3-907044-88-6

Joachim Krausse, 1943, Dr. phil, professor of design history at Hochschule Anhalt (FH) in Dessau, Germany. Publications on architecture, design and cultural history, e.g. *Gebaute Weltbilder* (Arch+). Frequent collaborator of the Arch+ magazine. Editor of RBF writings, e.g. *Bedienungsanleitung für das Raumschiff Erde* (Amsterdam/Dresden 1998). Lives in Berlin.

Claude Lichtenstein, 1949, Architect, curator at the Zurich Museum of Design (Museum für Gestaltung). Exhibitions and publications on architecture and design, e.g. *O.R. Salvisberg* (1985), *Anonymous Swiss Design* (1987), *Ferdinand Kramer* (Giessen 1991), *Streamlined* (Baden 1992/93, with F. Engler), *Bruno Munari* (Baden 1995, with A.W. Häberli), *Luigi Snozzi* (Basel/Boston 1997). Teaches history and theory of design at the Hochschule für Gestaltung Zurich.

Acknowledgements
The editors are greatly indebted to the following individuals and institutions for their advice, help and cooperation:

Edward J. Applewhite, Carlo Barck, Juliet Betita, Michael Burke, Joe Clinton, Tony DeVarco, Benedikt Erhard, Franz Engler, John Ferry, Rudolf Fluri, Allegra Fuller Snyder, Bonnie Goldstein DeVarco, Fiona Greenwood, Christian Guler, Tony Gwillian, Peter Hahn, Hardt-Waltherr Hämer, Hochschule der Künste Berlin, Reyer Kras, Hanna Krumteich, Nikolas Kuhnert, Kyoshi Kuromia, Klaus Lehmann, Jackie Lohrke, Wolfgang Meighörner, Hans Meyer, Gary Milliken, Lars Müller, Martin Pawley, James Peto, Urs Beat Roth, Shoji Sadao, Hubert Salden, Annette Schmidt, Caspar Schwabe, Robert Snyder, Jaime Snyder, Carl Solway, Paul Thompson, Wilfried Wang, Heidi und Peter Wenger, Thomas L. Zung

I would like to thank my wife, Lia, and my daughters, Hannah, Jelena und Vera, for their forbearance, indulgence and impatient patience. C. L.